AF560016

Environmental Biotechnology

Environmental Biotechnology

Editors
Prof. D.R. Khanna
Prof. A.K. Chopra
Dr. Gagan Matta
Dr. Vikas Singh
Dr. R. Bhutiani

Associate Editor
Dr. Chakresh Pathak

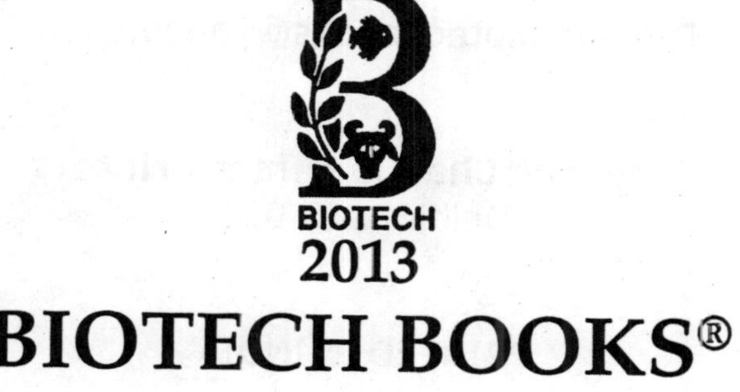

2013
BIOTECH BOOKS®

ISBN 978-81-7622-271-6

Published by: **BIOTECH BOOKS®**
4762-63/23, Ansari Road, Darya Ganj,
New Delhi - 110 002
Phone: +91-011-23262132
E-mail: biotechbooks@yahoo.co.in

Printed at: **Chawla Offset Printers**
Delhi - 110 052

PRINTED IN INDIA

Preface

Environmental Biotechnology is the field of applying principles of microbiology to solve environmental problems. There are various aspects in Environmental Biotechnology like treating industrial and municipal wastewaters biologically, restoring the pre existing flora and fauna and the contaminated and degraded land. It could be beneficial in protecting our surface water resources from environmental contaminants. The principles could be used in preventing various diseases related to air and water.

Environmental Biotechnology employs microorganisms to remove contaminants from wastewater. It is a vast and rapidly growing field with increasing relevance for sustainable development through environmental protection. Biotechnological approaches can be applied to assess and maintain the well being of the ecosystem, transform pollutants into harmless substances, generate biodegradable material from renewable resources and develop environmentally safe manufacturing and disposal processes. Regarding Environmental Biotechnology, its contributions span from environmentally-friendly and cost effective "end-of-the-pipe" solutions to environmental pollution and problems (bioremediation of soils and aquifers, biological waste treatment), to the development of sustainable alternatives for their prevention and alleviation, such as the replacement of fossil fuels by biohydrogen and methane from wastes and futuristic "bioreneries". Biotechnology has the potential of a reduction of operational and investment costs for the design and operation of more sustainable processes based on microbes and other living organisms as agents

Different articles delineates the current and prospective applications in these sub-areas of environmental biotechnology, and documents, case studies on environmental monitoring, restoration of environmental quality, resource recovery etc. We acknowledge special thanks to the authors for contributing their valuable work in the form of research papers in this book. This will definitely be helpful to the young researchers working in the concern field and hope best for this book in fulfilling the purpose of making awareness among social community.

This publication has been prepared as a tool for a better communication, information exchange and understanding of "WCMANU, 2011". Articles appeared in this Proceedings are basically *received based, non-reviewed* papers that have been submitted by the pre-registered persons to the "WCMANU, 2011". The reader who intends to refer the article in this Proceeding is recommended to contact directly to the author.

Editors

Contents

2013, Environmental Biotechnology Pages 1–9
Editors: D.R. Khanna, A.K. Chopra, Gagan Matta, Vikas Singh & Rakesh Bhutiani
Published by: BIOTECH BOOKS, NEW DELHI

Chapter 1

Antibacterial Activity of *Tridax procumbens, Calotropis gigantea* and *Calendulla officinalis* Plant Extracts Against Clinical Isolates from Wound Infection

P.M. Tumane and Durgesh D. Wasnik
Post Graduate Teaching Department of Microbiology, Rashtrasant Tukadoji Maharaj, Nagpur University, Nagpur – 440 033

Purpose: Wound infection is the severe problem among the individuals there may be alteration of function which interferes with the individual's ability to carry on with daily life activities. In the present investigation, antimicrobial activity of different medicinal plant extracts such as *Tridax procumbens, Calotropis gigantea* and *Calendulla officinalis* were tested against the isolated microorganisms from wound infection. *Material and Methods*: A total of 8 clinical isolates are isolated from wound infection such as *Escherichia coli, Staphylococcus aureus, Staphylococcus epidermidis, Proteus vulgaris, Proteus mirabilis, Pseudomonas aeruginosa, Klebsiella pneumoniae* and *Serratia marscensces* on the basis of morphological, biochemical, cultural characteristics and antibacterial activity of three medicinal plants against clinical isolates by using agar well diffusion method and their comparative study with the standard antibiotics using Kirby - Bauer disc diffusion technique was carried out. *Result*: All the plant extracts showed remarkable antimicrobial activity against all the isolated microorganisms. Among the plant extract tested, the aqueous leaf extract of *Calendulla officinilis* showed excellent zone of inhibition (18-22 mm) against all the isolated microorganisms specially against *Ps. aeruginosa* which have resistant against all antibiotics. Aqueous leaf extract of *Tridax procumbens* (10-13 mm) and milky white fluid from *Calotropis gigantea* (10-12 mm) had shown least inhibitory activity against all the clinical isolates from wound infection. Ethanolic and methanolic extracts of medicinal plants against all organisms shown very least inhibitory activity. *Conclusion*: The aqueous leaf extract of *Calendulla officinialis* show higher inhibitory action against all the test microorganisms, specially against *Pseudomonas aeruginosa* which is resistant to all the present antibiotics.

Keywords: *Medicinal plants, Antibacterial activity, Clinical isolates.*

Introduction

Wounds occur when the continuity of the skin or mucous membrane is broken. Injury to tissues results in bleeding (which may be life-threatening depending on the severity) with subsequent activation of acute inflammatory reactions:[1] Bleeding from damaged blood vessels in the injured tissue must be arrested through the process of haemostasis. The injury and associated acute inflammatory response result to necrosis of specialized cells and damage to the surrounding matrix and the host tissues must activate the healing process to replace dead tissues with healthy ones.[2] However, microbial infection of the wound impairs the healing process and may delay tissue repair. Besides the pain and general discomfort arising from injuries or wounds there may be alteration of function which interferes with the individual's ability to carry on with daily life activities. Consequently, there is an overriding need to stimulate healing and restore the normal functions of the affected part(s) of the body to ease the discomfort and pain associated with wounds by arresting bleeding from fresh wounds, preventing infection, and activating tissue repair processes. Wound healing occurs in three stages: inflammation, proliferation, and remodeling.[3,4,5] The proliferative phase is characterized by angiogenesis, collagen deposition, granulation tissue formation, epithelialization and wound contraction. In angiogenesis, new blood vessels grow from endothelial cells.[6,7,8] Medicinal plants are a source of great economic value in the Indian subcontinent.[9] Nature has bestowed on us a very rich botanical wealth and a large number of diverse types of plants grow in different parts of the country. India is rich in all the 3 levels of biodiversity, namely species diversity, genetic diversity and habitat diversity. In India thousands of species are known to have medicinal value and the use of different parts of several medicinal plants to cure specific ailments has been in vogue since ancient times. Herbal medicine is still the mainstay of about 75-80 per cent of the whole population, mainly in developing countries, for primary health care because of better cultural acceptability, better compatibility with the human body and fewer side effects. However, the last few years have seen a major increase in their use in the developed world. Now a day's multiple drug resistance has developed due to the indiscriminate use of commercial antimicrobial drugs commonly used in the treatment of infectious disease.[10-12]

Material and Methods

Plant Collection

1) For *Tridax procumbens* and *Calendula officinalis*

Fresh leaves of *Tridax procumbens* were collected and the fresh plant leaves were washed under running tap water, air dried and then homogenized to fine powder in pestle mortal and stored in airtight bottles.

2) For *Calotropis gigantean*

Fresh plants were collected and the fresh milky white fluids were directly collected from the leave. First leaves were cut, then the milky white fluid were collected directly in sterilize bottle. This fluid was stored in bottle for further study.

Preparation of Extracts

Aqueous Extraction

Leaves of *Tridax procumbens and Calendula officinalis* were collected from plant, wash with distilled water. These leaves are then transferred in clean and dry pestle mortal, chopped it well. The liquid extract was filtered through muslin cloth. This extract then filters three to four times through filter paper. The supernatant was collected, and use it as final extract. The milky while fluid collected directly from *Calotropis gigantea* and used for antimicrobial activity against clinical isolates from wound infection.

Solvent Extraction

1) For Ethanol Extraction

The leaves of *Tridax procumbens* and *Calendula officinalis* were air dried and then blended into powdery form. A 95 per cent Ethanol was the solvent used for extraction in which 7 grams of the powdered sample was weighted and poured in the 95 per cent ethanol and kept on rotary shaker for 12 hrs. It was sieved using muslin cloth and concentrated by evaporating and gets crude extract. For milky white fluid from *Calotropis gigantea,* take 2 ml of milky white fluid and dissolve in 2 ml of 95 per cent ethanol. The concentrated extract was stored in freeze until used.

2) For Methanol Extraction

The leaves of *Tridax procumbens* and *Calendula officinalis* were air dried and then blended into powdery form. A methanol was the solvent used for extraction in which 7 grams of the dried powdered sample was weighted and poured in 20 ml of methanol and kept on rotary shaker for 12 hrs. It was sieved using muslin cloth and concentrated by evaporating and gets crude extract. For milky while fluid from *Calotropis gigantea,* take 2 ml of milky white fluid and dissolve it in 2 ml of methanol. The concentrated extract was stored in freeze until used.

Isolation of Test Microorganisms

1. Initially, the pus sample was collected directly from the patient suffering an acute disorder of wound infection which is cause due to several factors.
2. The lesion was first clean with surgical soap or application of 70 per cent ethyl isopropyl alcohol.

3. If the specimen is collected as a swab with discharges then collect the excaudate.
4. If there is no discharge then sterile cotton swab socked in sterile physiological saline is extended deep into the wound.
5. Care should be taken not to touch adjacent skin margin.
6. If tissue is deeply ulcerated and nacrodate then the specimen is best collected by aspirating the correlate fluid from the depth of wound with sterile needle and syringe or whole inner dracing may be sent in a sterile jar.
7. The syringe should be kept in anaerobic jar, in case of delay in transporting specimen.
8. Prepare agar plates for isolation of microorganisms from wound infection, cool agar plates and solidify it.
9. Swabs are spread on nutrient agar plate and incubate these plates in incubator at 37°C for 24 hours.
10. In the next day, there was different colonies were observed, out of that, the colonies have certain characteristics are then transferred to the nutrient agar slant and it also incubate at 37°C and 24 hours.
11. All the isolated organisms used in the study were maintained on nutrient agar slant (HI, media, Mumbai) and stored in refrigerator at 4°C. Sub-culturing was done after every 15 days.

Bacterial Strains

Bacterial strains of *Escherichia coli, Staphylococcus aureus, Staphylococcus epidermidis, Proteus vulgaris, Proteus mirabilis, Klebsiella pneumoniae* and *Serratia marcescens* were isolated from wound infected patients and maintained on nutrient agar slant (HI, media, Mumbai) and stored in refrigerator at 4°C. Sub- culturing was done after every 15 days. Media components were purchased from HI Media Mumbai. All other chemicals used were of analytical grade.[13] The antimicrobial activity of all plant extracts against isolated microorganisms from wound infection were carried out by using agar well method (cup plate method) or Kirby-Baurer Method.[14]

Results

This was the comparative study, which plant extract gave large extent of relief against wound infection. The antibacterial activity of different plant extracts were presented in Table 1.1. All the plant extracts tested have exhibited different degrees of antibacterial activity against isolated microorganisms. The aqueous extract of *Tridax procumbens* gave the zone diameter of inhibitions: 13mm for *Escherichia coli*, 10mm for *Staphylococcus aureus* and no zone was found in remaining isolates. While the aqueous extract of *Calotropis gigantea* gave zones in the range of 10-12 mm in diameter of test organisms and no zone was found against *Pseudomonas aureuginosa*. But the aqueous extract of *Calendulla officinalis* gave better result against all isolated microorganisms in the range of 18-22 mm (Table 1.1). The methanolic extract of *Tridax procumbens* and

Calotropis gigantean gaves the least inhibitory action against all test microorganisms but the methanolic extract of *Calendula officinalis* gave the zone of inhibition in the range of 14-18 mm against all isolates (Table 1.1). The same results were found in case of ethanolic extract of *Tridax procumbens and Calotropis gigantean* gaves the least inhibitory action against all test microorganisms but the ethanolic extract of *Calendula officinalis* gave the zone of inhibition in the range of 10-21 mm of all isolates (Table 1.1). Antibacterial activities of antibiotic against all microorganisms from wound isolates were shown in Table 1.2.

Discussion

In our study, the antibacterial activity of antibiotics was found that the *Escherichia coli* were resistant to Amikacin (17 mm) and were intermediate to Streptomycin (18 mm) and Tobramycin (20 mm). No result was found against vancomycin. But the aqueous extracts of *Calendula officinalis* (Marigold) were sensitive against *E. coli* producing 21 mm zone and the aqueous extract of *Tridax procumbens*[15] and *Calotropis gigantea* were least effective to *Escherichia coli*. Methanolic and ethanolic extract doesn't show proper activity against *Escherichia coli* (Table 1.1). Matthew J. Leach, PhD, BN (Hons), ND, RN, MATMS who shows that the *Calendula officinalis* was found for the treatment of wound healing.[16] It is found that *Staphylococcus aureus* are resistant to all the antibiotics used (Table 1.2), but the aqueous extract of *Calendula officinalis* was found to be sensitive against *Staphylococcus aureus* (21 mm). The aqueous extract of *Tridax procumbens* and *Calotropis gigantea*[17] were resistant to *Staphylococcus aureus* (Table 1.1).

The methanolic and ethanolic extracts are resistant to *Staphylococcus aureus* (Table 1.1). *The Proteus vulgaris* and *Proteus mirabilis* show intermediate activity against Tobramycin and no result were obtained against Vancomycin. These organisms were sensitive against Amikacin (Table 1.2). Aqueous extract of plant leaves shows intermediate activity against clinical isolates (Table 1.1). The methanolic and ethanolic extract also shows intermediate activity against *Proteus vulgaris* and *Proteus mirabilis* (Table 1.1). *Klebsiella pneumonia* and *Serratia marcescens* were showed intermediate against Tobramycin and Strepomycin, but it was sensitive to Amikacin (Table 1.2). The aqueous extract of *Calendula officinalis* was sensitive against *Klebsiella pneumonia* and *Serratia marcescens*. Aqueous extract of *Tridax procumbens* and *Calotropis gigantea* was resistant to clinical isolates (Table 1.1). Methanolic and Ethanolic extract showed poor inhibitory action (Table 1.1). *Staphylococcus epidermidis* was resistant to Vancomycin and Tobramycin but these organisms were sensitive to Streptomycin and Amikacin (Table 1.2). In present study, the aqueous extract of *Calendula officinalis* was found that it was sensitive to *Staphylococcus epidermidis*. Other aqueous extracts were resistant to these organisms (Table 1.1). Methanolic and ethanolic extract doesn't show activity (Table 1.1). It was found that *Pseudomonas aeruginosa* were resistant to all the antibiotics present now a day (Table 1.2). But, it was found that aqueous extract of *Calendula officinalis* were sensitive to all clinical isolates from wound infection (Table 1.1). The other plant extract such as *Tridax procumbens* and *Calotropis gigantea* did not show any activity (Table 1.1). The methanolic and ethanolic extract did not show any activity against *Pseudomonas aeruginosa* (Table 1.1).

Table 1.1: Antibacterial Activity of Different Leaf Extracts Against Isolated Organisms from Wound Infection

Sl.No.	Wound Isolates	Diameter of Zone of Inhibition in mm								
		Aqueous			Ethanol			Methanol		
		Tridax procumbens	*Calotropis gigantea*	*Calendula officinalis*	*Tridax procumbens*	*Calotropis gigantea*	*Calendula officinalis*	*Tridax procumbens*	*Calotropis gigantea*	*Calendula officinalis*
1.	*Escherichia coli*	13	10	21	13	13	21	NZ	12	16
2.	*Staphylococcus aureus*	10	12	21	10	10	17	NZ	NZ	17
3.	*Staphylococcus epidermis*	NZ	10	19	NZ	NZ	15	NZ	12	18
4.	*Klebsiella pneuomonia*	NZ	11	18	NZ	NZ	NZ	10	10	17
5.	*Pseudomonas aureuginosa*	NZ	NZ	20	NZ	NZ	12	NZ	NZ	14
6.	*Proteus Vulgaris*	NZ	10	19	NZ	NZ	12	NZ	NZ	NZ
7.	*Proteus Mirabilis*	NZ	10	19	NZ	NZ	14	13	10	18
8.	*Serratia marcescens*	NZ	10	22	NZ	NZ	10	12	13	18

NZ: No Zone.

Table 1.2: Antibacterial Activity of Antibiotics against Isolated Organisms from Wound Infection

Sl.No.	*Wound Isolates*	*Diameter of Zone of Inhibition in mm*			
		Vancomycin	*Tobramycin*	*Streptomycin*	*Amikacin*
1.	*E. coli*	NZ	20 (I)	18 (I)	17 (R)
2.	*S. aureus*	17 (R)	18 (R)	17 (I)	17 (R)
3.	*S. epidermidis*	14 (R)	19 (I)	22 (S)	27 (S)
4.	*Klebsiella pneumonia*	NZ	19 (I)	18 (I)	20 (I)
5.	*Ps. aeruginosa*	NZ	NZ	NZ	NZ
6.	*P. vulgaris*	15 (NR)	22 (I)	22 (S)	21 (I)
7.	*P. mirabilis*	NZ	22 (I)	22 (S)	21 (I)
8.	*Serratia marcescens*	NZ	19 (I)	11 (R)	20 (I)

NZ: No Zone; NR: No Result; I: Intermediate; R: Resistant; S: Sensitive.

However it was similar to finding of M.K. Oladunmoye, Department of Microbiology, School of Sciences, Federal University of Technology, F.M.B 704, Akure, Nigeria who documented on Effects of Ethanolic Extract of *Tridax procumbens* on Swiss Albino Rats Orogastrically Dosed with *Pseudomonas aeruginosa* (NCIB 950).[18] In present study it was found that among all plant extracts, the most inhibitory effect was shown by *Calendula officinalis*. This plant extract are more effective against all clinical isolates from wound infection. One most important thing is that this plant extract shows sensitive (more inhibitory) action against *Pseudomonas aeruginosa*. All present antibiotics do not show any inhibitory activity against this organism. Other aqueous extract of *Tridax procumbens* and *Calotropis gigantea* show poor inhibitory action.[19]

CO Okoli, PA Akah, and AS Okoli (2007) who documented on evaluation of the potentials of *A. africana* in wound care showed that the leaves extract and fractions exhibited haemostatic, antimicrobial and wound healing activities suggesting that the constituents of the leaves may play a useful role in wound care.[20] Usually, microbial contaminations of wounds involve a variety of organisms such as *Pseudomonas aeruginosa, Staphylococcus aureus, Streptococcus faecalis, Escherichia coli, Clostridium perfringens, Clostridium tetani, Coliform bacilli* and *enterococcus. A. africana* leaves, is known to possess anti-inflammatory activity and may contribute to the wound healing activity by suppressing inflammatory reactions invoked by the injured tissues. Further phytochemical studies are required to determine the type of compounds responsible for the antimicrobial effect of theses medicinal plants. In addition, the results support the uses of these plants in traditional medicine for the treatment of infections. These medicinal plants products have great potential as antimicrobial compound and lot of research is going on to these plant products. The herbs and plants are easily available, cheaper and having no side effect. So we can use it directly and its formulation in making products which are effective, cheaper and safe to use.[21]

References

1. Ahmad I, Mehmood Z, Mohammad F. Screening of some Indian medicinal plants for their antimicrobial properties. J Ethnopharmacol 62: 183-193, 1998.
2. Giacometti A, Cirioni O, Schimizzi AM, Prete DM, Barchiesi F, Errico DM, *et al.*,. Epidemiology and microbiology of surgical wound infections. J Clin Microbiol 2000; 38:918-22.
3. Whaley K; Burt AD. Inflammation, healing and repair. In: MacSween RMN, Whaley K. editor. *Muir's Textbook of Pathology.* 13. London: Arnold; 1996. pp. 112–165.
4. Lawrence WT, Diegelmann RF. Growth factors in wound healing. *Clin Dermatol* (1994) 12:: 157–69.
5. Falanga V. The Chronic wound; impaired wound healing and solutions in the context of wound bed preparation. *Blood cells Mole dis* (2004) 32:: 88–94.
6. Ahmed S, Rahman A, Qadiruddin M, Qureshi S. Elemental analysis of Calendula officinalis plant and its probable therapeutic role in health. Pakistan J Sci Indian Res. 2003; 46(4):283–287.
7. Haque MM, Rafiq, Sherajee S, Ahmed Q, Hasan, Mostofa M. Treatment of external wounds by using indigenous medicinal plants and patent drugs in guinea pigs. *J Biol Sci* (2003) 3:: 1126–33.
8. Bowler PG, Duerden BI, Armstrong DG (2001). Wound microbiology and associated approaches to wound management. *Clin. Microbiol. Rev.*, 14: 244 – 269.
9. Nair R, Chanda SV. Antibacterial activity of some medicinal plants of Saurashtra region. J Tissue Res 4: 117-120, 2004.
10. De Boer HJ, Kool A, Broberg A *et al.*, Antifungal Antibacterial activity of some herbal remedies from Tanzania. J Ethnopharmacol 96: 461-469, 2005.
11. Nair R, Kalariya T, Chanda S. Antibacterial activity of some selected Indian medicinal flora. Turk J Biol 29: 41-47, 2005.
12. Chung PY, Chung LY, Ngeow YF *et al.*, Antimicrobial activities of Malaysian plant species. pharm Biol 42: 292-300, 2004.
13. Ananthanarayan R, Panikar CKJ. In text book of microbiology 5th edition.
14. Bauer AW., Kirby WMM., Sherris JC, Twick M. (1996), Antibiotic susceptibility testing by standardized disk method, American Journal of clinical Pathology 45: 493- 495.
15. Taddel A. and RAJ, Rosas, 2000. Bioactivity studies of extracts from *Tridax procumbens Phytomedicine*, 7: 3235-3238.
16. Leach Matthew J, PhD, BN (Hons), ND, RN, MATMS.
17. Kupchan S. Morris, Knox John R, Kelsey John E, and Saenz Renauld JA. Calotropin, a Cytotoxic Principle Isolated from *Asclepias curassavica* L. Science 25 December 1964.

18. Oladunmoye MK. Department of Microbiology, School of Sciences, Federal University of Technology, P.M.B 704, Akure, Nigeria who documented on Effects of Ethanolic Extract of *Tridax procumbens* on Swiss Albino Rats Orogastrically Dosed with *Pseudomonas aeruginosa* (NCIB 950).

19. Saxena VK and Albert S (2005). β-Sitosterol-3-O-β-D-xylopyranoside from the flowers of *Tridax procumbens* Linn. *J Chem Sci* 117:3 263-266.

20. Okoli CO, Akah PA, Nwafor SV, Anisiobi AI, Ibegbunam IN, Erojikwe O. Antiinflammatory activity of hexane leaf extract of *Aspilia africana* C.D. Adams. *J Ethnopharmacol.* 2006

21. Puri HS. (2003) RASAYANA: Ayurvedic Herbs for Longevity and Rejuvenation. Taylor and Francis, London.

Figure 8.1: Plots Between Rate vs Concentration [illegible] Ir-nano [illegible] for the Oxidation of Glycine

Figure 8.2: Plots Between Absorbance and Time for the Effect of Uncatalysed, PVP, Ir(III) Salt and Ir-nano on the Rate of Oxidation of Glycine

PVP plays an important role in the present system. The present colloidal dispersion was stable in air under the protective action of polymers. No precipitation had occurred after storing the dispersions 30 for month. Figure 8.3 shows a small effect of PVP concentration on the rate of oxidation.

2013, Environmental Biotechnology *Pages 11–17*
Editors: **D.R. Khanna, A.K. Chopra, Gagan Matta, Vikas Singh & Rakesh Bhutiani**
Published by: **BIOTECH BOOKS, NEW DELHI**

Chpater 2

Qualitative Determination of Phosphate Solubilization, Seed Germination, HCN and Ammonia Production by Rhizospheric Fungi of Soybean

Manish K. Sharma and D.M. Kumawat
School of studies in Environment Management,
Vikram University, Ujjain, M.P.

Rhizosphere is one of the most active zone of soil which expresses the interaction between microorganisms and plant roots. This study was based on the isolation of rhizospheric fungi and their evaluation for the phosphate solubilization, seed germination, HCN production and ammonia production. A total of 7 fungal isolates were isolated from soybean rhizospheric soil and two isolates of which *Aspergillus niger Aspn1* and *Aspn2* were found suitable for the above mentioned parameters. Results indicate that phosphate solubilization was positive by both fungal isolates on Pikovskaya's medium. Along with this HCN and ammonia production was also found positive, but seed germination was effective in other isolates as well. At this juncture of study it appears that presence of fungal population in the rhizopshere does affect the plant and other microbial growth and should not be excluded from the study when rhizosphere is taken into account. The presence of such important fungus in the rhizosphere may affect the plant growth directly (by solubilizing phosphate, ammonia production and seed germination) or indirectly (by secreting HCN).

Keywords: *Plant growth promoting microorganisms, Pikovskaya's medium, Potato Dextrose agar medium.*

Introduction

The rhizosphere may be defined as that portion of the soil which is adjacent to the root system of a plant and is influenced by the root exudates. The word rhizosphere is made up of two word (G) *Rhizo*= root and *sphere*= surrounding and the term 'Rhizosphere' was proposed by Hiltner (1904). The phenomenon of accumulation of microorganisms around the root zone was reported by a number of earlier workers (Jackson, 1960; Katznelson, 1946; Mukerji and Ranga Rao, 1968; Parkinson, 1957; Pinton *et al.*, 2001; Rouatt, 1959; Sadasivan, 1965; Sorensen *et al.*, 1997; Starkey, 1958). There is a need to learn more about the beneficial rhizosphere microorganisms. Ultimately the work on rhizosphere microbes has to aim at augmenting the biomass production (Yang-Ching *et al.*, 2000). While it is common knowledge that some plants grow only in special soils or in combination with microorganisms and the yield depends on the condition of the soil, nature and diversity of microbes in that area. The fact that the plant may strongly influence the soil in which it grows has not very often been taken into account. Many of these effects may be caused by specific ionic uptake or the exudation of specific organic compounds or other rhizosphere effects (Peters and Long, 1988). The ecological importance of such rhizosphere effects has been emphasized time and again (Mukerji, 2002). Microbes may interact with the same plant simultaneously independently, synergistically and/or antagonistically resulting sometimes in beneficial effect and at other times in harmful consequences and play a crucial role during seedling stage. This preliminary study was aimed to isolate and identify the rhizospheric fungi and their evaluation on the basis of the phosphate solubilization, seed germination, HCN and ammonia production. These parameters are few criteria for plant growth promoting activity and the ultimate goal of sustainable development.

Material and Methods

Isolation and Maintenance of Fungal Isolates

Serial Dilution and Plate Count for Fungi

Serial dilution and plate count method were used for isolation as per Johnson *et al.* (1959). Ten grams of soil sample was taken in the conical flask containing 95 ml distilled water and shaken vigorously to obtain a suspension for preparing the dilutions. Two dilutions *viz.*, 10^{-3} and 10^{-4} were used for fungal isolation. All isolates were maintained on the Potato Dextrose Agar (HiMedia Co.).

Screening of Fungal Isolates

Phosphate Solubilization

Pikovskaya's medium containing tricalcium phosphate was used for phosphate solubilization. Fungal isolates were transferred on plates and were incubated at 27°C for 7 days. Observation was based on the clear zone around the colonies.

Production of Ammonia

All seven fungal isolates were grown in peptone water (Dye 1962) and incubated at 27°C for 7 days. One ml of Nessler's reagent was added in each tube. Faint yellow color indicates small amount of ammonia and deep yellow to brownish colour indicates high production of ammonia.

Seed Germination Test

All the seven isolates were bio-assayed for their ability to promote/inhibit seedling growth using the method as described by Shende *et al.* (1977) with few modifications. Surface sterilization of seeds was done with 0.1 per cent $HgCl_2$ for 3 minutes followed by successive washing with sterilized distilled water. Seeds were placed in sterilized Petri-dishes containing two circles of cellulose filter papers. All the fungal isolates were grown in PDB and liquid suspension was added in the Petri-dishes containing seeds. Control plate was treated with distilled water. All the plates were kept for germination at 30°C for incubation. Observations were recorded as percent germination.

HCN Production

Sterilized PDA supplemented with 4.4 g 100^{-1} ml glycine. Twenty ml of the medium was poured in Petri-dishes. After solidification, spotting of fungal isolates was done. Filter paper circles (Whatman No.1) were soaked in 0.5 per cent picric acid and 2 per cent sodium carbonate. Each filter paper thus soaked was placed in the upper half of each Petri-dish. The Petri-dishes then were sealed using parafilm and incubated at 27°C for 4 days. Uninoculated (control) plate was also kept in triplicate for comparison. All the seven fungal isolates were tested for this test.

Identification and Maintenance of Isolates

Out of the seven isolates, two isolates were used for identification.

Colony Morphology

For this, fungal cultures were grown on the Sabourd-dextrose agar medium for proper incubation period and the colony characteristics were recorded.

Staining and Microscopic Observation of Isolates

After satisfactory growth of fungal isolates on the plates, small amount of growing mycelia was taken from periphery of the growth. Fungal cytoplasm was stained with Lacto-Phenol and Cotton blue. The mycelium content was transferred on the centre of a glass slide having drop of stain. Using two flamed-needles, fungal propagules were gently spread. After placing cover slip, slide was observed under the microscope.

Results

Phosphate Solubilization

Phosphate solubilization was observed on Pikovskaya's medium. All 7 isolates were tested for the zone of clearance around the colonies and the results are shown in

Table 2.1 and Plate 2.1. Out of seven rhizospheric fungi, only two *viz.*, F1 and F2 showed ability to solubilize the phosphate.

Table 2.1: Ammonia, Phosphate Solubilization and HCN Production by Fungal Isolates

Sl.No.	*Isolates*	*Ammonia Production*	*Phosphate Solubilization*	*HCN Production*
1.	F1	Positive	Positive	Positive
2.	F2	Positive	Positive	Negative
3.	F3	Negative	Negative	Negative
4.	F4	Negative	Positive	Negative
5.	F5	Negative	Negative	Negative
6.	F6	Positive	Negative	Negative
7.	F7	Negative	Negative	Negative

Ammonia Production

All seven isolates were tested for the production of Ammonia and results have been presented in Table 2.1 which indicated that only three isolates produced Ammonia.

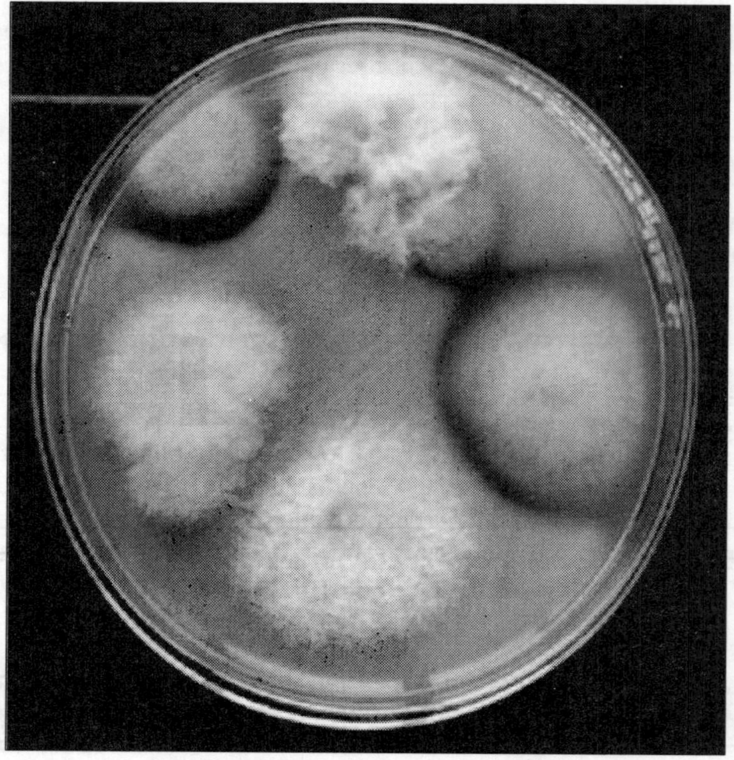

Plate 2.1: Pikovskaya's Medium Showing Zone of Clearance Around the Fungal Isolates

HCN Production

This attribute was analyzed following Bakker and Schipper (1987). Table 2.1 indicates that only F1 isolate was able to produce HCN.

Effect of Fungal Culture Filtrate on Seed Germination

Effect of culture filtrate on seed germination of soybean was tested. Culture filtrate of three fungal isolates *viz.*, F1, F2 and F3 showed 30-40 per cent increase in seed germination while in rest only minor differences over control could be noticed (Table 2.2).

Table 2: Effect of Fungal Isolates on Seed Germination (n=20)

Sl.No.	*Fungal Isolates*	*Germination (per cent)*
1	Control	50
2	F1	85
3	F2	90
4	F3	90
5	F4	65
6	F5	50
7	F6	60
8	F7	50

Identification of Isolates

The results were recorded using laboratory manual of Lilly and Barnet (1962). The following are the characteristics of the two isolates:

Colony Morphology

Colonies on Czapek's agar showed rapid growth with abundant submerged mycelia which were initially more or less white and after full growth turned to black in color. Arial hyphae were usually scanty.

Conidiophore

Conidiophores mostly arose directly from the substratum and were smooth, septate or non-septate, varying greatly in length and diameter from one to several μm. Conidial heads fuscous, blakish brown to carbonous black.

Conidia

Conidia globose, initially smooth but later spinulose with coloring substances, mostly 2.5–4 μm in length but less frequently 5 μm. Globose, superficial sclerotia produced in some strains, but were not common.

Discussion

Above mentioned results are the mere indications that few fungi can accelerate the plant growth directly or indirectly. It was studied that in addition to nitrogen, many soils are also deficient in available phosphorus nutrients and there is need of inoculation of both *Rhizobium* and phosphate solubilizing microorganisms (PSM). Pikovskaya's medium contains tri-calcium phosphate as a rock phosphate. Microbes use this phosphate and convert it into soluble form which appears on the plate as a clear zone. In the same manner these microbes help in solubilization of soil rock phosphate which may be transported to the plants and fulfill the requirement of phosphorus. Out of 7 rhizospheric fungi, only 3 were solubilizing phosphate. It has also been reported that PGPMOs are able to solubilize organic and inorganic phosphorus in soil (Liu *et al.*, 1992). Recently, there is a growing interest in PGPR due to their potential in biological control and growth promotion agents for many crops (Thakuria *et al.*, 2004). Production of HCN has been suggested as a mechanism by which PGPR may inhibit other microbes (Kloepper 1993) and only one isolate (F1) was showed HCN production out of seven fungal isolates. Plant growth promoting microorganisms (PGPMOs) may result either from indirect effects such as the biocontrol of soil-borne diseases through competition for nutrients, siderophore-mediated competition for iron, antibiosis or the induction of resistance in the plant host, or from direct effects such as the production of phytohormones or by providing the host plant with fixed nitrogen or the solubilization of soil phosphorus and iron (Glick 1995, Kinkel *et al.*, 2000, Shishido *et al.*, 1999, Sturz *et al.*, 2000). The present study though is a preliminary one but suggest a great potential for further deep research. Such plant growth promoting fungi (*A.niger Aspn1* and *Aspn2*) not only provide the nutrients in the available form to the plant but can also minimize the incidence of pathogenic microbes, which indirectly can promote the crop growth. Soybean being a major crop of Malwa region needs such study on a top priority basis which will certainly be beneficial for the farmers and can directly affect the economy of the region.

References

Dye DW (1962). The inadequacy of the usual determinative tests for identification of *Xanthomonas* spp. Nzt. Sci. 5: 393-416.

Glick BR (1995). The enhancement of plant growth by free living bacteria. Can. J. Microbiol. 41: 109-117.

Jackson TM (1960). Soil fungi stasis and rhizosphere. In: Parkinson D, Waid JS (eds) The ecology of soil fungi. Liverpool Univ Press, Liverpool, pp 168–176.

Johnson LF, Curl EA, Bond JH, Friburg HA (1959). Microorganism in the plant rhizosphere. In: Johnson LF, Curl LF (Eds.) Methods for Studying Soil Microflora-Plant Disease Relationship. Burgess Publi. Minn. pp: 29-37.

Katznelson H (1946). The Rhizosphere effect of mangles on certain groups of soil microorganisms. Soil Sci 62: 343–354.

Kinkel LL, Wilson M and Lindow SE (2000). Plant species and plant incubation conditions influence variability in epiphytic bacterial population size. Microb. Ecol. 39: 1-11.

Kloepper JW (1993). Plant growth-promoting rhizobacteria as biological control agents. In: Metting FBJ (eds) Soil microbial ecology. Marcel Dekker, New York, pp 255–274.

Liu ST, Lee LY, Tai CY, Hung CH, Chang YS, Wolfram JH, Rogers R, Goldstein AH (1992). Cloning of an *Erwinia herbicola* gene necessary for gluconic acid production and enhanced mineral phosphate solubilization in *Escherichia coli* HB101. J. Bacteriol. 174: 5814-5819.

Mukerji KG (2002). Rhizosphere biology. In: Mukerji KG *et al.* (eds) Techniques in Mycorrhizal studies. Kluwer Academic Publishers, Netherlands, pp 87–101.

Parkinson D (1957). New methods for qualitative and quantitative study of fungi in the rhizosphere. Pedologi Gand 7:146–154.

Peters NK, Long SR (1988). Alfalfa root exudates and compounds which promote or inhibit induction of *Rhizobium meliloti* nodulation genes. Plant Physiol. 88:396–400.

Pinton R, Varanini Z, Nannipieri P (2001). The rhizosphere-biochemistry and organic subtances at the soil-plant interface. Marcel Dekker. New York Basel Rouatt JW (1959) Initiation of Rhizosphere effect. Can J Microbiol 5:67–71.

Ranga Rao V and Mukerji KG (1969). Fungi of Delhi. IX. Additions to our knowledge of soil fungi. J Ind Bot Soc 48:258–261.

Rouatt JW (1959). Initiation of Rhizosphere effect. Can J Microbiol 5:67–71.

Sadasivan TS (1965). Root and its environment. J Sci Ind Res 24:11–113.

Shishido M, Breuil C and Chanway CP (1999). Endophytic colonization of spruce by plant growth-promoting rhizobacteria. *FEMS Microbiol. Ecol.* 29: 191-196.

Sorensen J, van Elsas JD, Trevors JT (1997). The rhizosphere as a habitat for soil microorganisms. In: Wellington EMH (ed) Modern soil microbiology. Marcel Dekker, New York, pp 21–45.

Starkey RL (1958). Inter-relation between microorganisms and plant roots in rhizosphere. Bot Rev 22:154–172.

Sturz AV, Christie BR and Nowak J (2000). Bacterial endophytes: potential role in developing sustainable systems of crop production. Crit. Rev. Plant Sci. 19: 1-30.

Yang-Ching H, Crowley DE, Yang CH (2000). Rhizosphere microbial community structure in relation to root location and Plant iron nutritional status. Appl Environ Microbiol. 66:345–351.

Introduction

2013, Environmental Biotechnology *Pages* **19–27**
Editors: **D.R. Khanna, A.K. Chopra, Gagan Matta, Vikas Singh & Rakesh Bhutiani**
Published by: **BIOTECH BOOKS, NEW DELHI**

Chapter 3

Effect of Molasses Effluent on Yielding Characteristics of Rice (*Oryza sativa*) ADT-43

S. Gershome Daniel Rajadurai and S. Dawood Sharief
School of Environmental Sciences, Department of Zoology,
The New College, Chennai – 14

Water is one of the most important requirements of all living beings for performing essential life functions and is considered as precious natural resources. It is one of the most important commodities, which man has exploited more than any other resources for the substances of his life. But with the rapid growth of industries in the country, pollution of natural water by industrial wastes has increased tremendously leading to an *"ecological disaster"*.

Waste management in the industries has always been a complex problem due to large amount of waste. Wastewater produced by the molasses industries are generally discharged into land or water sources. It contains high amount of organic matter, Biochemical Oxygen Demand (BOD), Chemical Oxygen Demand (COD) and Total free sugars. Molasses is thick syrup a byproduct of sugar cane detained while processing for sugar. It is used in industries like food beverages, lactic acid and acetic acid manufacturing units. For the present study, the effluent samples were collected from lactic acid manufacturing unit which uses molasses as a raw material at SIDCO industrial estate, Thiruvellour, Chennai.

Investigation was carried out to study the biometric yield parameters such as plant height, effective tillers, panicle length and weight, total number of grains in the panicle, number of filled grains and spikelet fertility percentage in Paddy variety ADT-43 at farmers land located near Chennai. Paddy seeds were irrigated with various concentration of untreated and industry treated molasses effluent

(25 per cent, 50 per cent, 75 per cent and 100 per cent). At lower dilutions, the Paddy showed favorable effects on seed germination, plant growth and yield. Among them 100 per cent of untreated effluent showed inhibitory effect on the paddy yield, which may be due to pollutional load in the effluent. The results are discussed in detail, both individually and comparatively, in the light of recent literature

Keywords: *Molasses effluent, Seed germination percentage, Growth parameters, Kernel elongation ratio.*

Introduction

Water is one of the most important requirements of all living beings for performing essential life functions and is considered as precious natural resources. It is one of the most important commodities, which man has exploited more than any other resources for the sustenance of his life. But with the rapid growth of industries in the country, pollution of natural water by industrial waste has increased tremendously leading to an "*ecological disaster*".

Waste management in the industries has always been a complex problem due to large amount of waste. Wastewater produced by the industries are generally discharged into land or water sources. Industrial effluent rich in organic matter and plant nutrients are finding its way in to agricultural use as a cheaper way of disposal. The evaluation of toxicity of these wastes by biological testing is therefore extremely important for screening the suitability of waste for land crop production by toxic pollutants is global agricultural problem (Nagda, 2010). Plants adapt to pollution stress by different mechanisms, including changes in morphological and developmental stress as well as physiological and biochemical process (Zhu, 2001). The evaluation of organic waste toxicity by biological testing is therefore extremely important for screening the suitability of waste for land application (Fuentes *et al.*, 2004).

Molasses is thick syrup byproduct from the processing of the sugarcane or sugar beet in to sugar. The conversion of molasses by the processes of fermentation and distillation is extensively used in the manufacture of methyl, ethyl alcohol and potable spirit. Due to its unique physical and chemical properties, molasses has been traditionally used as a major compound for feeds, live stock feeds and silage additives.

However, wastewater from factories using molasses contains not only high concentration of organic matters, BOD, COD but also containing a large amount of dark brown pigments called melanoidin (Sirianuntapiboon *et al.*, 1988). Commonly molasses effluents are easily biodegrable since it mainly consists of unfermented sugars, soluble starch and volatile fatty acids as well as solids. Organic and inorganic nutrients has significant effect on crop yield (Pathak *et al.*, 1998 and Ramana *et al.*, 2001). Against this back drop, the study that follows was conducted to find out the biometric parameters of Paddy (*Oryza sativa*) ADT- 43 crop against untreated and industry treated molasses effluent.

Material and Methods

The effluent was collected from a lactic acid manufacturing company which uses molasses as a raw material at Kakkalur SIDCO Industrial Estate, Tiruvallur, Chennai. 100 litres of effluents (untreated and industry treated) was collected every-week for field experiments. Physicochemical characters such as colour, odour, pH, BOD, COD TSS, TDS, were studied following standard methods (APHA, 2005). Field experiments were conducted in farmer's field at Ikkadu village, Tiruvallur near Chennai, Tamil Nadu. The experiment setup was laid out in strip plot designed with 9 replications (Figure 2.1).

Figure 2.1: Experimental Field Setup

a) Control plot; b) 25 per cent Industry treated effluent irrigated plot; c) 50 per cent Industry treated effluent irrigated plot; d) 75 per cent Industry treated effluent irrigated plot; e) 100 per cent Industry treated effluent irrigated plot; f) 25 per cent Untreated effluent irrigated plot; g) 50 per cent Untreated effluent irrigated plot; h) 75 per cent Untreated effluent irrigated plot; i) 100 per cent Untreated effluent irrigated plot

Paddy seeds were obtained from Tamil Nadu Govt. agriculture department (TNAD), Tiruvallur. Different concentration of the effluent (untreated and industry treated) were prepared using native lake water (25 per cent, 50 per cent, 75 per cent, 100 per cent) and used for experiments in the field.

The seeds were sowed and irrigated with different concentration of molasses effluent. At the time of maturity, paddy were harvested and different biometric yield parameters such as plant height, effective tillers, panicle length and weight, total number of grains in the panicle, number of filled grains, spikelet fertility percentage were studied both in untreated and industry treated effluent (25 per cent, 50 per cent, 75 per cent, 100 per cent).

Standard Evaluation System for Rice (SES) has been prepared to enable rice scientists from around the world to speak common language on evaluation of rice characters. SES remains most popular method used in mass evaluation in breeding lines (RRI, 2002).

Plant Height

The plant height was measured from soil surface to tip of the tallest panicle (awns excluded), in centimeters (cm)

Spikelet Fertility (SP Fert)

The fertile spikelet's were identified by pressing the spikelets with the fingers and noted those that do not have grains.

Spikelet Fertility Percentage

The spikelet fertility was calculated by the following standard formula

$$= \frac{\text{Total number of filled grains in the panicle}}{\text{Total number of grains present in the panicle}} \times 100$$

Kernel Length Measurement

On the basis of average length, kernels were classified as following

Size	*Length (mm)*
Extra long	> 7.50
Long	6.61–7.50
Medium	5.51–6.50
Short	< 5.50

Kernel Length/Breadth Ratio

Shape	*L/B ratio*
Slender	> 3.0
Medium	2.1–3.0
Bold	1.1–2.0
Round	< 1.0

Results and Discussion

The colour of the untreated effluent was chocolate brown while the industry treated effluent was colourless. The odour of the untreated effluent was burnt sugar smell and industry treated effluent was odourless. pH of untreated effluent ranged from 7.6 - 8.9 and industry treated effluent ranged from 6.4 - 7.9 during the period of study, and these results are within the permissible limit of CPCB norms (5.5- 9.5). BOD level of untreated molasses effluent ranged between 998 - 1414 mg/L and industry treated effluent ranged between 663 - 1026 mg/L which was above the permissible limit prescribed by the CPCB (30mg/L). COD levels of untreated and industry treated effluents ranged between 3536 - 4329 and 3000 to 3990 mg/L respectively. COD level of untreated effluent was very high when compared to industry treated effluent (CPCB permissible limit is 250 mg/L). TSS of untreated and industry treated samples ranged between 481 to 676 mg/L and 367 to 684 mg/L and the TDS of untreated effluent ranged between 163 - 272 mg/L and industry treated effluent ranged between 107 - 210 mg/L.

It is evident from the data (Table 3.1) the plants height was found to reduce significantly due to the usage of untreated and industry treated molasses effluent at different concentration. The plants treated with control water (lake water) the plant height was 46.3 ± 0.12 cm. In case of 25 per cent, 50 per cent, 75 per cent and 100 per cent of untreated molasses effluent the plant height declined to 43.66 ± 0.19, 42.54 ± 0.26, 43.42 ± 0.03 and 38.64 ± 0.38 cm respectively. Different concentration of (25 per cent, 50 per cent, 75 per cent, 100 per cent) industry treated molasses effluent was used for the experimental study, and the height of the plants slightly reduced when compared with the height of control plants *i.e.* 44.48 ± 0.59, 44.68 ± 0.14, 43.44 ± 0.37 and 39.28 ± 0.08 respectively.

The height of the plants varied depending up on the concentration of untreated and industry treated effluents. In lower concentration the height of the plants was more or less equal to control plants (plants treated with nature lake water) but as the concentration increased the height of the plant was slightly reduced. According to Rani *et al.* (1990) diluted molasses effluent increases the growth of shoot length, leaf number per plant and chlorophyll content of leaf. Increased concentration of molasses effluent causes decreased seed germination, seedling growth and it could safely used for irrigation purpose at lower concentration (Rajendran, 1990 and Ramana *et al.*, 2001). This could be due to maximum absorption of NPK by the plants at lower dilutions, which is clearly depicted in our studies, indicating lower concentration of effluents help in promoting growth and higher concentration suppresses the growth.

Number of grains present in the panicle is one of the important components of yielding. The control plants (treated with lake water) showed 78 ± 1.58 numbers of grains per panicle. In plants treated with 25 per cent of untreated molasses effluent the yield (number of grains) were 60.8 ± 1.3 per panicle, in the same way experimental plants treated with 50 per cent and 75 per cent of untreated molasses effluent the yield was recorded as 51.8 ± 1.78 and 47.4 ± 1.14 per panicle respectively. There was no yield (zero number of grains) in the plants which were treated with 100 per cent of untreated molasses effluent. The plants treated with 25 per cent, 50 per cent, 75 per cent and 100 per cent of industry treated molasses effluents, the number of grains (yield) per panicle were decreased as mentioned 67 ± 0.7, 64.6 ± 0.8, 69.2 ± 0.4 and 67.8 ± 1.3 respectively (Table 3.1).

Table 2.1: Biometric and Yield Characters of Rice (*Oryza sativa*) ADT-43 in Different Concentration of Untreated and Industry Treated Molasses Effluent

Sl.No.	Percentage of Effluent	Nature of the Effluent	Plant Height (cm)	Effective Tillers	Panical Length (cm)	Panical Weight (gm)	Total Number of Grains in the Panical	Fleld Grains	Spikelet Fertility Percentage
1.	Control		**46.3** (± 0.12)	**7.2** (±0.44)	**19.14** (± 0.37)	**3.82** (±0.08)	**78** (± 1.58)	**75.5** (± 1.30)	**97.1** (± 1.05)
2.	25%	Untreated	**43.66** (±0.19)	**5** (± 0.70)	**15.16** (± 0.26)	**2.6** (± 0.07)	**60.8** (± 1.30)	**51.8** (± 1.30)	**85.19** (± 0.31)
		Industry Treated	**44.48** (± 0.59)	**5** (± 0.70)	**17.98** (± 0.42)	**3.3** (± 0.15)	**67** (± 0.70)	**64** (± 0.70)	**95.53** (± 1.80)
3.	50%	Untreated	**42.54** (± 0.26)	**4.2** (± 0.44)	**16.14** (± 0.13)	**2.16** (± 0.08)	**51.8** (± 1.78)	**23.8** (± 1.64)	**46.02** (± 3.99)
		Industry Treated	**44.68** (± 0.14)	**4** (± 0.70)	**18.4** (± 0.41)	**3.16** (± 0.15)	**64.6** (± 0.89)	**56.2** (± 0.44)	**87.01** (± 1.58)
4.	75%	Untreated	**43.42** (± 0.33)	**2.8** (± 0.83)	**16.18** (± 0.51)	**1.72** (± 0.08)	**47.4** (± 1.14)	**20.6** (± 1.94)	**43.41** (± 3.25)
		Industry Treated	**43.44** (± 0.37)	**6** (± 0.70)	**17.34** (± 0.31)	**3.14** (± 0.19)	**69.2** (± 0.44)	**58.6** (± 0.89)	**84.68** (± 1.27)
5.	100%	Untreated	**38.64** (± 0.38)	–	–	–	–	–	–
		Industry Treated	**39.28** (± 0.08)	**4.4** (± 0.89)	**16.44** (± 0.13)	**2.84** (± 0.05)	**67.8** (± 1.30)	**54** (± 2.34)	**79.66** (± 3.73)

Each value represents mean of 5 observations.

Values in parenthesis shows the Standard Deviation (S.D) value.

Table 3.2: ANOVA between Biometric and Yield Characters of Rice (*Oryza sativa*) ADT-43

Sl.No.	*Yield Parameters*		*Sum of Squares*	*df*	*Mean Square*	*F*	*Sig.*
1.	Plant height	Between Groups	250.594	8	31.324	367.474 *	.000
		Within Groups	3.069	36	.085		
		Total	253.663	44			
2.	Effective tillers	Between Groups	165.644	8	20.706	47.782 *	.000
		Within Groups	15.600	36	.433		
		Total	181.244	44			
3.	Panicle length	Between Groups	1362.474	8	170.309	1606.691 *	.000
		Within Groups	3.816	36	.106		
		Total	1366.290	44			
4.	Panicle weight	Between Groups	51.554	8	6.444	547.151 *	.000
		Within Groups	.424	36	.012		
		Total	51.978	44			
5.	Spikelet fertility per cent	Between Groups	41942.173	8	5242.772	962.036 *	.000
		Within Groups	196.188	36	5.450		
		Total	42138.361	44			

Table value - 3.07 at 0.01 level or 1 per cent level.

*: Significant.

Spikelet fertility percentage is another important key factor to evaluate the yield of the crops. In this present study when the plants were treated with lake (control) water, the spikelet fertility percentage was 97.1 ± 1.05. Plants treated with 25 per cent of untreated molasses effluent, spikelet fertility percentage were reduced to 85.19 ± 0.31. 50 per cent and 75 per cent of untreated molasses effluents when irrigated with the experimental plants they showed the spikelet fertility percentage as 46.02 ± 3.99 and 43.41 ± 3.25. But the plants treated with 100 per cent of untreated molasses effluent the spikelet fertility was zero (No yield). In the same way plants treated with different concentration of (25 per cent, 50 per cent, 75 per cent and 100 per cent) industry treated molasses effluent the spikelet fertility percentage shows 95.53 ± 1.80, 87 ± 1.58, 84.68 ± 1.27, and 79.66 ± 3.73 respectively during the period of the field study. It may be noted that lower concentration of molasses effluent is found to yield better number of rice growth than the higher concentration treated.

It is well documented that, nitrogen and phosphorus play important roles in regulating plant growth and development (Wang *et al.*, 2005; Yan *et al.*, 2005). According to Hu and Wang (2004), potassium fertilizer could improve rice nutrient-absorbing capacity from soil and promote transfer of nitrogen form leaf and straw to panicle. As a physiological response to potassium, the number of filled spikelet and grain weight increased, so grain yield increased.

The present study revealed that out at 8 concentrations (25 per cent, 50 per cent, 75 per cent and 100 per cent of both untreated and industry treated) analyzed, two concentrations (25 per cent and 50 per cent) of industry treated effluent showed desirable yield. These concentrations of industry treated molasses effluent could be effectively utilized in rice improvement programmes which could be helpful to reduce the pollutional load of the environment as well as reduce the food scarcity and save groundwater directly and saving electricity indirectly.

References

APHA.2005. Standard methods for the examination of water and wastewater American Public Health Association, Washington, D.C.

Hu H, and Wang GH (2004). Nutrient Uptake and Use Efficiency of Irrigated Rice in Response to N uptake and groundwater in taihu region. Acta Pedol. Sin., 42: 440-446 (in Chinese).

Nagda, G.K., Diwan, A.M. and Ghole, V.S. 2006. Seed germination bioassays to assess toxicity of molasses fermentation based bulk during industry effluent. *European J. Environment and Agriculture food chemistry*, 5(6): 1598-1603.

Patnaik, H.K., Garg, S.S., Pradeep Sing and Sudhirdixit. 1998. A review on design consideration, constriction, operation and application of Root Zone Treatment System in tropical and sub-tropical countries. *J. Industrial pollution control* 18 (2): 223-230.

Rajendran, K. 1990. Effect of distillery effluent on the seed germination, seedling growth, chlorophyll content and mitosis in *Helianthus annus* – Indian botanical contactor, 7: 139-144.

Ramana, S., Biswas, A.K., Kundu, S., Saha, J.K. and Yadav, R.B.R. 2001. Effect on distillery effluent on seed germination in some vegetable crops. Bio-resource Tech., 82, 3: 273-275.

Rani, R. and Sri Vastava, M.M. 1990. Eco physical response of *Pisum sativa* and *Citrus maxima* to distillery effluents. *Int.J. Eco. Environ.Sci.*, 16- 23.

Wang, J.H., Liu, J.S and Li YF (2005). The effect of nitrogenous and phosphorus fertilizers on accumulation and movement of nitrogen in shallow layer of dark soil. Chin. J. Soil Sci., 36: 341-344 (in Chinese with English abstract).

Yan, D.Z., Wang, D.J and Lin, J.H (2005). Effects of fertilizer-N application rate on soil N supply, rice N uptake and groundwater in Taihu region. Acta Pedol. Sin., 42: 440-446.

Zhu, Y.L., Shen, Q.R and Duan, Y.H (2004). Physiological effects of different nitrogen forms on rice. Nanjing Agr. Univ. J. Nat. Sci., 27: 130-135 (in Chinese with English abstract).

2013, Environmental Biotechnology *Pages* **29–40**
Editors: **D.R. Khanna, A.K. Chopra, Gagan Matta, Vikas Singh & Rakesh Bhutiani**
Published by: **BIOTECH BOOKS, NEW DELHI**

Chapter 4

Impact of Environmental Factors on Population Dynamics of Pests of Safflower, *Carthamus tinctorius*

C.K. Deshmukh and S.M. Sardar
P.G. Department of Zoology, S.G.B. Amravati University, Amravati – 444 602, M.S.

The aphid, *Uroleucon carthami* and the borer, *Helicoverpa armigera* recorded after seven days of interval during 2006- 07 and 2007-08.The polyphagous semi looper larval stages of *Achaea janata* were also found while remaining pests were not found but good number of lady bird beetles was found along with it. *U. carthami* was found prevalently than that of *H.armigera.* The population of *U. carthami* and *H. armigera* was at peak when temperature and humidity were 31°C, RH -56 per cent in 2006-07 for both and 26.7°C, RH -71 per cent and 29.5°C, RH-80 per cent in 2007-08 respectively. The temperature, relative humidity, stage of the plant and rain fall affects their population.

Thus, it is concluded that environmental abiotic factors affect their population of both the pests. The environment was not suitable for the incidence of the remaining pests as they are not found during the study. The late sowing showed late incidence of both the pests but complete their life cycle in due period. Thus the incidence of both the pest is at different stages of the plant and at different time with respect to their abiotic factors. The variation in the abiotic factors causes change in the incidence of the pests and temperature plays key role in the multiplication of their population.

Keywords: *Aphid Uroleucon carthami, Helicoverpa armigera, Abiotic factors.*

Introduction

The incidence of insect pests considerably reduces both the yield and quality of production. The incidence and development of all the insect pests are much dependent upon the prevailing environmental factors such as temperature, relative humidity and precipitation (Aheer *et al.*, 1994 and Nath *et al.*, 2000). Insects are cold blooded therefore; temperature is probably the single most important environmental factor influencing insect behavior, distribution, development, survival, and reproduction. Insect life stage predictions are most often calculated by using accumulated degree days from a base temperature and biofix point (Bale *et al.*, 2002). One of the oilseed rabi crops Safflower, *Carthamus tinctorius* grown all over India.India ranks first amongst the safflower growing countries of the world covering about 75 per cent of the world's acreage. Maharashtra and Karnataka are the major states accounting for nearly 90 per cent of the area and production of safflower (Anonymous, 2004–2005, Singhal, 1999). Safflower is used for dyes, contains 78 per cent linolic acid and besides it has many medicinal properties (Weiss, 1971; Wang Guimiao and Li Yili, 1995).),

The reason for low productivity of safflower is due to biotic and abiotic stresses during crop growth period. Safflower crop is known to be infested by 22 insect's pests. Aphids Uroleucon (*Dactynotus*) *carthami H.R.L.* locally called Mava, infestation threatens the cultivation of the crop (Deshmukh *et al.*, 1981). Losses ranging from 15 to 80 per cent have been reported to be infected by this pest in different parts of the country. Indirectly photosynthesis is also adversely affected by this pest (Reddy, 1968, Rathore, 1983, Thontadarya and Hiremath, 1983) and further important because they act as vector for 50 different virus diseases of plants (Blackman and Eastop, 2000). Both apterous and alate forms are presents throughout the season. This pest is active from second fortnight of November to February and causing 35 per cent crop losses (Rathore and Pathak, 1983). The capsules fly *Acanthiophilus helianthi* (Rossi) is also considered to be important insect pests of safflower and others are semi looper, *Plusia orichalcea* (Hb) and Bihar hairy caterpillar, *Diaorisia oblique* Walker (Chavan, 1961).

Helicoverpa armigera is considered as a national key pest due to it's damaging potential and wide range of host plants including safflower (Rai, 1976, Bilapte, *et al.*, 1980). Another pest of safflower is caterpillar of *Perigea capensis* (Guinnee) which is a major defoliating pest throughout India (Singh *et al.*, 1994; Singh *et al.*, 1999). Enormous yield losses of 62.6 to 100 per cent have been encountered due to excessive foliage feeding by a large number of larvae (Sekhar and Rai, 1989). Weather conditions play the most favorable role for rapid multiplication of aphids (Sinha *et al.*, 1989; Rana *et al.*, 1993; Aheer, *et al.*, 1994; Singh and Malik, 1998).

The present work has therefore been focused on to find out the impact of environment factors and incidence of these four pests on safflower a) *Acanthiophilus helianthi* (Diptera: Tephritidae) b) *Perigea capensis* (Lepidoptera: Noctuidae) c) *Helicovera armigera* (Lepidoptera: Noctuidae) and d) *Uroleucon carthami* (H.R.L.) (Hemiptera: Aphididae)

Materials and Methods

A healthy safflower plot having medium black cotton soil was selected for the study during rabi season 2006-2007 and 2007- 2008 in the Oil Seed Department of Dr. Panjabrao Deshmukh Krishi Vidyapeeth College of Agriculture, Akola (M.S.).The aphid infections counted per ten number of plants per shoot in which the distance is 5cm on the shoot of each plant and also per twig and per leaves on upper and lower surfaces. The larvae of *H. armigera* counted on per plant in marked plot. The intensity and the population dynamics were calculated.

Method for Study of Population Dynamics of Aphid, *Uroleucon carthami* and Others

Bhima (unspiny) variety of safflower seed was sown by seed drilling with the help of tiffan (wooden drill) method on 10 th October and 2nd November 2006, 2007 respectively, in the Randomized Block Designing plot, are in three replications of a plot size: 3.30 x 2.60 = 8.5800 m^2. The row to row and plant to plant distance was 45 cm, 15 cm respectively. Inter replication spacing was 1.0 m, total 30 plots observed in which 6 rows were per plot. Total plants per plots were 132. The data on weather parameters *viz.*, maximum, minimum temperature, relative humidity and rain fall were collected from the weather department, P.K.V. Akola. Recommended agronomic practices like fertilizer application, weeding and irrigation etc. were taken up. The crop was raised under unprotected conditions. Data pertaining to the age of the crop and pests and disease incidence was taken at weekly interval till the harvest of the crop.

Statistical Analysis

The data analyzed statistically with weather parameters using factor analysis by Kaiser Meyer Olkin and Bartle Test, correlation coefficient value is determined by Principal component analysis (factor analysis) with the help of SPSS software to work out the significance relationship between weather parameters and pest incidence. Mean and ±S.D values calculated. Integer figure shows only adult forms in which unwinged and winged forms are involved each reading is the mean of observations of 10 plants per shoot in which distance is 5 cm on the shoot of each plant (Tables 4.1 and 4.2).

Collection Method for Aphid

The aphids feeding on plants were carefully and slowly touched with the hair point of brush as a result the disturbed aphids pulled out their beaks from plant tissue and distributed on plant surface. This was further collected with the wet tip of the brush and were preserved in 70 per cent alcohol in insect bottle (Bhat, *et al.*, 1989) later identified with key character given by Blackman and Eastop (2000) and reported as *Uroleucon carthami.* The larvae of *H. armigera* were also collected for identification.

Result

The Study of Population Dynamics of Aphid *Uroleucon carthami* during Season 2006-2007 and 2007-2008 (Tables 4.1 and 4.2 and Figure 4.1–4.3)

The first incidence of *U. carthami* was recorded in last week of November, 2006 on very tender plants but soon after seven days interval the population was reached by more than five times to that of the first appearance at that time the maximum and minimum temperature and relative humidity were 29.3° and 10.7°C, 64 per cent respectively. The population started to build up and reached to the peak level in second week of January and second week of February. In the first week of February the maximum number was 96.50 aphids found per 5cm length of shoot. At that time the maximum and minimum temperature was 31°C-11.1°C respectively. The relative humidity was 56 per cent. The aphid population was 30.15 more than that of initial population. It was the highest level population in the season. At this stage the plant was young and in the flower condition. The aphid population falls down due to the rain fall in the third week of February later it declined in the first week of March.

Table 4.1: The Population Dynamics of *U. carthami* and *H. armigera* on per Ten Number of Plants on Safflower with Respect to Meteorological Data

Sl.No.	*Date of Observation*	*Number of Pests Population*		*Meteorological Data*			
		Aphid	*Shoot borer*	*Temperature °C*		*RH%*	*Rainfall mm*
				Maximum	*Minimum*	*Morning*	
1.	27.11.2006	3.2	0	29.3	10.7	64	0.0
2.	4.12.2006	16.95	1	30.2	10.2	63	0.0
3.	11.12.2006	29.55	2	29.2	11.2	75	0.0
4.	18.12.2006	35.15	2	27.6	8.0	73	0.0
5.	25.12.2006	43.80	2	28.2	8.6	67	0.0
6.	01.01.2007	68.05	3	29.6	11.1	73	0.0
7.	08.01.2007	79.50	4	28.4	9.6	71	0.0
8.	15.01.2007	90.20	4	30.4	11.2	67	0.0
9.	22.01.2007	84.20	5	31.2	13.7	65	0.0
10.	29.01.2007	83.20	6	30.8	11.9	65	0.0
11.	05.02.2007	96.50	6	31.0	11.0	56	0.0
12.	12.02.2007	94.40	4	30..0	12.7	56	0.0
13.	19.02.2007	63.80	2	30.0	14.0	53	0.0
14.	26.02.2007	21.35	1	29.4	13.3	73	25.2
15.	05.03.2007	4.93	0	32.9	15.3	56	0.0
	Mean	54.32	2.8	29.54	11.5	65.13	1.68
	S.D.	±33.74	±1.97	±1.94	±2.03	±7.21	±6.51

Mean and ± S.D are the total 15 number readings.

Table 4.2: The Population Dynamics of *U. carthami* and *H. armigera* on per Plants of Safflower with Respect to Meteorological Data

Sl.No.	Date of Observation	Meteorological Data				Number of Aphids for 5 cm per Plant					No. of Larvae Observed of Shoot Borer		
		Temperature °C											
		Max	Min.	RH per cent	Rainfall mm	Twig	Flower	Upper	Lower	Total	Flower	Shoot	Total
1	16.12.2007	30.2	17	72	0.0	3	2	1	1	7	0	1	1
2	23.12.2007	30.9	17.1	78	0.0	4	2	1	2	9	0	2	2
3	30.12.2007	30.3	15.1	69	0.6	4	3	2	1	10	1	3	4
4	06.1.2008	29.1	15.5	71	0.0	5	4	1	2	12	2	2	4
5	13.12008	28.1	11.0	68	0.0	11	2	5	3	21	2	2	4
6	20.1.2008	26.8	12.1	70	0.0	12	8	6	3	27	3	2	5
7	27.1.2008	27.6	9.2	66	0.0	12	8	4	5	31	3	3	6
8	03.2.2008	26.7	8.2	71	0.0	20	12	14	8	54	3	4	7
9	10.2.2008	25.8	8.2	75	20.0	5	4	4	3	16	4	3	7
10	17.2.2008	28.8	11.8	83	0.0	4	3	3	2	12	5	4	9
11	24.2.2008	29.5	14.0	80	0.0	4	3	2	2	11	7	6	13
12	02.3.2008	25.0	11.4	80	0.0	4	3	1	2	10	6	5	11
13	09.3.2008	30.8	15.2	86	0.0	2	1	1	2	6	4	2	6
14	16.3.2008	28.8	13.0	75	0.0	1	–	–	–	2	1	–	1
15	23.3.2008	29.3	12.2	77	0.0	–	–	–	–	–	–	–	–
	Mean	28.51	12.73	74.73	1.37	6.50	4.23	3.46	2.64	16.29	2.93	3.00	5.71
	S.D.	±1.81	±2.91	±5.85	±5.16	±5.27	±3.17	±3.60	±1.86	±13.50	±2.13	±1.41	±3.54

Each value is the mean of 5 observations. Mean and ±S.D are the total 15 number readings.

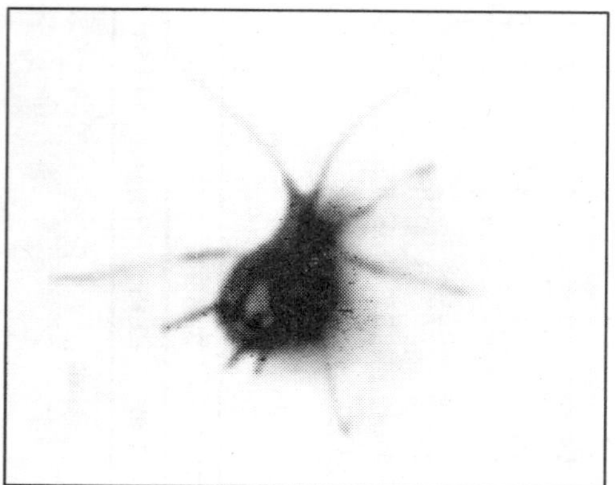

Figure 4.1: Adult Aphid *U. carthami*, Winged Form

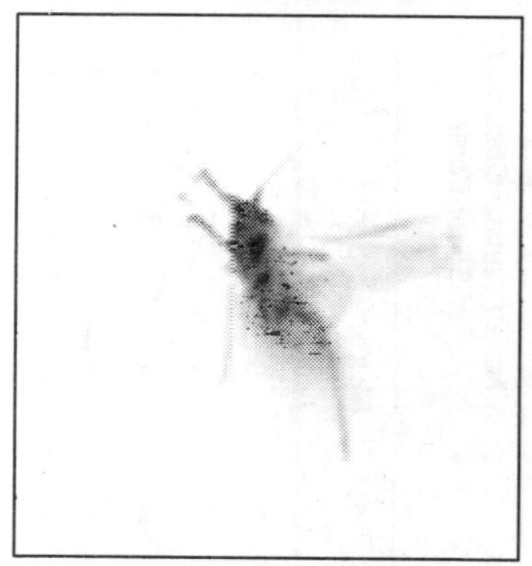

Figure 4.2: Winged Form

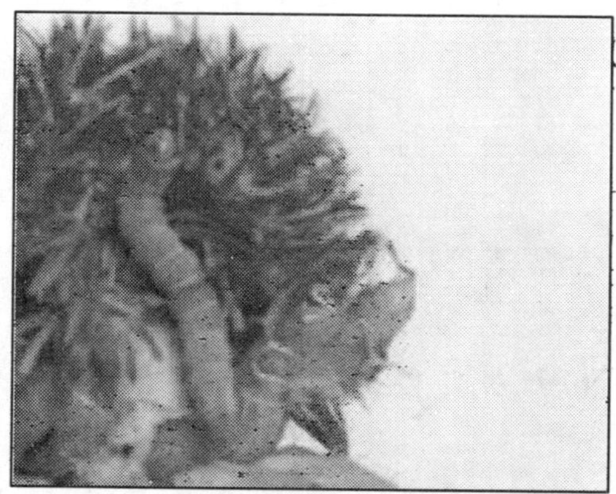

Figure 4.3: Sowing Larva of *H. armigera* on Flower Head

Figure 4.4: Heavy Incidence of *U. carthami* on Plant

In the second week of December, 2007 the first incidence of *U.carthami* was recorded when maximum and minimum temperature were 30.2°C and 17°C respectively. The relative humidity was 72 per cent. The pest incidence was gradually increased in the following two weeks of December when maximum and minimum temperature and relative humidity were 28.10C, 11°C and 68 per cent respectively. In the third week of January, the pest population was increased. At that time the maximum and minimum temperature were 27.6°C and 9.2°C respectively. Along with the relative humidity was recorded 66 per cent. When the population was counted per plant of parts of safflower, found more on twigs than flower and more on upper than lower surfaces of leaves. In the first week of February, the population was not only highest on the twig but also on all parts of the plants. Therefore the total population was highest also *i.e.* 54 in number on per plant at 5cm distance during the season. The population was 6.6, 6, 14 and 8 times more than that of initial population on twig, flower, upper and lower surfaces of leaves respectively. At that time, the maximum and minimum temperature and relative humidity were 26.7°C, 8.2°C71 per cent respectively.

In the second week of February, due to the rain fall, the population suddenly dropped. Later population was still remained decreased up to the end of February and then declined gradually in the month of March. The population of aphid was nil in the fourth week of March when the maximum and minimum temperature and relative humidity were 29.3°C, 12.2°C and 77 per cent respectively.

Population Dynamics of the Stem Borer,. *Helicoverpa armigera* during the Season 2006-2007 and 2007-2008 (Tables 4.1 and 4.2, Figure 4.4)

The first incidence of larvae of *H. armigera* was in the first week of December2006-2007 on the tender shoot of the plant when the maximum and minimum temperature and relative humidity were 30. 2°C, 10.2°C and 63 per cent respectively. In the last week of January the population was highest at that time the maximum and minimum temperature and relative humidity were 30.8°C, 11.9°C and 65 per cent respectively. The last incidence was observed in the last week of February when the maximum and minimum temperature and relative humidity were 29.4°C, 13.3°C and 73 per cent respectively.

In 2007-2008 the first incidence of the larvae *H. armigera* appeared on the shoot of the plant on the second week of December when the maximum and minimum temperature were 30.2°C, 17°C respectively. The relative humidity was recorded 72 per cent. At that time the plant was of medium height. In the last week of December the larval incidence was recorded on flower and shoot when maximum and minimum temperature were 30.3°C,15.1°C respectively the relative humidity was 69 per cent.

In the fourth week of February, the infestation was at the peak on both the flower and shoot. It was 6 and 7 times more than that initial level when temperature was 4 29.5°C, 14°C respectively. The relative humidity was 80 per cent. In the first week of March, the infestation was increased and later on it was decreased in second week and third week gradually. When the maximum and minimum temperature were 30°C, 28. 2°C, 15.2°C-13°C and relative humidity was 86,75.5 per cent respectively.

The infestation of larvae was nil when there was increased in temperature *i.e.* Maximum and minimum temperature were 29.3°C 12°C respectively. The relative humidity was 77 per cent.

Comparative Study of Pests

A pilot survey was carried out in the year 2004 - 2005 all the four proposed pests were found in the field but when the study was conducted during the season 2006-2007, 2007-2008 in the field the aphid *U. carthami* was observed predominantly on twigs, leaves and flower of safflower. The caterpillars of *H. armigera* were present on shoot and flower of the plant but appearance of *A. helianthi* and *P. capensis* were not observed. On the leaf the lady bird beetles were found in 1-2 numbers per plant. In 2006 -2007 the larvae of semilooper, *Achea janata* were also observed on leaves.

Discussion (Table 4.3)

Acanthiophilus helianthi, Perigia capensis, Helicoverpa armigera and *Uroleucon carthemi* are the serious pests of safflower (Ricci and Ciriciofolo, 1983; Sekhar and Rai, 1991; Akashe *et al.*, 1997). Plant protection schedule are generally formulated for the convenience of the farmers who are not aware of the pests and their occurrence schedule, incidences of pests, the stages of crops but they are slightly aware about climatic conditions.

Table 4.3: Correlation Matrix for Both Pests

	Twig	*Flower*	*Upper*	*Lower*	*Total*	*Flowers*	*Shoot*	*Totals*
Maximum	–0.470	–0.522	–0.443	–0.480	–0.489	–0.446	–0.367	–0.421
Minimum	–0.635*	–0.635*	–0.724**	–0.711**	–0.678**	–0.420	–0.378	–0.403
Humidity	–0.563*	–0.493	–0.407	–0.343	–0.486	0.506	0.347	0.433
Rainfall	–0.086	–0.025	0.041	0.048	–0.010	0.137	0.000	0.101

*: Significant at 5 per cent level; **: Significant at 1 per cent level.

Raja Rau (1954) observed that wet and drizzly weather was necessary for the rapid multiplication of aphids. At the repining stage of the plants, *i.e.* the seed formation stage, the aphids either perished or migrated to younger plants. Bindra and Rathore, (1967) studied the population dynamics of aphids. According to them that the aphids were active from early December to the end of April while Lanjewar (1976) reported that the aphids develop rapidly before the plants develop flower buds and sudden decreased in summer when the temperature was maximum 37°C and minimum at 18.7°C with relative humidity 61 per cent during the first week of April.

In the present study the aphids *U. carthemi* and the stem borer *H armigera* were active November to February and decreased in the first and second week of March and when maximum and minimum temperature were 32.9°C and 15.3 °C, 25.0°C, 11.4°C with relative humidity 56 per cent and80 per cent during the season 2007 and in 2008 respectively. The present observations support the earlier workers views that rise in the temperature decrease the population (Rathore, 1983, Lanjewar, 1967;

Rathore and Pathak, 1983, Sarode, 1999; Jallow and Matsumura, 2001; Rochesteret *et al.*, 2002 and Jalali *et al.*, 2009). The sowing of safflower usually done in early October and intends late sowing up to early November. So in the months of December and January generally the plants become two months old with respective sowing.

Upadhya *et al.* (1980) observed that the aphid was first appeared in the first week of December but in the present study the first incidence of aphid *U. carthami* were observed in the last week of November in 2006 second week of December in 2007 but stem borer *H. armigera* appeared in first and second week of December 2006 and 2007respectively. When maximum and minimum temperature 29.3°C and 10.7°C with respective humidity 64 per cent while in 2007-2008 season, In the present study the aphid and stem borer incidence were delayed in the field due to delay in sowing of safflower (Aheer *et al.*, 1993, Sarode, 1999; Shuaib, *et al.*, 2008, Saleha, *et al.*, 2009). Thus, incidence of the pests must be related to the age of the plant and weather factor (Aheer *et al.*, 2007). During the second week of February, the plant was in young stage and the population was at the peak level. At that time, usually the seed formation was completed in plant. Late planting of safflower makes the crop prone to pest infestation. However, the sowing in October carried lower incidence of capsule fly (Singh *et al.*, 2007). Besides these, other factors like availability of enough green foods in the forms of tender leaves and succulent shoots also influence the growth and multiplication of the population so the age, condition and variety of the plants and have direct relation with the growth and development of pests a populations (Atwal *et al.*, 1971 and Sinha *et al.*, 1989).

The diagonal correlation is found to all parameters but the correlation coefficient value -0.678 shows(Table 4.3) that minimum temperature and population of *U. cartaemi* (Total) are inversely related which implies that the increase in minimum temperature, the population of *U. carthami* was significantly reduced at 5 per cent and 1 per cent level of significance. Sing and Singh (1994) concluded that abiotic factors shares 77.69 per cent impact on aphid population while in the present study the abiotic factors like maximum temperature, rainfall and humidity does not effect on the population of *U. carthami* except relative humidity on twig. In the present study larvae of *H.armigera* do not significantly correlated with the components (Table 4.3). Thus, present observations concluded that the change in intensity in population of the pests directly depends up on the climatic conditions which varies from region to region agree with the reports of Singh and Malik (1998) and Ansari (2007) that the temperature is significantly conducive for aphid multiplication.

Ackowledgement

We are grateful Dr. R.K. Rathod, Department of Oilseed, Dr. Panjabrao Krishi Vidyapith, Agriculture College, Akola for providing facilities and also thankful of Dr. Sushil Agrekar, Statistical Institute And Research Lab, Amravati.

References

Aheer, G.M., I. Haq, M. Ulfat, K.J. Ahmad and A. Ali. (1993). Effects of sowing dates on aphids and grain yield in wheat. *J. Agric., Res.*, 31(1): 75-79.

Aheer, G.M., K.J. Ahmed and A. Ali (1994) Role of weather in fluctuating aphid density in wheat crop. J. Agric. Res., 32: 295–301.

Aheer, G.M., K.J. Ahmad and A. Ali (2007). Impact of weather factors on population of wheat aphids at Mandi Baha-ud-din District. *J. Agric. Res.*, 45(1): 61-68.

Akashe, V., S. Mehetre, B. Coli and D. Veer (1997) Biology of *Helicoverpa armigera* (Hb) on safflower. The P. K. V. Research Journal 21: 230- 231.

Anonymous (2004–2005.) Annual Progress Report. Safflower. Directorate of Oilseeds Research, Rajendranagar,Hyderabad-500 030, India, 181 pp.

Ansari, M.S. Barkat Hussain and N.A. Qazi (2007) Influence of Abiotic Environment on the Population Dynamics of Mustard Aphid, *Lipaphis erysimi* (Kalt.) On *Brassica* Germplasm Journal of Biological Sciences Volume: (7): 6 Pp: 993- 996

Atwal, A. S., Chaudhary J. P. and Ramzan M. (1971). Mortality factor on the natural population of cabbage aphid, *Lipaphis erysimi* (Kalt.) (Aphididae: Homoptera:Hemiptera) in relation to parasites, predators and weather conditions. Ind. J. Agric. Sci., 41: 507-510.

Bale, J. S. G. J. Masters, I. D. Hodkinson, C. Awmack, T. M. Bezemer, V. K. Brown, J Butterfield, A. Buse, J.C. Coulson, J. Farrar, J. E. G. Good, R. Harrington, S. Hartley, T. H. Jones. R. L. Lindroth, M. C. Press, I. Symrnioudis, A. D. Watt, and J. B. Whittaker. (2002.) Herbivory in global climate change research: direct effects of rising temperatures on insect herbivores. Global Change Biology 8:1-16.

Bhat, N. S. M. Manjunath and G.T.T. Raju (1989) Sampling of *Uroleucon compositiae* Theobald (Homoptera: Aphidae) on safflower IMARC.6(1)175-176

Bilapate, C.G.; A.K. Raodeo and V.M. Pawar(1980). Investigations on *Helothis armijera* (Hubner) in Marathwada –II. Entomology 5(4):319-326)

Bindra O. S. and Y.S. Rathore (1967) Investigation on the biology and control of safflower aphid at Jabalpur. JN KVV Res. J. 1:60-63

Blackman R.L., Eastop V.F. (2000). Aphids on the World's Crops: An Identification and Information Guide. John Wiley and Sons Ltd.: Chichester.

Chavan, V.M. (1961) Niger and safflower.Indian Central oilseed Committee Publ; Hyrdabad, India pp57-150.

Deshmukh, S.D. M.D. Dakhare and M.N. Borle (1981) Chemical control of aphid on safflower Vidharbha Pestology,11(6):9-10

Jalali,M.A. Luc Tirry, Abbas Arbab,Patrick De Clercq. (2009) Temperature dependent development of the two spotted lady beetle *Adalia bipunctata* on the green peach aphid, *Myzus persicae* and facctitious food under constant temperature,Journal of Insect Science (10)24,1-10

Jallow and M. Matsumura (2001) Influence of temperature on the rate of development of *Helicoverpa armijera* (Hubner)(Lepidoptera: Noctuidae). Applied Entomology and Zoology 36: 427-430.

Lanjewar, A. K. (1976) Bionomics and control of safflower aphid *Dactynotus carthemi* H.R. L. unpublished M. Sc. Thesis P. K. V. Akola.

Nath, P., O.P. Chaudhary, P.D. Sharma and H.D. Kaushik,(2000) The studies on the incidence of important insect pests of cotton with special reference to *Gossypium arborium* (Desi) cotton. Indian J. Entomol., 62: 391

Rana, J. S., Khokar K. S., H. Singh and Sucheta,(1993). Influence of abiotic environment on the population dynamics mustard aphid. *Lipaphis erysimi* (kalt.) Crop. Res., 6:116- 119.

Raja Rau (1954) Bionomics and life history of aphids sp. A parasite on *Aphis gossyil.* Glov.. on Brinjal. Indian J. Ent. 16(4): 362-371.

Rai, B. K. (1976) Pests of Oil seed crops in India and their control. Published by Indian Council of Agricultural Research, New Delhi, pp89-97.

Rathore, V.S. (1983) Bionomics and management of safflower Aphid Ind. Farming 33 (8):32-33.

Rathore, V.S. and S.C. Pathak, (1983) Seasonal Incidence of safflower aphid *Uroleucon carthemi* Theobald at Jabalpur.Indian Journal.Plant Prot. IX (2): 162-168.

Reddy, D.B. (1968) Plant Protection In India, Allied Publishers Bombay pp213-215.

Ricci, C. and E. Circiofolo (1983) Observations on *Acanthophilus helianthi* Rossi (dipteral: Tephritidae) injurious to safflower in central Italy. Redida 66: 577-592.

Rochesteret, W. M: Zaluci A. W.; M. Miles and D. Murry (2002) Testing insects movement theory empirical analysis of pest data routinely collected from agricultural crops. Computers and Electronics in Agriculture. 35:139-149.

Sarode, S. V. (1999) Sustainable management of *Helicoverpa armigera* (Hub). Proceeding of Second Asia-pacific crop Protection Conference. Pestology.22(2):pp279-283.

Saleha, S., Siddiqui, M. Naeem and Nadeem A. Abbasi (2009). Effect of planting dates on aphids and their natural enemies in cauliflower varieties Pak. *J. Bot.*, 41(6): 3253-3259.

Sekhar, P. R.; Rai, P.S. (1989) Incidence of different caterpillars on safflower and assessment of grain loss due to *Prospalta conducta* Walker (*Lepidoptera:Noctuidae*) in Karnataka, India. Tropical Pest Management 37 (1)

Singh, D. and H. Singh, (1994.) Correlation coefficient between abiotic biotic factors (Predator and parasitoid) and mustard aphid *Lipaphis erysimi* (kalt.) population on rapeseed-mustard. J. Aphidol., 8:102-104

Singh, S. V. and Malik Y. P. (1998). Population dynamics and economic threshold of *Lipaphis erysimi* (Katenbach) on mustard. Ind. J. Ent., 60: 43-49.

Singh, H. Ahlawat, D.S. Rohilla,H.R. (1994) Perspective of Insects Pests Management. G. S. Dhaliwal and Arora common Wealth Publishers, New Delhi pp 265-306.

Singh,V.; Prasad Y.G.,Laxminarayan,M. (1999) Insects pests of safflower and their management In:IPM system in Agriculture vol(5) Oil seeds. Aditya Books Pvt. Ltd. New Delhi, pp 553.

Singh, R., Diwan Singh and V.U.M. Rao (2007) Effect of abiotic factors on mustard aphid (*Lipaphis erysimi* Kaly) on Indian Brassica Indian J. Agric. Res., 41 (1): 67 - 70.

Singhal, V. (1999) Indian Agriculture, Indian Economic data Research centre, New, Delhi. Pp 300-303.

Shuaib, M., Khan, S. H., Mulghani, N. A (2008.) Effect of temperature and relative humidity on population dynamics of sucking insect pests of cotton (*Gossypium hirsutu*L. Endure International Conference Diversifying crop protection, 12-15October La Grande-Motte, France

Sinha, R. P., Yazdani S. S. and Verma G. D. (1989) Population dynamics of mustard aphid, *Lipaphis erysimi* (kalt.) in relation to ecological parameters. Ind. J. Ent., 51:334-339.

Thontadarya, T. S. and I.G. Hiremath, (1983) Losses due to insect's pests in Karnataka crop Special Issue Indian J. Ent. 1:43-58.

Upadhya, V. R.; C.L. Kaul and G. M. Jalati (1983) Seasonal incidence of the aphid *Dactynotus carthemi* and coccinellids in relation to weather conditions. Indian J. Plant Prot. 8(2): 117-121.

Wang Guimiao and Li Yili, (1995) Clinical application of safflower Carthamus *tinctorius* Zhejiang Traditional Chinese Medical Science.Journal 20(1): 42-43.

Weiss, E. A. (1971) Castor sesame and safflower Barnes and Noble Inc. New York, pp 529- 744

2013, Environmental Biotechnology *Pages* **41–53**
Editors: **D.R. Khanna, A.K. Chopra, Gagan Matta, Vikas Singh & Rakesh Bhutiani**
Published by: **BIOTECH BOOKS, NEW DELHI**

Chapter 5

Changes in Germination, Seedling Growth and Peroxidase Activity in *Pisum sativum* and *Phaseolus vulgaris* in Response to Arsenic Stress

Sayan Bhattacharya[1], *Riazuddin Ahammad*[1], *Moumit Roy Goswami*[1], *Subarno Mukherjee*[1], *Arundhati Poddar*[1], *Dhrubajyoti Chattopadhyay*[2] *and Aniruddha Mukhopadhyay*[1]

[1]*Department of Environmental Science, University of Calcutta, 51/2, Hazra Road, Kolkata – 700019*

[2]*B.C. Guha Centre for Genetic Engineering and Biotechnology, University of Calcutta, 35, Ballygunge Circular Road, Kolkata – 700 019*

Arsenic is a metalloid which has significant environmental health effects because of its detrimental toxicity and colossal abundance. At high concentrations, arsenic is toxic to most plants. It interferes with metabolic processes and inhibits plant growth and development. Seed germination rate and the early seedling growth are sensitive to metal toxicity. Arsenic accumulated in the plant tissues can also stimulates peroxidase enzyme activity during the early phases of plant development. The objective of this study was to investigate the *in vitro* effects of different arsenic concentrations on germination and change in peroxidase activity in pea (*Pisum sativum*) and kidney bean (*Phaseolus vulgaris*) seeds and seedlings. Pea and kidney bean seeds were germinated in Petri plates treated with sodium arsenate solutions (0.5 ppm, 1 ppm, 1.5 ppm, 2 ppm and 2.5 ppm concentrations) for 96 hours. In order to assess the oxidative stress, Guiacol peroxidase (GPX) activities were estimated in different parts of the seedlings (seeds, roots and shoots). Germination of the seeds of both species deceased with increase in sodium arsenate concentrations. Interestingly, the lengths of the roots and shoots increased with the increase in concentrations of sodium arsenate. On the other hand, peroxidase activities in the seeds and seedlings increased significantly with time in response to arsenic stress. Interestingly, maximum increase in enzyme activity was observed in the shoot parts of both species. Therefore, under natural field conditions, application of arsenic contaminated irrigation water may have adverse effect on seed germination and growth. The experiment, therefore, indicates that arsenic has the potential to affect both the quantity and quality of pea and kidney bean seed production in arsenic affected areas, which can also affect the agricultural economy.

Keywords: *Arsenic, Germination, Enzyme activity, Phytotoxicity.*

Introduction

Arsenic is a metalloid (atomic no. 33) of great environmental concern because of its highly toxic carcinogenic properties to human beings (Nordstrom, 2002). Arsenic is a potent endocrine disruptor and can alter hormone mediated cell signaling process at extremely low concentration (Kaltreider *et al.*, 2001). It ranks 20th in abundance in the earth's crust, 14th in the seawater and 12th in the human body (Mondal and Suzuki, 2002). Arsenic naturally occurs in over 200 different mineral forms, of which around 60 per cent are arsenates, 20 per cent are sulfides and sulfosalts and the rest 20 per cent are arsenides, arsenites, oxides, silicates and elemental arsenic (Onishi, 1969). The source of arsenic is mainly geological, but anthropological activities like mining, burning of fossil fuels and uses of pesticides also cause arsenic contamination (Bissen and Frimmel, 2003). Arsenic contamination in groundwater has been reported in Bangladesh, India, China, Taiwan, Vietnam, USA, Argentina, Chile and Mexico. In many of these places the concentration has exceeded the permissible limit of 50 ppb recommended by EPA and WHO (WHO, 2001).

The Bengal basin is regarded to be the most acutely arsenic affected geological province in the world (Mukherjee and Fryar, 2008). Groundwater is regularly used for agricultural and household purposes in these areas. The use of arsenic contaminated groundwater for irrigation purpose in crop fields elevates arsenic concentration in surface soil and in the plants grown in those areas (Meharg and Rahman, 2003). Soil arsenic levels are very much related with local well water arsenic concentration, which suggests that the source of soil contamination is the irrigation water (Bhattacharya *et al.*, 2009). The absorption of arsenic by plants is influenced by the concentration of arsenic in the soil.

The arsenic concentrations in the edible parts of a plant depend on the availability of the soil arsenic and the accumulation and translocation ability of a plant (Huang *et al.*, 2006). The arsenic detoxification in plants involves arsenic mobilization from roots to aerial parts of the plant (translocation). This movement is controlled by the external arsenic concentration (Zao *et al.*, 2010).

The arsenic concentrations in the edible parts of a plant depend on the availability of the soil arsenic and the accumulation and translocation ability of a plant (Huang *et al.*, 2006). In general, plants uptake and metabolize As(V) through the phosphate transport channels (Tripathi *et al.*, 2007). Because of their chemical similarity, arsenic competes with phosphate for root uptake and interferes with metabolic process like ATP synthesis and oxidative phosphorylation (Tripathi *et al.*, 2007). Arsenate is taken up by phosphate transporter in plants grown on aerobic soils. Plants generally have a low efficiency of arsenic translocation from roots to shoots, may be due to the formation of complexes of arsenite, less toxic organic compounds and thiol compounds and subsequent sequestration in the root vacuoles, or because of the strong efflux of arsenite to the external medium (Zhao *et al.*, 2009).

At a higher concentration, arsenic is toxic to most plants. It interferes with metabolic processes and inhibits plant growth and development through arsenic induced phytotoxicity (Marin *et al.*, 1993). When plants are exposed to excess arsenic either in soil or in solution culture, they exhibit toxicity symptoms such as inhibition of seed germination (Abedin and Meharg, 2002); decrease in plant height (Marin *et al.*, 1993; Abedin *et al.*, 2002); decrease in tillering (Kang *et al.*, 1996; Rahman *et al.*, 2004); reduction in root growth (Abedin and Meharg, 2002); decrease in shoot growth (Cox *et al.*, 1996); lower fruit and grain yield (Kang *et al.*, 1996; Abedin *et al.*, 2002); and sometimes, leads to death (Baker *et al.*, 1976; Marin *et al.*, 1993). However, visible injuries and significant changes in growth inhibition and poor yield become apparent only after the plants are exposed to relatively high levels of pollutants or after a certain growth period.

Seeds are a vital component of the world's diet. Cereal grains alone, which comprise 90 per cent of all cultivated seeds, contribute up to half of the global per capita energy intake. Seed biology is one of the most extensively researched areas in plant physiology. With the seed, the independence of the next generation of plants begins. The seed, containing the embryo as the new plant in miniature, is structurally and physiologically equipped for its role as a dispersal unit and is well provided with food reserves to sustain the growing seedling until it establishes itself as a self-sufficient, autotrophic organism. In comparison, seed germination frequency and the early seedling growth are more sensitive to metal toxicity than the mature plants because some of the plants' defense mechanisms have not yet developed and hence effects at early stages of plant development can be very useful for toxicity assessment (Xiaoli *et al.*, 2007). Seed germination is the first physiological process affected by metals (Shanker *et al.*, 2005).

Heavy metals including arsenic have been reported to stimulate the formation of free radicals and reactive oxygen species leading to oxidative stress (Zhang *et al.*, 2003; Azevedo *et al.*, 2005). Interestingly, some recent reports have provided experimental evidences that arsenic-induced generation of free radicals can cause cell damage and death through activation of oxidative sensitive signaling pathways (Kamat *et al.*, 2005). Requejo and Tena (2005) proved that the induction of oxidative stress is the main process underlying arsenic toxicity in plants. Plants respond to oxidative stress by increasing the production of antioxidant enzymes, *e.g.* peroxidase (POD). The overexpression in peroxidase (POD) activity was correlated with increasing arsenic stress in plants (Miteva and Peycheva, 1999). According to the studies, arsenic accumulated in the plant tissue stimulates peroxidase synthesis during the early phases of plant development, long before the visible changes take place.

The objective of this study was to investigate the *in vitro* effects of different concentrations of sodium arsenate solution (pentavalant arsenic) on germination and change in peroxidase activity in pea (*Pisum sativum*) and kidney bean (*Phaseolus vulgaris*) seeds and seedlings.

Materials and methods

Germination Assay

Prior to germination, pea and kidney bean seeds were surface-sterilized in 3 per cent H_2O_2 and then rinsed with distilled water. Seed germination was tested on moist filter papers. Pieces of filter papers were placed on petri plates and then moistened with 10 ml aqueous solutions of sodium arsenate (0.5 ppm, 1 ppm, 1.5 ppm, 2 ppm and 2.5 ppm concentrations in each petri plate). The arsenic solutions were freshly prepared by dissolving sodium arsenate in deionized water and adjusting their pH to 5.8 with 2mM Mes-Tris buffer (Liu *et al.*, 2007). Controls were maintained by moistening the filter papers with 10 ml deionized water. Thirty seeds were placed in each plate, covered by lid, and incubated at 27°C. Geminated seeds were counted 5 days after initiation. Seeds were considered germinated when both the plumule and radicle were extended from their junction. Each treatment was replicated three times. The fresh masses of the shoots and roots were taken and shoot height was measured from culms base to the tip of the longest leaf and root length was measured from the root-shoot junction to the tip of the longest root. The root and shoot lengths of all the seeds under of each set were measured and mean values were calculated accordingly.

Peroxidase Assay

Shoots, roots and seeds were separated carefully by cutting the junctions and were rinsed thoroughly with distilled water to remove the arsenic solutions on the surfaces of the plant parts. For determination of peroxidase activity, 0.5 gm. of roots, shoots and seeds were homogenized in 0.5 ml. of respective extraction buffer in a pre-chilled mortar and pestle. The homogenate was filtered and centrifuged at 22,000g for 20 mins. At 4°C. Peroxidase (POD) activity was determined with guaiacol by spectrophotometry (Lagrimini, 1991). In the presence of H_2O_2, POD catalyzes the transformation of guaiacol to tetraguaiacol. This reaction can be recorded at 470 nm. The reaction mixture contained 100mM phosphate buffer (pH 6.0), 33mM guaiacol and 0.3mM H_2O_2. Enzyme activity was measured in spectrophotometer in 470 nm. wavelength in every 10 second interval. Six readings were taken for each sample. The enzyme activities were calculated by converting the readings in OD/min./gm. fresh tissue.

Results and Discussions

As a preliminary experiment, pea and kidney bean seeds were exposed to different concentrations of As(V) to determine their effects on germination. In high concentrations of sodium arsenate, specially in 2.5 ppm. concentration, some seeds turned black coloured and not germinated at all. Germination percentages were reduced in response to sodium arsenate stress (90 per cent, 85 per cent, 75 per cent, 75 per cent and 60 per cent germination with respect to the control in pea seeds in response to 0.5, 1, 1.5, 2, 2.5 ppm. sodium arsenate concentrations respectively; 95 per cent, 85 per cent, 85 per cent, 75 per cent and 65 per cent germination with respect to the control in kidney bean seeds in response to 0.5, 1, 1.5, 2, 2.5 ppm. sodium arsenate concentrations respectively). Interestingly, in the germinated seeds, there was a positive change in growth in both pea and kidney bean seedlings after 5 days of incubation. With the increase in sodium arsenate concentrations, the lengths of the roots and shoots increased accordingly. Most of the seeds showed increase in growth with no morphological indication of toxicity. However, in pea seedlings, the effects of sodium arsenate on growth were more prominent compared to the kidney bean seeds, and there was a significant increase in root and shoot lengths in response to sodium arsenate stress. In pea seeds, the root lengths increased from 3.38 cm. (control) to 5.29 cm. (2.5 ppm. sodium arsenate stress), whereas the shoot lengths increased from 1.21 cm. to 3.11 cm. respectively (Figures 5.1 and 5.2). In kidney bean seeds, the average lengths of all the roots and shoots were higher than that of pea seeds, but the rate of increase in length were lesser than that of pea seeds. Kidney bean root lengths varied between 6.21cm.-8.18 cm. and the shoot lengths varied between 7.69 cm.-9.97 cm. (Figures 5.3 and 5.4). Previous works have also been done different species, and some of them found increase in root and shoot lengths in response to arsenic stress, as our

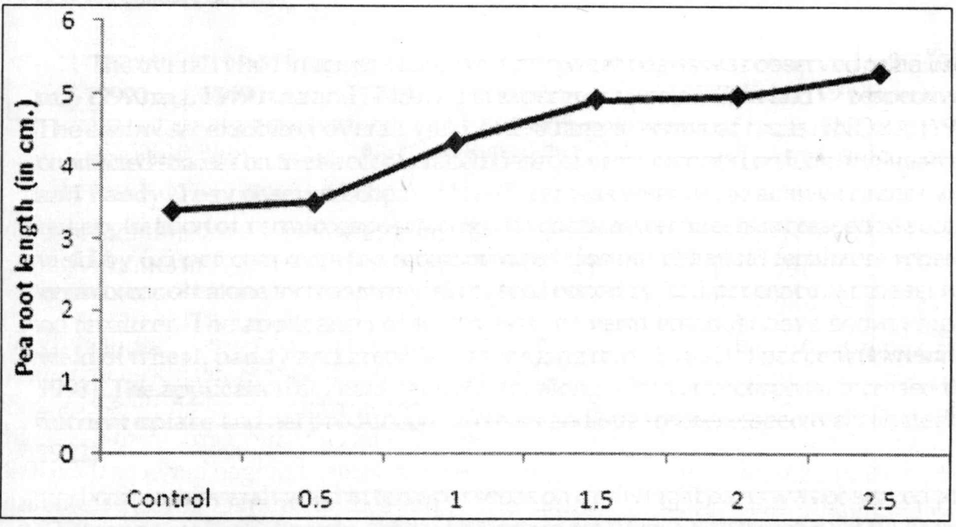

Figure 5.1: Peaseed Samples Treated with Sodium Arsenate (Conc. in ppm)

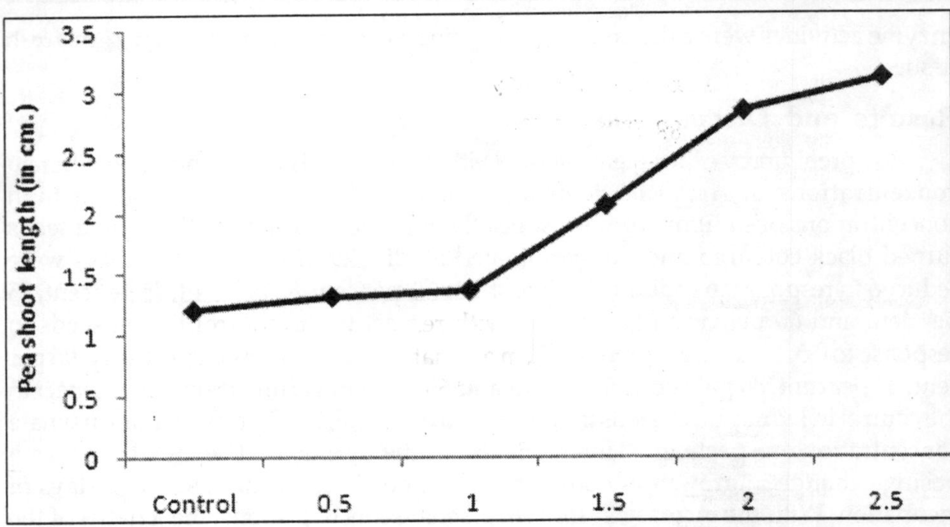

Figure 5.2: Peaseed Samples Treated with Sodium Arsenate (Conc. in ppm)

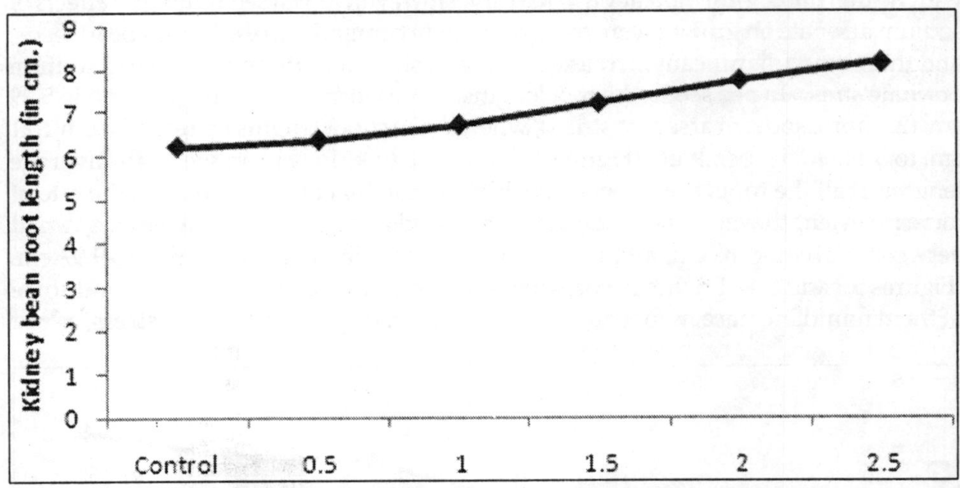

Figure 5.3: Kidney Bean Seed Samples Treated with Sodium Arsenate (Conc. in ppm)

study showed (Liu *et al.*, 2007). Some researchers, however, also found that arsenic stress can increase the growth of roots and shoots upto a certain concentration, after that can execute negative effects (Chun-xi *et al.*, 2007).

Peroxidase activity in the shoots, roots and seeds increased significantly in response to sodium arsenate stress (0.5 ppm.-2.5 ppm.). In kidney bean seeds, changes in peroxidase expressions were more prominent in roots, followed by seeds and shoots. In roots, the changes varied between 0.7-20.8 OD/min./gm. fresh tissue, in shoots 6-21 OD/min./gm. fresh tissue, whereas in seeds, it ranged between 0.75-10.3 OD/min./gm. fresh tissue. In presence of 2.5 ppm. sodium arsenate stress, peroxidase expression was almost 30 times greater in kidney bean root samples compared to the

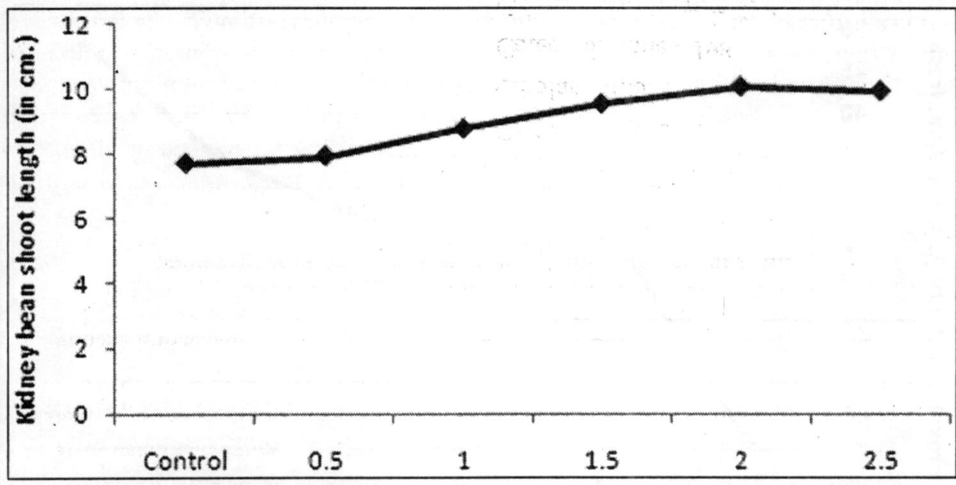

Figure 5.4: Kidney Bean Seed Samples Treated with Sodium Arsenate (Conc. in ppm)

control. In case of seeds and shoots, they were almost 14 and 4 times higher than the controls, respectively (Figures 5.5–5.7). In pea seeds, however, more pronounced effects were observed in roots, followed by shoots and seeds. In the roots, the changes varies between 0.1-7.8 OD/min./gm. fresh tissue, in shoots 1.1-47.5 OD/min./gm. fresh tissue, whereas in the seeds, it ranged between 0.5-18.5 OD/min./gm. fresh tissue. Peroxidase levels of pea roots were increased by 78 times at 2.5 ppm. sodium arsenate concentration with respect to the control, whereas, in shoots and seeds, they were 43 and 37 times higher than the controls, respectively (Figures 5.8–5.10). The results indicated that arsenic has more pronounced effect on peroxidase activities in pea seeds than kidney bean seeds. It is the first ever reported arsenic induced

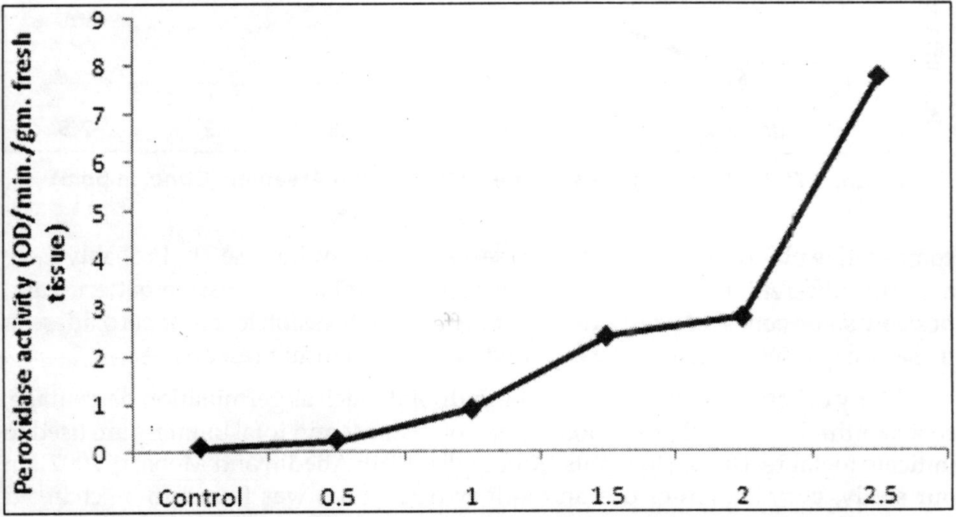

Figure 5.5: Pea Root Samples Treated with Sodium Arsenate (Conc. in ppm)

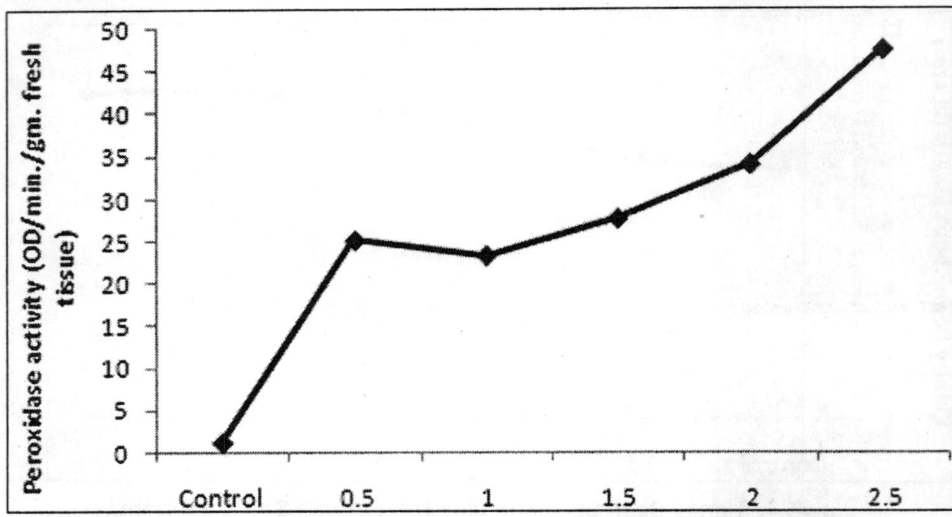

Figure 5.6: Pea Root Samples Treated with Sodium Arsenate (Conc. in ppm)

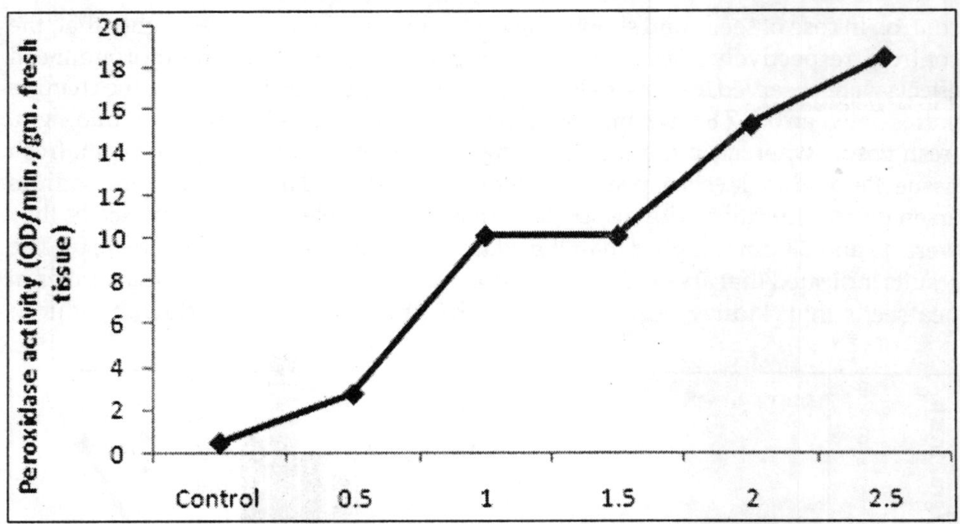

Figure 5.7: Pea Seed Samples Treated with Sodium Arsenate (Conc. in ppm)

comparative phytotoxicity study of pea seeds and kidney bean seeds. Probably there are some different molecular mechanisms for such unique expression patterns, and possibly some genes may responsible for different expression levels of peroxidase in these two species, which can be explored in details in future researches.

Many different vegetative response endpoints such as germination percentage, root length, shoot length; root biomass, shoot biomass and total biomass are used to indicate metal resistance to plants (Karataglis, 1980; Abedin and Meharg, 2002). In our study, germination of pea and kidney bean seeds was found to be changed considerably in response to As(V). However, reduced root length growth in response

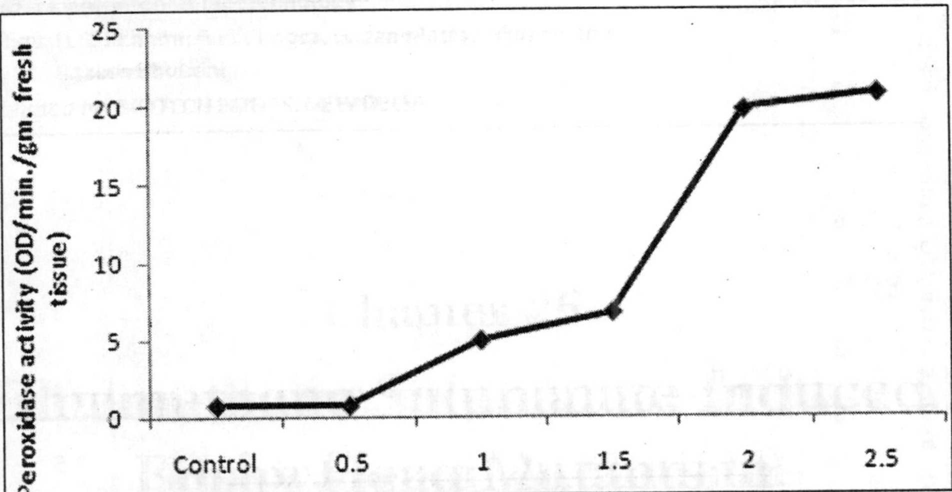

Figure 5.8: Kidney Bean Seed Samples Treated with Sodium Arsenate (Conc. in ppm)

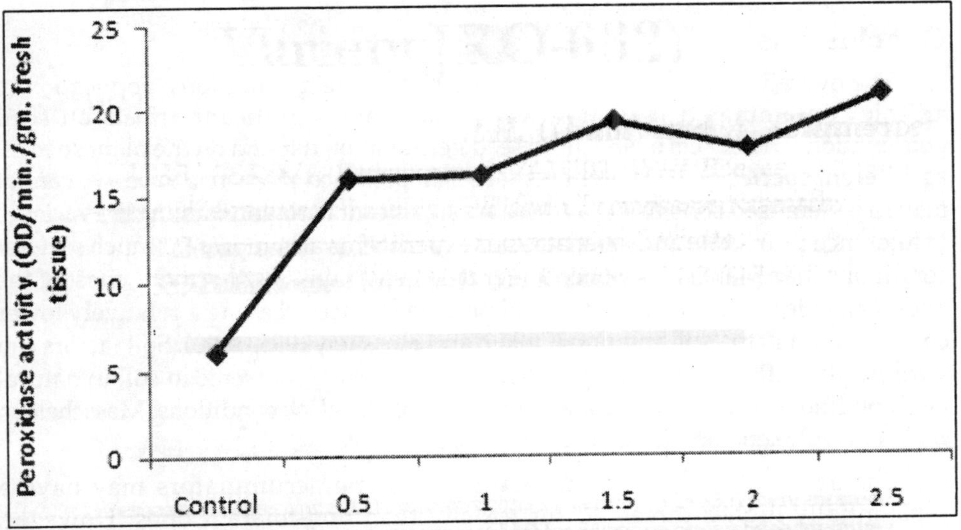

Figure 5.9: Kidney Bean Seed Samples Treated with Sodium Arsenate (Conc. in ppm)

to arsenic exposure has been reported by a number of investigators in different species (Hartley-Whitaker *et al.*, 2001; Sneller *et al.*, 2000). Differences of responses in terms of growth were possibly due to variation of species, because of which arsenic can stimulate the growth in some species, whereas negative effects can also be observed simultaneously in others. The up-regulation of peroxidase enzyme activities found in the seeds in response to arsenic stress indicated that excess arsenic can generate oxidative stress. According to our experimental results, it is also clear that peroxidase can be considered as an effective indicator for arsenic stress in germinating seeds.

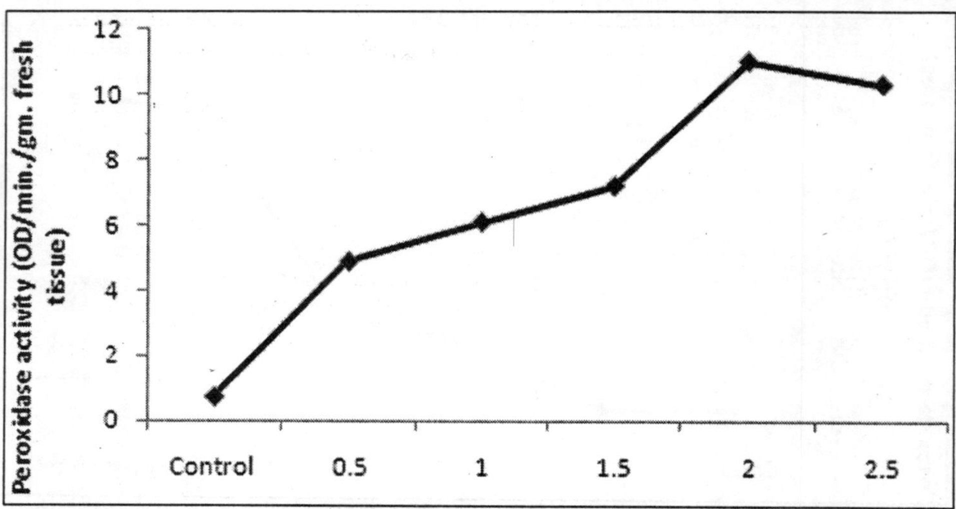

Figure 5.10: Kidney Bean Seed Samples Treated with Sodium Arsenate (Conc. in ppm)

Conclusions

The overall study revealed that under normal field conditions, application of arsenic contaminated irrigation water may have significant effects on seed germination. The effects of arsenic in seed germination depend on the plant species, as different species give different responses in presence of arsenic. Sure, we cannot make a generalized conclusion, as there are significant variations among the varieties in tolerance to arsenite and arsenate. Our experiments also represent much simpler conditions than plants experience under field conditions. Furthermore, most of the added arsenic will be adsorbed on soil particle surfaces, leaving a relatively lower concentration in the soil solution (Onken and Hossner, 1995, 1996). Soil factors can significantly influence solubility, mobility and toxicity of arsenic in soil in natural field conditions, and the effects may differ from artificial lab conditions (Masscheleyn *et al.*, 1991; Onken and Hossner, 1995).

The seeds of metal tolerant plants and hyperaccumulators may have a substantially higher threshold for toxicity than non-tolerant ones. However, knowledge of the effects of metals on seeds is only starting to emerge, in particular on the mechanisms of damage. In addition, more studies are needed that investigate the effects of metals on seed germination and also how they affect the species composition in soil seed banks. Moreover, it is largely unknown where metals are deposited in developing seeds and which levels are toxic to the embryo as compared to the endosperm or cotyledons. More research is needed to understand how seed longevity is affected by metals, both in soil seed banks and in seed banks used for agricultural or conservation purposes. Importantly, for a more comprehensive understanding of the effects of metals on seed viability and dormancy, the signalling networks need to be explored including the interaction of ROS, RNS and seed hormones with metals.

References

Abedin, M.J.; Meharg, A.A. 2002. *Relative toxicity of arsenite and arsenate on germination and early seedling growth of rice (Oryza sativa L.).* Plant and Soil, 243, 57–66.

Abedin, M.J.; Cottep-Howells, J.; Meharg, A.A. 2002. *Arsenic uptake and accumulation in rice (Oryza sativa L.) irrigated with contaminated water.* Plant and Soil, 240, 311–319.

Azevedo, H.; Gomes, C.; Fernandes, J.; Loureiro, S.; Santos, C. 2005. *Cadmium effects on sunflower growth and photosynthesis.* Journal of Plant Nutrition, 28, 2211–2220.

Baker, S.; Barrentine, W.L.; Bowmaan, D.H.; Haawthorne, W.L.; Pettiet, J.V. 1976. *Crop response and arsenic uptake following soil incorporation of MSMA.* Weed Science, 24, 322–326.

Bhattacharya, P.; Samal, A.C.; Majumdar, J.; Santra, S.C. 2009. *Transfer of Arsenic from Groundwater and Paddy Soil to Rice Plant (Oryza sativa L.): A Micro Level Study in West Bengal, India.* World Journal of Agricultural Sciences, 5, 425-431.

Bissen, M.; Frimmel, F.H. 2003. Arsenic – A Review. Part I: occurrence, toxicity, speciation and mobility. *Acta hydrochimica et hydrobiologica, 31, 9-18.*

Chun-xi, L.I.; Shu-li, F.; Yun, S.; Li-na, J.; Xu-yang, L.U.; Xiao-li, H.O.U. 2007. *Effects of arsenic on seed germination and physiological activities of wheat seedlings.* Journal of Environmental Sciences, 19, 725–732.

Cox, M.S.; Bell, P.F.; Kovar, J.L. 1996. *Different tolerance of canola to arsenic when grown hydroponically or in soil.* Journal of Plant Nutrition, 19, 1599–1610.

Hartley-Whitaker, J.; Ainsworth, G.; Meharg, A.A. 2001. *Copper and arsenate induced oxidative stress in Holcus lanatus L. clones with differential sensitivity.* Plant, Cell and Environment, 24, 713–722.

Huang, R.; Gao, S.; Wang, W.; Staunton, S.; Wang, G. 2006. *Soil arsenic availability and the transfer of soil arsenic to crops in suburban areas in Fujian Province, southeast China.* Science of The Total Environment, 368, 531-541.

Kabata-Pendias, A.; Pendias, H. 1984. *Trace Elements in Soils and Plants.* CRC Press, Boca Raton, Florida, USA. pp. 64.

Kaltreider, R.C.; David, M.A.; Lariviere, J.P.; Hamilton, J.W. 2001. *Arsenic alters the function of glucorticoid receptor as a transcription factor.* Environmental Health Perspectives, 109, 245-251.

Kamat, C.D.; Green, D.E.; Curilla, S.; Warnke, L.; Hamilton, J.W.; Sturup, S.; Clark, C.; Ihnat, M.A. 2005. *Role of HIF signaling on tumorigenesis in response to chronic low-dose arsenic administration.* Toxicological Sciences, 86, 248–257.

Kang, L.J.; Li, X.D.; Liu, J.H.; Zhang, X.Y. 1996. *The effect of arsenic on the growth of rice and residues in a loam paddy soil.* Journal of Jilin Agricultural University, 18, 58–61.

Karataglis, S.S. 1980. *Zinc and copper effects on metal tolerant and non-tolerant clones of agrostis tenuis.* Plant Systematics and Evolution, 134, 173–182.

Lagrimini, L.M. 1991. *Wound-induced deposition of polyphenols in transgenic plants overexpressing peroxidase.* Plant Physiology, 96, 577–583.

Liu, X.; Zhang, S.; Shan, X.; Christie, P. 2007. *Combined toxicity of cadmium and arsenate to wheat seedlings and plant uptake and antioxidative enzyme responses to cadmium and arsenate co-contamination.* Ecotoxicology and Environmental Safety, 68, 305–313.

Ma, L.Q.; Komar, K.M.; Tu, C.; Zhang, W.; Cai, Y.; Kennelley, E.D. 2001. *A fern that hyperaccumulates arsenic.* Nature, 409, 579.

Marin, A.R.; Pezeshki, S.R.; Masscheleyn, P.H.; Choi, H.S. 1993. *Effect of dimethylarsinic acid (DMAA) on growth, tissue arsenic and photosynthesis of rice plants.* Journal of Plant Nutrition, 16, 865–880.

Masscheleyn, P.H.; Dlaune, R.D.; Patrick, W.H. 1991. *Effect of redox potential and pH on arsenic speciation and solubility in a contaminated soil.* Environmental Science and Technology, 25, 1414–1418.

Meharg, A.A.; Rahman, M.M. 2003. *Arsenic contamination in Bangladesh paddy field soils: implication for rice contribution to arsenic consumption.* Environmental Science and Technology, 37, 229-234.

Miteva, E.; Peycheva, S. 1999. *Arsenic accumulation and effect on peroxidase activity in green bean and tomatoes.* Bulgarian Journal of agricultural Science, 5, 737-740.

Mondal, B.K.; Suzuki K.T. 2002. *Arsenic round the world: a review.* Talanta, 58, 201–235.

Mukherjee, A.; Fryar, F.E. 2008. *Deeper groundwater chemistry and geochemical modeling of the arsenic affected western Bengal basin, West Bengal, India.* Applied Geochemistry, 23, 863–894.

Nordstrom, D.K. 2002. *Worldwide occurrences of arsenic in groundwaters.* Science, 296, 2143–2144.

Onishi, H. 1969. *Arsenic, In Handbook of Geochemistry.* Springer- Verlag: New York, pp. 39.

Onken, B.M.; Hossner, L.R. 1995. *Plant uptake and determination of arsenic species in soil solution under flooded conditions.* Journal of Environmental Quality, 24, 373–381.

Onken, B.M.; Hossner, L.R. 1996. *Determination of arsenic species in soil solution under flooded conditions.* Soil Science Society of America Journal, 60, 1385–1392.

Rahman, M.A.; Rahman, M.M.; Miah, M.A.M.; Khaled, H.M. 2004. *Influence of soil arsenic concentrations in rice (Oryza sativaL.).* Journal of Subtropical Agricultural Research and Development, 2, 24–31.

Requejo, R.; Tena, M. 2005. *Proteome analysis of maize roots reveals that oxidative stress is a main contributing factor to plant arsenic toxicity.* Phytochemistry, 66, 1519–1528.

Shanker, A.K.; Cervantes, C.; Loza-Tavera, H. 2005. *Chromium toxicity in plants.* Environment International, 31, 739–753.

Sneller, E.F.C.; Van Heerwaarden, L.M.; Schat, H.; Verkleij, J.A.C. 2000. *Toxicity, metal uptake, and accumulation of phytochelatins in Silene vulgaris exposed to mixtures of cadmium and arsenate.* Environmental Toxicology and Chemistry, 19, 2982–2986.

Tripathi, R.D.; Srivastava, S.; Mishra, S.; Singh, N.; Tuli, R.; Gupta, D.K.; Maathuis, F.J.M. 2007. *Arsenic hazards: strategies for tolerance and remediation by plants.* Trends in Biotechnology, 25, 158-165.

WHO. 2001. *United Nations synthesis report on arsenic drinking water.* World Health Organization.

Zhang, F.; Shi, W.; Jin, Z.; Shen, Z. 2003. *Response of antioxidative enzymes in cucumber chloroplasts to toxicity.* Journal of Plant Nutrition, 26, 1779–1788.

Zhao, F.J.; McGrath, S.P.; Meharg A.A. 2010. *Arsenic as a Food Chain Contaminant: Mechanisms of Plant Uptake and Metabolism and Mitigation Strategies.* Annual Review of Plant Biology, 61, 535–59.

Zhao, F.J.; Ma, J.F.; Meharg, A.A.; McGrath, S.P. 2009. *Arsenic uptake and metabolism in plants.* New Phytologist, 181, 777–794.

2013, Environmental Biotechnology *Pages* **55–66**
Editors: **D.R. Khanna, A.K. Chopra, Gagan Matta, Vikas Singh & Rakesh Bhutiani**
Published by: **BIOTECH BOOKS, NEW DELHI**

Chapter 6

Kinetic Studies on Nanocatalysis by Iridium Nanoclusters in some Oxidation Reactions

Anjali Goel and Ranjana
Department of Chemistry, KGM, Gurukul Kangri University, Hardwar, U.K.

The use of a catalyst is a common way to minimize energy use and less material waste formation by lowering the energy of activation for a reaction. With catalysts reactions can be more efficient and selective. This provides an enormous energy savings while reducing the risks of explosion or other high temperature hazards. The field of nanocatalysis has undergone an explosive growth during the past decades, both in homogeneous and heterogeneous catalysis. Since nanoparticles have a large surface-to-volume ratio compared to bulk materials, they are attractive to use as catalysts. With the improved developments in nanochemistry, it is now possible to prepare soluble analogues of heterogeneous catalysts, materials that might have properties intermediate between those of the bulk metal and single metal-particle (homogeneous) catalysts. Noble metal nanoparticles with high specific catalytic activity are ubiquitous in modern synthetic organic chemistry during the recent decades. However how to reduce their dosage is one of the most exiting challenges due to the limited reserves of noble metals.

In this study colloidal dispersion of iridium nanoparticles have been synthesized by wet reduction method using different polymers as protecting agent after the reduction of precursor salt by alcohol. The particle size has been effectively controlled in the range of 10nm by the variation of solvent, reductant, temperature and polymer concentration. The synthesized iridium nanoparticles were characterized by UV-vis, XRD, FT-IR, TEM etc. techniques. The catalytic activity of colloidal iridium nanoparticles was evaluated in some oxidation reactions like amino acids-hexacyanoferrate(III) redox system in alkaline medium. These nanoparticles show a better catalytic activity than an equal amount of iridium precursor. Easy recovery of the catalyst from the reaction mixture shows another positive significance. Superior catalytic activities of nanoparticle-based catalytic systems than the corresponding bulk materials and recycling of metal nano particles are helpful in reducing the raw material costs and engineering a greener process via limiting the amount of waste chemicals for disposal.

Keywords: *Nanocatalyst, Iridium, oxidation reaction.*

Introduction

Catalytic reactions are preferred in environmental friendly green chemistry due to the reduced amount of waste generated. The field of nanocatalysis has undergone an explosive growth during the past decades, both in homogeneous and heterogeneous catalysis (Bradley *et al.*, 1994, Thomas *et al.*, 2003). Since nanoparticles have a large surface-to-volume ratio compared to bulk materials, they are attractive to use as catalysts (Bruss *et al.*, 2006, Firooz *et al.*, 2011). With the improved developments in nano-chemistry, it is now possible to prepare soluble analogues of heterogeneous catalysts that might have properties intermediate between those of the bulk metal and single metal-particle (homogeneous) catalysts. Noble metal nanoparticles with high specific catalytic activity are ubiquitous in modern synthetic organic chemistry during the recent decades. (Jansat *et al.*, 2004). However how to reduce their dosage is one of the most exiting challenges due to the limited reserves of noble metals.

In this study colloidal dispersion of iridium nanoclusters have been synthesized by wet reduction method using polymer as protecting agent after the reduction of precursor salt by alcohol. (Faraday *et al.*, 1857, Turkevich *et al.*, 1985, Hashemipour *et al.*, 2011). The particle size has been effectively controlled in the range of 0- 10 nm by the variation of solvent, reductant, temperature and polymer concentration used in synthesis. The synthesized Ir-nano were characterized by UV-vis, XRD, FT-IR, TEM etc techniques.

The catalytic activity of colloidal Ir- nano was evaluated in the oxidation of some amino acids/dyes- hexacyanoferrate(III) redox system in alkaline medium. These nanoparticles show a better catalytic activity than an equal amount of iridium precursor. Easy recovery of the catalyst from the reaction mixture shows another positive significance. Superior catalytic activities of nanoparticle-based catalytic systems than the corresponding bulk materials and recycling of metal nano particles are helpful in reducing the raw material costs and engineering a greener process via limiting the amount of waste chemicals for disposal.

Method

To make the kinetic studies on nanocatalysis in some oxidation reactions nanoparticles of iridium below size 10 nm were synthesized by wet chemical reduction method. The synthesized particles are used as catalyst in the oxidation of some amino acids and dyes.

All the chemicals and reagents used were of AR grade. All solutions were prepared with doubly distilled water. $K_3[Fe(CN)_6]$ was recrystallized before use. Its solutions were kept in amber colored bottles to prevent photodecomposition. KCl (Merck) and NaOH (Merck) were used to provide the required ionic strength and alkalinity respectively of the reaction mixture. Some stabilizer as polyvinylpyrrolidone is used to stabilize the particles.

Preparation of Colloidal Iridium Nanoparticles

In the present work Ir-nano were prepared by the reduction of $IrCl_3.3H_2O$or $H_2IrCl_6.6H_2O$(Merck) precursor. Methanol (Merck) was used as reducing agent. NaOH (Merck) was used to make medium alkaline. Some stabilizer as polyvinylpyrrolidone, is used to stabilize the particles. For their preparation required amount of precursor was dissolved in 25 ml of methanol. A solution of stabilizer in water was mixed slowly at room temperature by stirring magnetically. After that 1.6 ml of an aqueous solution of sodium hydroxide (0.2 M) was added drop wise with vigorous stirring. Now the solution (in a three necked round bottom flask) was heated in oil bath over 1 hr with a heating rate of 4°/min. Polymer stabilized nanoparticles were obtained. These nanoparticles were dried at 50°C and analyzed by UV-vis, IR, XRD and TEM methods of analysis. Size controlled synthesis was carried out by varying the amount of precursor, NaOH, stabilizer and methanol as described in Table 6.1.

Eavaluation of Catalytic Activity

The kinetic study of the nanocatalysis by the Ir-nano synthesized by the method 'a' and 'c' has been carried out in the oxidation of (i) glycine (amino acid) by alkaline hexacyanoferrate (III) ion (ii) of methyl orange (dye) using hexacyanoferrate (III) ions respectively.

The kinetic measurements were performed at constant at constant temperature The requisite amount of each reactant was thermostated to attain thermal equilibrium. The appropriate quantities of reactants were mixed in a 100 ml iodine flask. The progress of reaction was measured by injecting a solution of glycine/methyl orange into the reaction mixture at λ_{max} 420/465 nm corresponding to the $[Fe(CN)_6]^{-3}$/methyl orange respectively. It was verified that there is negligible interference from other species present in the reaction mixture at these wavelength.

Results and Discussion

For Characterization of Colloidal Iridium Nanoparticles

UV-vis spectroscopy is a convenient technique for monitoring the progress of metal colloid formation. A PVP- H_2IrCl_6- methanol-water solution kept in an oil bath was sampled at different time and the samples were characterized by UV-vis spectrophotometer. Initially $IrCl_6^{-2}$ and $IrCl_3$ show an absorption peak at about 487 nm. The intensity of this peak decreases with the increase of time. After about 25 minutes refluxing this absorption peak disappeared completely and a new peak

Table 6.1

Method	Precursor (mol dm^{-3})		Stabilizer (mol dm^{-3})		Reductant (ml)	NaOH (ml)	Diameter (nm)	
	Iridium	Concenteration	Polymers	Concentration	Methanol	(0.2M)	XRD	TEM
a.	$H_2IrCl_{6.}6H_2O$	$2.83 *10^{-5}$	PVP	$1.4*10^{-4}$	25 ml	1.6 ml	10.55	10.00
b	$H_2IrCl_{6.}6H_2O$	$1.54*10^{-5}$	PVP	$1.4*10^{-4}$	25 ml	1.6 ml	42.50	50.00
c.	$IrCl_3.3H_2O$	$7.07*10^{-4}$	PVP	$8.69*10^{-5}$	25 ml	1 ml	3.91,4.46	4.00

appeared at about 235 nm indicating the formation of Ir(0) nanoparticles. No further change was observed. The colour of solution changes from yellow to black-brown.

The XRD spectra of the Ir-nano shows two broad peaks at 2-theta about 40° and 48.5° at reflection plane Ir(111), Ir(200) respectively Figure 6.5. Which are the characteristic of isolated Ir(0) nanoparticles. Analysis of the peaks broadening using the Scherrer equation gives an estimate of particle diameter. Diameters of particles calculated by Scherrer equation are given in Table 6.1. The particle size ranges between 10.55 nm by method 'a', and 42.50nm and 3.91to 4.46nm by method 'a' to 'c' respectively. The X- ray diffractogram of these metallic particles indicate that they are amorphous and show broad peak characterization of materials with a small size.

The IR spectra of Ir-nano show a new band near the frequency 2015 cm^{-1} indicates the formation of Ir(0) in method a, b and c.The red shift of resonance peak of pure PVP at 1700 cm^{-1} to 1664 indicates the interaction of >C=O group of PVP with Ir-nano. The presence of same bands at about 1500 cm^{-1}, 1461 cm^{-1} (aromatic C-C str, N-H bending) in spectra shows that these groups do not interact with Ir-nano.

TEM micrographs confirms the particle size as 10, 50 and 4.5 nm by method a, b and c respectively Figure 6.6.

For Eavaluation of Catalytic Activity

Nanocataltsis by Ir-Nano in the Oxidation of Glycine (α-amino Acids) by Hexacyanoferrate(III) in Alkaline Medium

In the first series of experiments Ir-nano are used as catalyst for studying its catalytic effect on the rate of oxidation of amino acids namely glycine kinetically. All kinetic runs were followed as described in foregoing section. The concentration of Ir-nano used in the experiments is of the order of 10^{-6} mol dm^{-3}. The effect of Ir-nano concentration on the rate of oxidation of glycine has been studied by varying its concentration from 1.23×10^{-6} mol dm^{-3} to 6.14×10^{-6} mol dm^{-3}. The concentration of other reactants was as follows:

[HCF(III)] = 3×10^{-4} mol dm^{-3}, [Glycine] = 3×10^{-3} mol dm^{-3}, [NaOH] = 0.4 mol dm^{-3}, μ = 0.5 mol dm^{-3}, Temp. = 35 ± 0.1°C, λ_{max} = 420 nm.

The experimental results are presented in Figures 6.1 and 6.2 for the oxidation of glycine. The initial rate of the reaction has been calculated by the slope of the curve between absorbance vs time (Figure 6.2). The graphical representation between reaction rate and [Ir-nano], (Figure 6.1), $r \geq 0.96$, S.D. ≤ 0.008 and 02, ≥ 0.97, S.D. ≤ 0.006) clearly reveals that the reaction follows first order kinetics with respect to Ir-nano concentration. The effect of particle size on oxidation rate was studied by studying the effect of precursor iridium (H_2IrCl_6), [Ir-nano] (≈50nm) and [Ir-nano] (≈10nm) as catalyst keeping the concentration of all other reactants constant. A graph plotted between rate vs concentration of precursor iridium/Ir-nano (≈50nm)/Ir-nano (≈10nm) shows that reaction rate of Ir-nano (≈10 nm) catalyzed reaction are more than that of precursor iridium and Ir-nano (≈50 nm) (Tsunoyama *et al.*, 2006 Somorjai *et al.*, 2006). This shows that the catalytic activity of Ir-nano is two to three times more than that of precursor iridium due to large surface area to volume ratio. It also shows that the catalytic activity of small particles is more than that of larger particles.

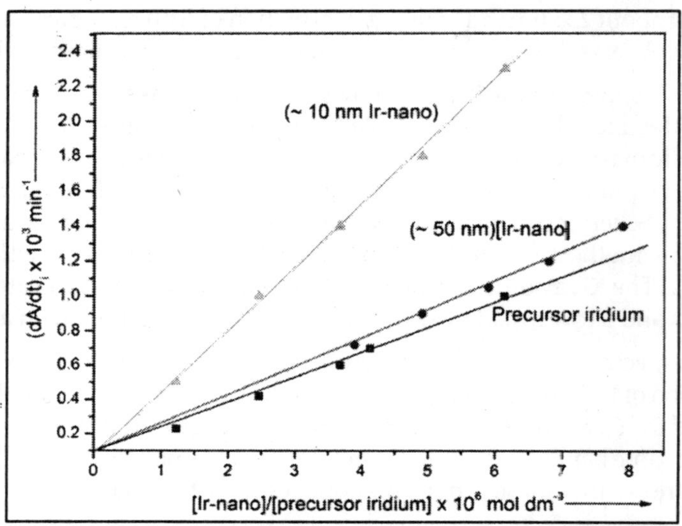

Figure 6.1: Plots Between Rate vs Concentration of Precursor Iridium/≈10nm Ir-nano/≈50nm Ir-nano for the Oxidation of Glycine

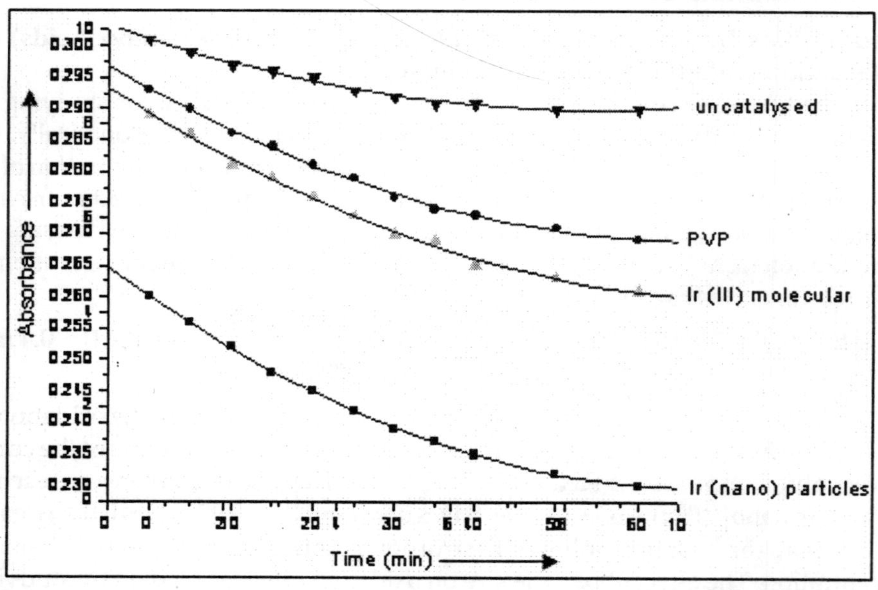

Figure 6.2: Plots Between Absorbance and Time for the Effect of Uncatalysed, PVP, Ir (III) Bulk and Ir-nano on the Rate of Oxidation of Glycine

PVP plays an important role in the present system. The present colloidal dispersion was stable in air under the protective action of polymers. No precipitation had occurred after storing the dispersion in air for month. Figure 6.2 shows a small effect of PVP concentration on the rate of oxidation.

To calculate the thermodynamic parameters the effect of temperature on the oxidation of glycine with alkaline Hexacyanoferrate(III) has been studied in the presence of Ir-nano particles at four temperatures (30°C, 35°C, 40°C and 45°C). The data presented in Table 6.2 show that presence of Ir- nano decreases the activation energy as compared to bulk iridium making the reaction rate fast in presence of nanoparticles. Large negative entropy values suggest formation of more ordered and polar complex formation during the reaction (Goel *et al.*, 2010).

Table 6.2: Kinetic Parameters for Oxidation of Glycine and Methyl Orange by HCF (III) in the Presence of Ir-Nano Catalyst

Parameter/Substrate	E_a *(k cal mol^{-1})*	*A* *(l mol^{-1} sec^{-1})*	$-\Delta S^{\#}$ *(E.U.)*	$\Delta H^{\#}$ *kcal/mol*	$\Delta F^{\#}$ *kcal/mol*
Glycine(Ir-precursor)	13.8	1.65×10^{-8}	23.03	13.17	20.44
Glycine(Ir-nano catalysed)	12.2	6.89×10^{-6}	29.32	11.59	20.69
Methyl Orange(uncatalysed)	18.9	7.36×10^{-12}	11.4	18.3	14.6
Methyl Orange (catalysed)	8.6	1.56×10^{-6}	20.02	7.96	1.55

Stoiciometry of the reaction was determined by making different sets of reaction mixtures containing an excess [Hexacyanoferrate(III)] over [substrate] at 0.4 mol dm^{-3} NaOH, same concentration of Ir-nano at constant ionic strength was allowed to react. This mixture was analyzed by titrating against Ce(IV) solution. The results indicated that two moles of hexacyanoferrate (III) were reacted with one mole

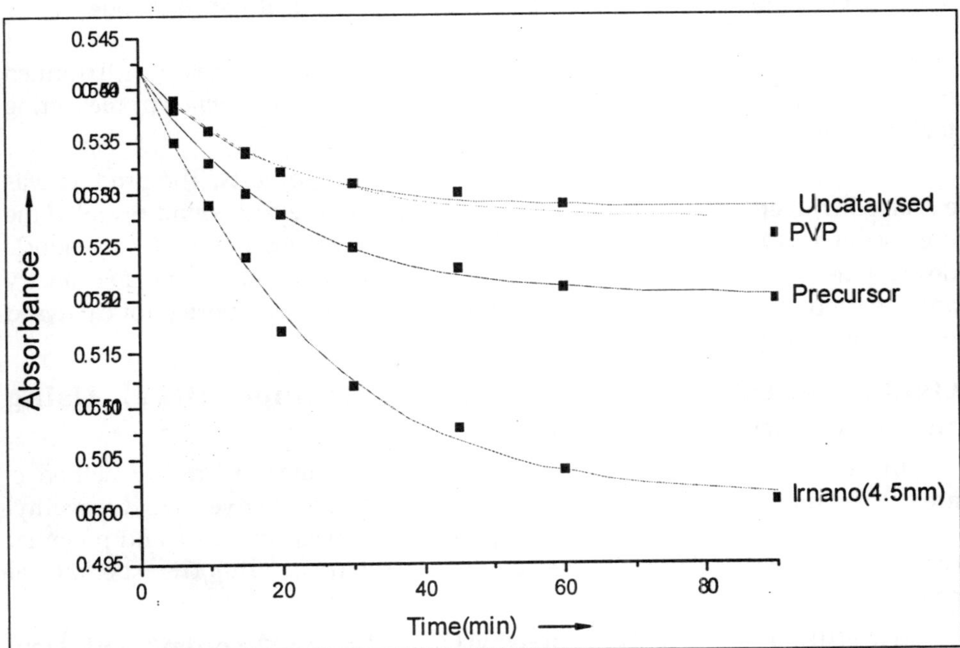

Figure 6.3: Plot between Absorbance and Time for the Effect of Uncatalysed, PVP, Ir (III) bulk and Ir-Nano on the Rate of Oxidation of Methyl Orange

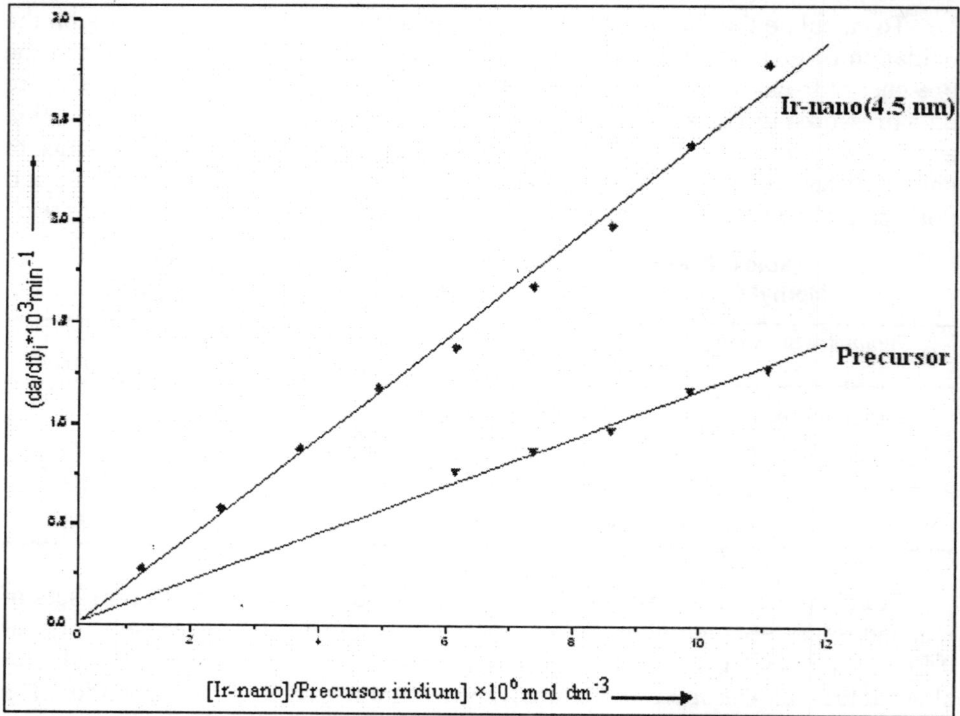

Figure 6.4: Plots between Absorbance and Time for the Effect of Nncatalysed, PVP, Ir (III) bulk and Ir-Nano on the Rate of Oxidation of Methyl Orange

of glycine. After completion of the reaction, the amount of hexacyanoferrate(II) formed (which is equal to the amount of ferrate(III) consumed) was determined by measuring the absorbance of the reaction mixture.

The main oxidative products were submitted to spot tests. The product was extracted with ether and purified. The IR spectrum of the compound revealed the presence of one carboxylic group and one keto group. It was observed that, the products do not undergo further oxidation under prevailing kinetic conditions. The stoiciometry and the products of Ir-nano catalysed oxidation are same as reported for catalysed reaction by bulk iridium(Goel *et al.*, 2008).

Oxidative Degradation of Methyl Orange (Dye) Using Hexacyanoferrate (III) Ions

In the second series of experiments the effect of Ir- nano prepared by method 'c' are studied on the rate of oxidation of methyl orange an azo dye. Azo dye methyl orange creats environmental pollution problems by releasing toxic and potential carcinogenic substances in to the aqueous phase (Meena *et al.*, 2010). The experimental conditions set are as follow:

[HCF(III)] = 3×10^{-6} mol dm^{-3}, [Methyl Orange] = 3×10^{-5} mol dm^{-3}, $\mu = 0.3$ mol dm^{-3}, [NaOH] =0.25 mol dm^{-3} λmax = 465, Temp. = 40 ± 0.1°C.

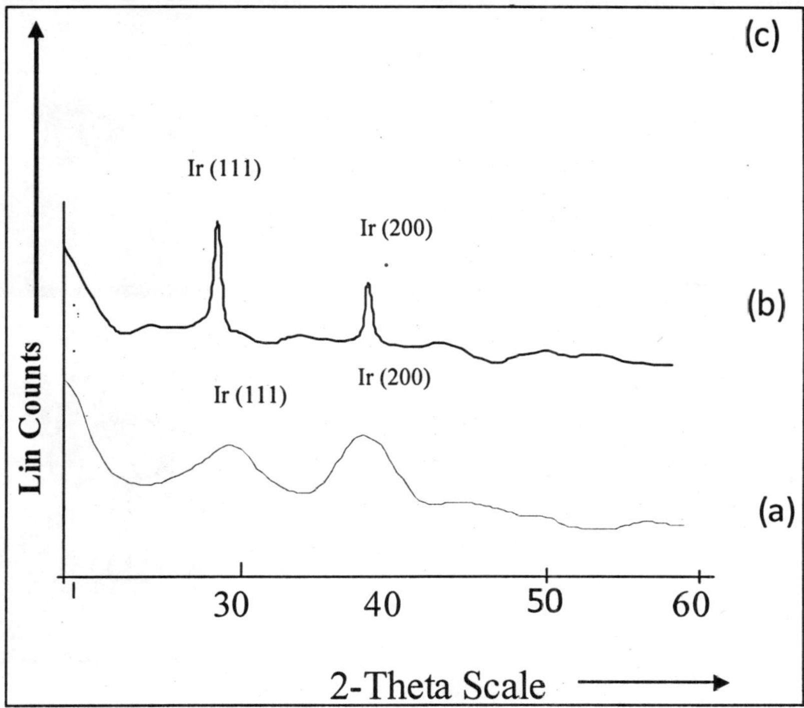

Figure 6.5: XRD Spectra of Ir-Nano Prepared by Method a, b and c

Under the above experimental conditions the reaction rate of uncatalysed and catalysed oxidation was determined. For catalysis iridium nanoparticles of size 4.5nm are used as catalyst. Figures 6.3 and 6.4 show that rate of catalysed reaction is more than the rate of uncatalysed reaction. The effect of PVP on the oxidation is also studied. Figure 6.3 show a negligible effect of PVP on the reaction rate. Thus the overall effect is only due the Ir-nano. The reaction was also studied at four different temperatures to calculate the thermodynamic parameters. The results in table 02 show that the presence of Ir-nano increases the oxidation rate many times by decreasing the energy of activation from 18.9 to 8.6 kcal mol^{-1}. The stoichiometry and oxidation product are studied by the above mentioned method as for glycine oxidation. Both are same as reported earlier for uncatalyzed reaction(Narayan *et al.*, 2003).

Table 6.3

Sl.No.	*Diameter of the Particles (nm) Before Starting of the Reaction*	*Diameter of the particles (nm) After Completion of the Reaction*
1.	10.55	12.63 ± 0.34

For recycling of Ir-nano from reaction mixture after kinetic study of reaction, solution was carried out for centrifugation. The obtained residue was dried at 100°C temperature and analyzed by XRD technique. XRD pattern was presented in Figure 6.7. This shows that the size of particles increases from 10.55 to 12.63 due to the

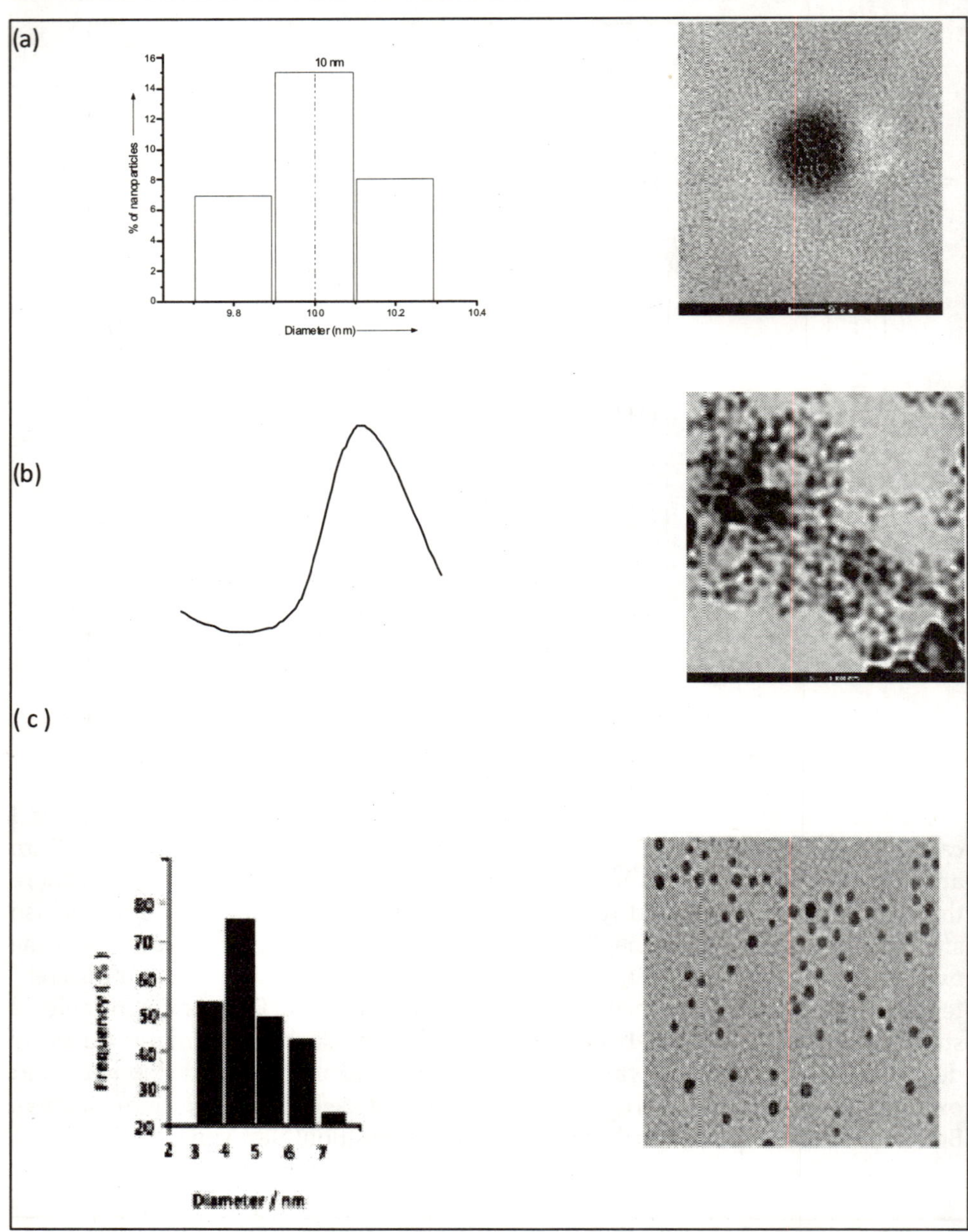

Figure 6.6: TEM Micrograph (a), (b) and (c) of Ir-Nano Synthesized by Method a, b and c Respectively

distortion and agglomeration of the particles. The recovery of Ir-nano is good though the colloidal PVP-Pd nanoparticle were found to have poor recycling potential (Narayan *et al.*, 2003).

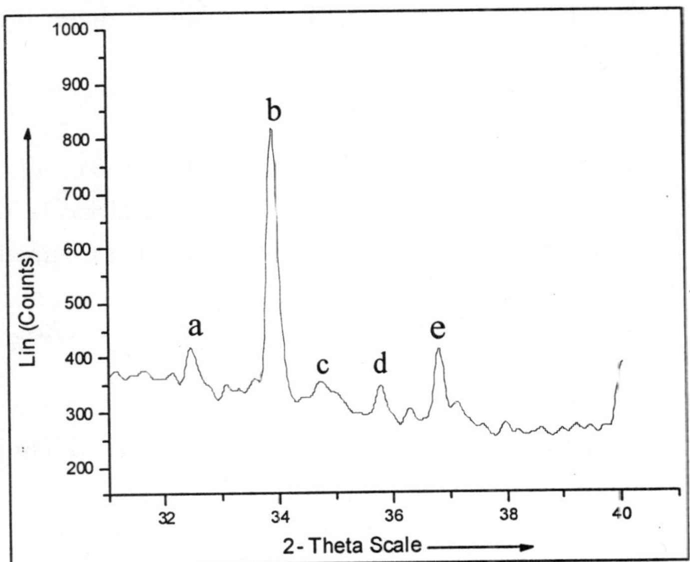

Figure 6.7: XRD Spectra of Ir-Nano Particles after Recovery

Conclusion

Thus from the above study it can be concluded Ir-nano, which were prepared by reduction of $H_2IrCl_6/IrCl_3.3H_2O$ remains stable even after several months. The kinetic study on nano catalysis show that iridium nano clusters act as good catalyst than the bulk Ir(III). It shows a remarkable effect on degradation of methyl orange azo dyes giving less hazardous oxidation products.Thus the study shall be helpful in making the environment green.

Abbreviations

- ☆ Ir-nano: Iridium nano particles
- ☆ PVP: Poly vinyl pyrrolidone

References

Bradley, J. S., Scmided, G., Clusters and Colloids: From Theory to Application VCH, New York *Cluster Colloids,* (1994) 459.

Bruss, A.S., Gelesky, M.A., Machado, G., Dupont, J., J. of Molecular Catalysis A; 252 (2006) 212-218

Faraday, M., Philos. Trans. R. Soc. London, 147 (1857) 145 – 153.

Firooz, A.A., Mahjoub, A.R., and Khodadadi, A.A., World Academy of Science, Engineering and Technology 76, 2011, 138-140.

Goel, A., Sharma, S., Trans. Met. Chem, 35 (2010) 549-554.

Goel, A., Sharma, S., Shakunj., Int. J. Chem. Sci.: 6(4), 2008, 1891-1899.

Hashemipour, H., Zadeh, M.E., Pourakbari, R. and Rahimi, P., International Journal of the physical sciences, 6(18), pp-4331-4336, 2011

Jansat, S., Gomez, M., Philippot, K., Muller, G., Guiu, E., Claver, C., Castillon, S., Chaudret, B., J. Am. Chem. Soc., 126 (2004) 1592.

Meena, V.K., Meena, R.C, J.Ind.Council Chem,27(2) 2010, 180-184.

Narayan, R., and El-Sayed, M. A.(2003), J, American Chemical Soc. 125(27) 8340.

Somorjai, G.A., Bratlie, K.M., Montano, M.O., J.Y. Park, J. Phys.Chem. *B*, 110 (2006) 2014.

Thomas, J.M., Johnson, B.F.G., Raja, R., Sankar, G., Midgley, P.A., Acc. Chem. Res., 36(1) (2003) 20.

Turkevich, J., Gold Bull., 18 (1985) 86-91.

Tsunoyama, H., Sakurai, H., Tsukuda, T., Chem. Phys. Lett., 429 (2006) 528.

2013, Environmental Biotechnology *Pages* **67–77**
Editors: **D.R. Khanna, A.K. Chopra, Gagan Matta, Vikas Singh & Rakesh Bhutiani**
Published by: **BIOTECH BOOKS, NEW DELHI**

Chapter 7

Suitability of Curcumin Pigment in the Coating of Triphla Guggul Ayurvedic Tablets

***Anjali Goel*[1], *Neetu Rani*[1] *and M.K Bhardwaj*[2]**
[1]*Department of Chemistry, KGM, Gurukul Kangri University, Haridwar*
[2]*Patanjali Ayurved Limited, Haridwar*

Curcumin [1,7 bis (4 hydroxy 3 methoxy phenyl)-1,6 heptadiene, -3,5 dione] is the major yellow pigment, extracted from turmeric, a commonly used spice derived from the rhizome of plant *Curcuma longa* L., which belongs to *Zingiberaceae* family. The aim of the present work is to extract curcumin pigment from turmeric rhizomes and use it in the film coating of triphala guggul ayurvedic tablet (TGA). For coating of ayurvedic tablets the work has been completed into two parts. Optimization of extraction conditions and coating of ayurvedic tablets with curcumin and acetylated curcumin. The optimum extraction conditions for extraction are 60°C and 75 minute stirring with ethanol (95 per cent) as solvent but from industrial point of view it is more economical to use ethanol-water (1:1) mixture. Separation and purification of curcumin from oleoresin was carried out by crystallization using selective solvent. The purified curcumin was characterized by UV-Vis, FT-IR, and TLC techniques.

Film coating protect the tablets from light, temperature and moisture and improve their appearance. In the present study core tablets of triphla guggul are coated with pure curcumin and acetylated curcumin. To study the effect of environmental factors on coated tablets, the tablets were packed in polypropylene pouches and subjected to storage at room temperature (19-40°C) and 65±5 per cent relative humidity (RH) for three months during June to August 2011. The colour of pure curcumin coated tablets start fading after two months as curcumin is sensitive to heat and light but acetylated tablets remain same even after three months. The coated tablets were evaluated in accordance with Indian Pharmacopeia or other pharmaceutical references for their physico-chemical parameters. The parameters studied for core tablets of triphla guggul, curcumin coated tablets and acetylated curcumin coated tablets are weigth variation, friability, disintegration time, hardness and loss on drying etc. The values of all the parameters are within the prescribed limits of Indian Pharmacopea. Thus the results show that pigment curcumin can be used as colourant in the film coating of triphala guggul tablets after acetylation.

Keywords: *Curcumin, Extraction, Coating, Turmeric rhizome.*

Introduction

There is an increasing demand for the use of natural colours as they are preferred in food over synthetic colours (Sowbhagya *et al.,* 2005). Accordimg to the prevention of food adulteration act (PFA-1994)) in India, eight synthetic pigments and 11 natural pigments are permitted in food. Curcumin is an important permitted natural colourant, isolated from the rhizomes of *Curcuma longa* L, used in food, nutritious and pharmaceutical preprations among others (Zebib *et al.,* 2010, Gouin., 2004, Sowbhagya *et al.,* 1998). The rhizome of *Curcuma longa* L., usually contains 1-5 per cent of curcumin and two related demethoxy compound i.e demethoxcy curcumin (DMC) and bisdemethoxy curcumin (BDMC) which are together known as curcuminoids (Govindarjan., 1980, Revathy *et al.,* 2011). The active ingredient of turmeric is curcumin (1E, 6E)-1, 7 bis (4 hydroxy-3-methoxy phenyl)-1, 6 heptadiene-3, 5-dione. (Rouhani *et al.,* 2009, Zaibunisha *et al.,* 2009) which is insoluble in water under acidic or neutral conditions but dissolves in alkaline conditions (Tonnesen *et al.,* 2002, Sowbhagya *et al.,* 2005). The maximum acceptable daily intake for curcumin is 0.1 mg/kg body weight (FAO, 2000).

All drugs have their own characterstic properties but few properties make the drug unacceptable. These are bitter in taste, unpleasant odour, sensitive to light or oxidation or hygroscopic in nature (Cole *et al.,* 1998, Porter *et al.,* 1980). Because of these reasons tablet coating is the choice of option to solve such problems. Film coating of tablet is a multivariate process with many different factors. Film coating masks the undesirable taste or odour, improve appearance, provide tablet identity, facilitate swallowing and control or modify release of drug. It also protect tablet from moisture, light and temperature effects (Sheth *et al.,* 2009). Some tablets may contain moisture sensitive ingredients. The use of water may creat problem of physical and chemical stability of coated tablets because moisture content shall also increase the degradation of drug (Ruotsalainen *et al.,* 2003, Porter *et al.,* 1997). Film coating are relatively thin, small differences in film thickness on tablets may result in significant colour vibration. There has been some success in using pacified dye systems, however these systems have been shown to have poor light stability than pigmented coating. In addition to providing colour, pigments have been reported to reduced, moisture diffusion through the film and improve light stability as compared with dye (Allam *et al.*, 2011, Stanley *et al.*, 2006). The objective of present study is to extract curcumin pigment from turmeric rhizome and use it in film coating of triphala guggle ayurvedic tablets (TGA). Also to study the stability of curcumin coated tablets under different environmental conditions.

Materials and Methods

The rhizomes of *Curcuma longa* L., were collected from Haridwar market and identified by quality control department of Patanjali Ayurved Limited, Haridwar. All chemicals and reagents used were of AR grade and obtained from E.Merck and Loba chemicals. Double distilled water was used throughout the experimental analysis.

Systronic -117, UV-Vis spectrophotometer was used for the determination of λ_{max} of curcumin. The measurement of colour intensity of extracted dye solution was made by Tintometer-F-Lovibond. The coating were applied on laboratory scale in a coating pan apparatus (Harison pharma pvt.ltd.). A high shear mixer (Remi eqipment pvt. Ltd.) is used to make coating solution.

For coating of ayurvedic tablets the work has been completed in two parts.

1. Optimization of extraction conditions and purification of curcumin.
2. Coating of ayurvedic tablets with curcumin and acetylated curcumin.

Optimization of Extraction Conditions and Purification of Curcumin

Rhizomes of *Curcuma longa L.*, were washed thoroughly with distilled water and dried. These were then crushed in the coarse powder of particle size 0.50 mm. In order to determine the optimum extraction conditions, a number of experiments were carried out at various temperature and different stirring time (Umbreen *et al.*, 2008). Powdred raw material (50 gm) was soaked in (500 ml) of solvent (water, ethanol, water-ethanol mixture) and kept overnight. The resulted coloured solution was subjected to warming at 60°C with continuous stirring for 75 min. The extracted solution is filtered and resulted coloured solution was concentrated under reduced pressure using rotavapour at 60-80°C. The identification and separation of curcumin was carried out by TLC using solvent chloroform:ethanol:glacial acetic acid (94:5:1) and visualized by UV light. The identification was confirmed by comparing their R_f values (0.75) with the reported values (0.70 - 0.80) (Rajpal., 1990) (Figure 7.3).

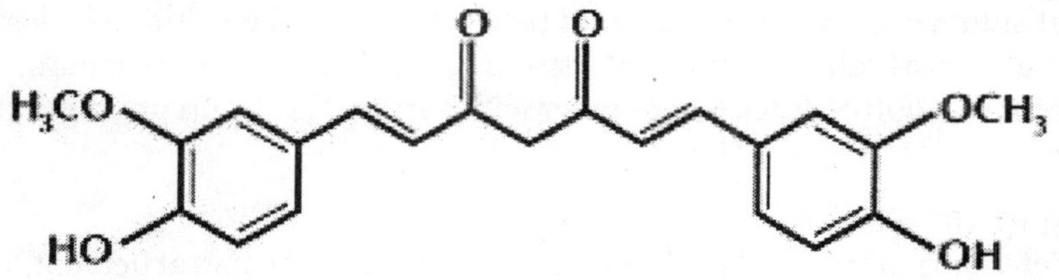

H$_3$CO OCH$_3$ O O HO OH

Figure 7.1: Structure of Curcumin 1, 7 bis (4 hydroxy, 3 methoxy phenyl)-1,6 heptadiene -3,5 dione

Further crystallization, (Rao *et al.*, 1970) of the separated sample was carried out to get curcumin of 90-95 per cent purity.

The extracted curcumin was characterized by the following methods.

By UV-vis Spectroscopy

The UV- visible spectra was measured in ethanol using systronic -117, UV - vis spectrophotometer in the range of 200-700 nm.

By TLC

In the present experiment ethanolic extract were tested in TLC for presence of curcumin. The TLC pre-coated silica gel (Merk-60 F 254, 0.25 mm thick) plate were developed using a Camag twin-trough glass tank which was pre-saturated with the mobile phase for 1- hour and each plate was developed to a height of about 10 cm.The composition of mobile phase was optimized by using different solvent of varying polarity. After development, the plate was removed and dried and spots were visualized in UV light.

By FT-IR Spectroscopy

The FTIR spectra of curcumin and acetylated curcumin in the range (4000-400) cm^{-1} were recorded as KBr disc on FT-IR, SHIMADZU Sperctrophotometer.

Coating of Ayurvedic Tablets with Curcumin and Acetylated Curcumin

Curcumin is unstable under rapid hydrolytic degradation in neutral or alkaline conditions. In aqueous medium it has poor stability towards oxidation, light, enzyme and heat. Thus to increase the stability of curcumin acetylation of curcumin has been carried out.

Acetylation

Take 5 gm of curcumin powder in a 100 ml round bottom flask. Add 25ml acetic acid glacial and 2-5 drops of conc. H_2SO_4. Reflux it for four hours at 40-50°C. After 4 hours refluxing, pour this mixture into ice-water and filter the precipitates. After filtration wash the precipitates with distilled water and then dry in vacuum oven.

Now the the coating of TGA tablets was carried out using curcumin and acetylated curcumin as follows-

Prepration of Coating Solution

Prepare 32 per cent solution of titanium dioxide in water and 2.5 per cent of instacoat solution in iso propyl alcohol (IPA). Mix both the solution in a stainless steel vessel with a high shear mixer at a speed of 250-300 rpm for 20-30 min. Add 2.0-2.5 per cent solution of curcumin powder with stirring till all curcumin get attached properly.

Coating of Tablets

A defenite quantity of the tablets were kept in a coating pan at bed temperature 45-50°C and the pan was rotated continuously. The various parameters like spray rate, inlet air temperature and rotating speed of pan were adjusted. After completion of coating, tablets were dried at 60°C for 20 minute. Then tablets were removed from the pan and evaluated for their physico-chemical parameter like average weight by Essae balance modal (DS-852J), diameter and thickness by Vernier calliper digital (MITUTOYO), friability by friabilator digital (TANCAO), hardness by (MONSATO) type tablet hardness tester and disintegration time by (HARISON PHARMA MACHINERY LTD.)

Results and Discussion

For Optimization of Extraction of Curcumin

In order to determine the optimum extraction conditions different sets of experiments were performed. The optimization of extraction conditions has been reported by (Goel *et al.*, 2011) elsewhere. The results show that the yield of dye depends a lot on extraction conditions i.e temperature, solvent, stirring time and solvent to meal ratio etc. It was observed that the dye yield first increased and then decreased with the increase of temperature. It might be due to increase in solubility with increase in temperature but beyond 60°C thermal degradation reduced the yield. So the extraction is carried out at 60°C (Sogi *et al.*, 2010). Thus it can be concluded that 60°C and 75 min stirring with ethanol (95 per cent) to meal ratio (1:10) is the best extraction conditions. Mandal *et al.* (2007) reported the high yield of curcumin in ethanol as compared to acetone.

The UV- visible spectra of separated curcumin was studied in ethanol. Curcumin exhibit absorption maxima at 420 nm which corresponds to the reported value (Singh., 1985). A bathochromic shift of absorption peak from 420 to 280 and 381nm confirms the acetylation of curcumin, (Figures 7.2 and 7.3). These were also characterized by their melting point 178°C and 182°C for curcumin and acetylated curcumin respectively (Andrew *et al.*, 2000).

Figure 7.4 shows the presence of three components curcumin and two curcuminoids ((DMC&BDMC) in the sample with the R_f values 0.75, 0.34, 0.48 respectively.

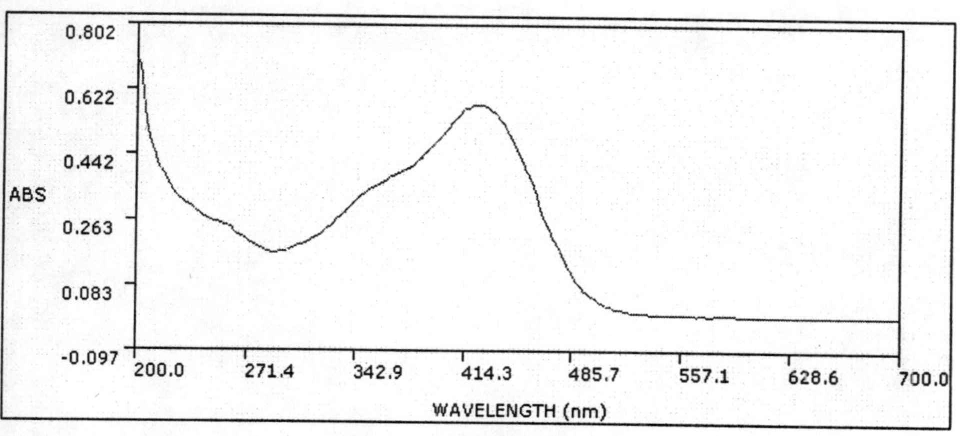

Absoption peak = 422
UV-Vis spectra of curcumin

Figure 7.2

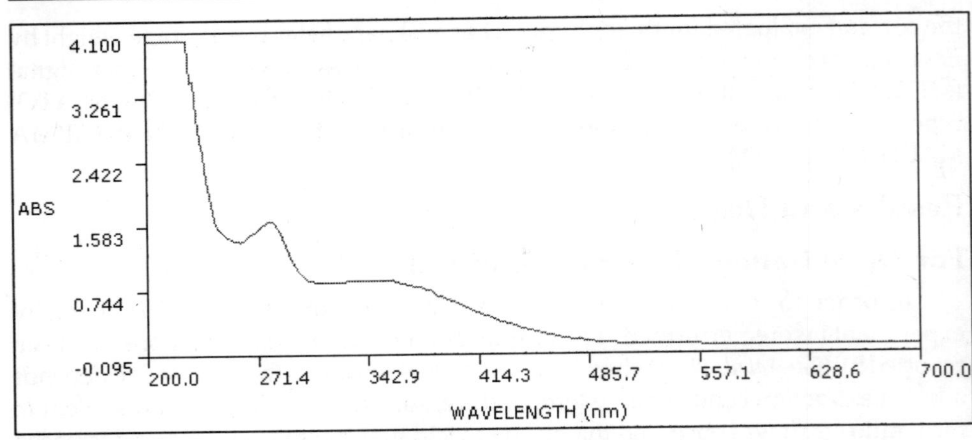

Absoption peak = 280
UV-Vis spectra of acetylated curcumin

Figure 7.3

Figure 7.4: TLC Profile of Curcumin (*Curcuma longa* L.)

Figure 7.5, compares the IR spectra of curcumin and acetylated curcumin. The FTIR spectra of Curcumin (Figure 7.5a) show bands at frequency 3200-3550 cm^{-1} due to O-H stretching and at 1550 cm^{-1} for C=C olefienic stretching. Other significant

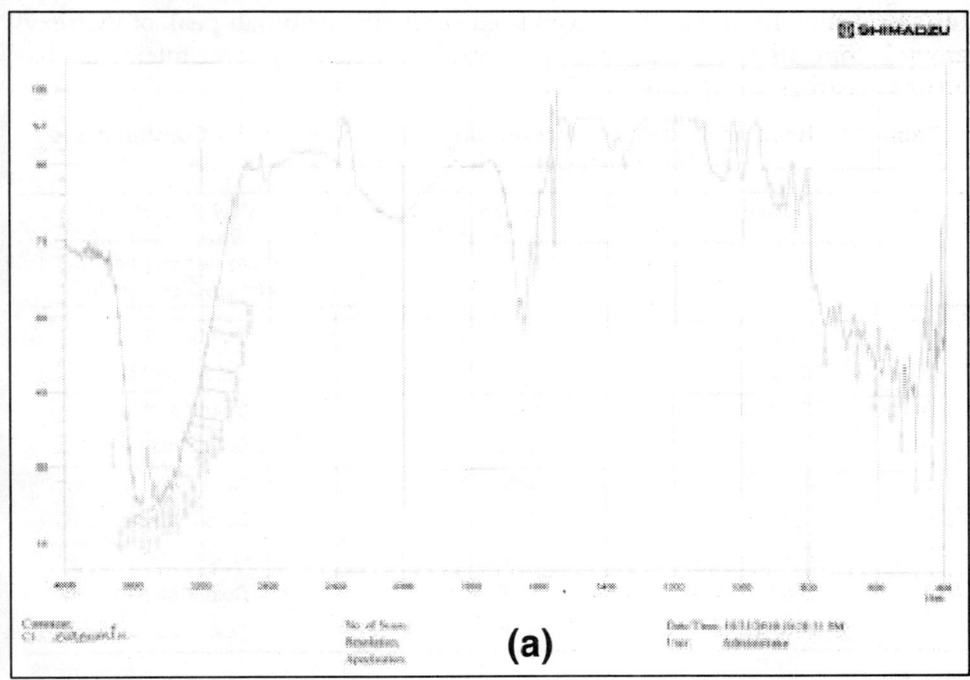

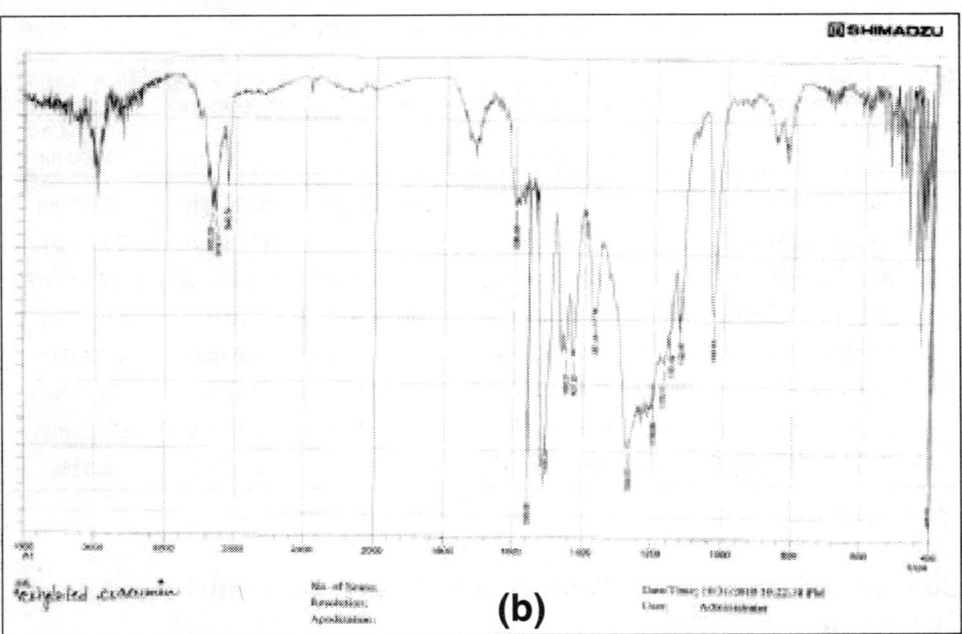

Figures 7.5(a) and (b)

bands were observed at 1625 cm^{-1} being assigned to the C=O stretching and at about 700,756, 840 cm^{-1} being assigned to the C=C-H stretching (Naama *et al.*, 2010). In acetylated curcumin (Figure 7.5b) a new peak appears at 1730 cm^{-1} showing strong carbonyl absorption. Due to C-O streatching two strong bands appears in the region 1450-1029 cm^{-1}. In IR spectra of acetylated curcumin the broad peak of hydroxyl group is converted in a sharp short peak which indicate that some interaction has occur at hydroxyl group (Bong, 1999).

Table 7.1: Results for Effect of Temperature and Humidity on the Curcumin and Acetylated Curcumin Coated Tablets

Sl.No.	*Time (Jun to Aug 2011)*	*Tablet Coated with Curcumin (Room Temp-19-40°C, RH-65±5 per cent)*	*Tablet Coated with Acetylated Curcumin (Room Temp-19-40°C, RH-65±5 per cent)*
1.	0-10 days	Colour stable	Colour stable
2.	10-20 days	Colour stable	Colour stable
3.	20-30 days	Colour stable	Colour stable
4.	30-40 days	Colour stable	Colour stable
5.	40-50 days	Colour stable	Colour stable
6.	50-60 days	Colour stable	Colour stable
7.	60-70 days	Fading	Colour stable
8.	70-80 days	Fading	Colour stable
9.	80-90 days	Fading	Colour stable

Table 7.2: Results for Coated Tablets

Sl.No.	*Physico-chemical Parameter*	*Standard (I.P. and Inhouse of Divya Pharmacy, Haridwar)*	*Before Coating*	*After Coating with Curcumin*	*After Coating with Acetylated Curcumin*
1.	20 Tablets weight	10.0±5%	10.05 gm	10.26 gm	10.28 gm
2.	Average weight	0.500±5%	0.502 gm	0.513 gm	0.514 gm
3.	Average Hardness	Not less than 1 kg/cm²	1.51kg/cm²	2.43 kg/cm²	2.25 kg/cm²
4.	Average diameter	10.40mm-10.80 mm	10.41mm	10.68 mm	10.69 mm
5.	Average Thickness	5.60 mm-6.20mm	5.63 mm	5.88 mm	5.89 mm
6.	Average Friability	Not more then 1%	0.89%	0.48%	0.19%
7.	Disintegration time	Not more then 60 min	38.5 6min	51.17 min	59.23 min
8.	Loss on drying	Not more them 8%	7.61%	4.75%	4.83%

I.P.: Indian Pharmacopea.

Coating of Ayurvedic Tablets with Curcumin and Acetylated Curcumin

The coated tablets are useful to extend the shelf life of components that are sensitive to moisture or oxidation and to cover up the bed smell of the active ingredients of tablets (Anwar *et al.*, 2007). The TGA tablets coated with curcumin and acetylated curcumin were exposed to environmental conditions. To study the effect of environmental factors on coated tablets, the tablets were packed in propylene pouches and subjected to storage at room temperature (19-40°C) and 65 ± 5 per cent RH for three months *i.e.* June to August 2011. The observations presented in Table 7.1 show that between room temperature (19–40°C) and 65±5 per cent RH condition, the curcumin coated tablets are stable up to two months and after that fading starts. It may be due to fast degradation of curcumin (Tonneson *et al.*, 1985, Khurana *et al.*, 1988), while acetylated curcumin coated tablets are stable up to three to four months. Further the suitability of coating was checked by studying different physico- chemical parameters. The important properties of the core tablets such as hardness, friability, must be of particular attention for preventing tablets fragmentation during coating process (Porter *et al.*, 1990). The observations of curcumin coated tablets (CCT) and acetylated curcumin coated tablets (ACCT) were shown in the Table 7.2. The data show that the results of all parameters are within the prescribe limits. Average friability of coated tablets decreased from 0.89 per cent to 0.48 per cent and 0.19 per cent respectively, while disintegeration time increased from 38.56 to (without coated tablets) 51.17 min (CCT) and 59.23 min (ACCT) respectively. The loss on drying was 7.61 per cent and after coating it decreases to 4.75 per cent, 4.83 per cent for CCT and ACCT respectively. Percentage LOD is a measure of the moisture content of the tablets. It can be extremely important to both for core tablet and drug stability. (Patel *et al.*, 2009). Initially hardness of uncoated tablets is 1.51 kg/cm^2 while after coating it increases from 2.43 kg/cm^2 to 2.25 kg/cm^2. The decreases in LOD and friability and increase in disintegration time and hardness are favourable for coated tablets. All the results like disintegration time, friability, hardness, LOD etc are with in the limits. But ACCT show lower solubility in water and 0.1N acid.

Conclusion

Thus from the above results it can be concluded that at 60°C and 75 min stirring time with ethanol to meal ratio (1:10) the extraction of dye is good. After coating, the stability of curcumin coated tablets is only for two months, while for acetylated curcumin coated tablets it is three to four months, but on disintegration testing the ACC tablets, has low solubility. Thus the coating of TGA tablets with CCT and ACCT show that both are suitable for coating of TGA but ACCT are more stable than CCT.

Abbreviations

TGA: Triphala guggle ayurvedic tablet

C.C.T: Curcumin coated tablet

A.C.C.T: Acetylated Curcumin Coated tablet

RH: Relative humidity

References

Anderson A.M., Mitchell M.S., Mohan R.S., Isolation of curcumin from Turmeric, Journal of Chemical Education, (2000); 77(3), 359-360.

Allam K V., Kumar G.P., COLORANTS - The cosmetics for the Phamaceuticle dosage forms International Journal of Pharmacy and Pharmaceuticle science, (2011); 3(3), 13-21.

Anwar E., Arsyadi., and Kardono L.B.S., Study of coating tablet extract noni fruit (*Morinda citrifolia*,L.) with maltodextrin as a subcoating material. J. Med. Sci. (2007); 7(5) 762-768.

Bong P., Spectral and photophysical behavior of curcumin and curcuminoids. Bulletin Korean Chemical Society, (1999); 21(1):81-86.

Cole G.C., Pharmaceutical Coating Technology, Taylor and Francis Ltd. (1998); 6-52.

FAO, Evaluation of certain food additives. FAO/WHO Expert Committee Report on Food Additives, WHO Technical Report Series 891, WHO Geneva. (2000); 26-27.

Govindarjan V.S., Turmeric-Chemistry, Technology, and Quality. CRC, Crit Rev Food Sci Nutr. (1980); 12: 199-301.

Goel Anjali., Bhardwaj M.K., Rani Neetu., Application of turmeric dye in coating of Ayurvedic tablets, Journal of applied and natural science, (2011); In press.

Indian Pharmacopia, (2007); 7: 2041-2047.

Khurana A., and Ho T.Chi., High performance liquid chromatographic analysis of curcuminoids and their photo oxidative decomposition compounds in *Curcuma longa* L., Journal of Liquid Chromatography, (1988); 11(11) 2295-2304.

Mandal V., Mohan Y., Hemlata S., Optimization of curcumin extraction by microwave assisted *in-vitro* plant bursting by orthogonal array designed extraction process and HPTLC analysis, Pharmacognosy Magzine (2007); 3(11): 132-138.

Naama J. H., Ali. A., Temimi Al, and Ahmad. A., Hussain Al. Amiery (2010). Study the anticancer activities of ethanolic curcumin extract, African Journal of Pure and Applied Chemistry, (2010); 4(5), 63-73.

Porter S.C., and Bruno C.H., Coating of Pharmaceutical solid-dosage form, Pharm. Tech. (1980); 4(3), 66-69.

Porter S.C., Verseput R.P., and Cunningham *C.R.*, Process optimization using design of experiments, Pharm.Techol, (1997); 21, 60-70.

Porter S.C., Bruno C.H., Coating of Pharmaceutical solid–dosage form. In: Liberman, H.A.,L, Lachman and J.B Schwatz (Eds.) Pharmaceutical dosage form, Marcel. Dekker Inc., Newyork, (1990); 77-88.

Patel J.K., Shah A. M., Sheth. N.R., Aqueous-based Film coating of Tablets: Study the effect of critical process parameters, International Journal of Pharm Tech Research, (2009); 1(2) 235-240.

PFA Prevention of Food Adulteration Act 1954.as amended Sixteenth Edition with supplement 1955 Lucknow, India: Eastern Book Company. (1994); 53.

Rouhani Sh., Alizadeh., N, Salimi.Sh., Haji-Ghasemi,T, Ultrasonic Assisted Extraction of Natural pigment From Rhizome of *Curcuma Longa,L.*, 2 (2009); 103-113.

Revathy S., Elumalai, S., Benny Merina., Isolation purefication and Identification of Curcuminoids from Turmeric (Curcuma longa L.,) by Coloum Chromatography, Journal of Experimental science (2011), 2(7) 21-25.

Rajpal. V., STANDARDIZATION OF BOTANICALS. (Testing and extraction methods of medicinal herbs) (2005); (II), 128-129.

Ruotsalainen. M., Studies on aqueous film coating of tablets in side vented perforated pan coater, (2003); ISBN 352-10-1041-X Helsinki.

Rao D.B, Sekhera N.C, Satyanarayan M. N, Srinivasan M, Effect of curcumin on serum and liver Cholesterol level in the rat, J.Nutr. (1970); 100: 1307-1316.

Singh A, Enhancement of turmeric (*Curcuma longa*) growth, rhizome yield and its yellow pigment content by processing chilling treatment to rhizomes, J India Bot Soc, 64, (1985), 210-213.

Showbagya H.B., Samhita, S. Krishnamurty, S.R. Bhattacharya, N and S. Stability of water soluble turmeric colorant in an extruded food product during storage. (2005); Journal of food Engineering 67(3): 367-371.

Sheth N., Potdar Arti., and Shah. Aanand., Shah. Sunny., Studies in optimization of Aquous film Coating parameters, International Journal of Pharmaceuticle Science and nanotechnology, (2009); 2(3), 621-626.

Stanley J. Tucker., Arnold E. Nicholson., Herbert E ngelbert., Tablet color coating with Pigment Journal of American Pharmaceutical Association. (2006); 47(12), 849–850.

S Gouin, "Microencapsulation: industrial appraisal of existing technologies and trends" Trends in Food Science and Technology, (2004); 15(7-8), 330–347,

Sowbhagya, H. B. Sampathu, S. R. Vatsala, C. N. and Krishnamurthy, N. "Stability of curcumin, a natural yellow colourant during processing and storage of fruit bread," Beverage Food World, (1998); 25(4), 40–43,

Sogi D.S., Sharma. S., Oberoi. P.S., Wani. I.A., Effect of extraction parameters on curcumin yield from turmeric. J. Food Sci Technol (2010); 47(3):300-304

Tonnesen H.H., and Karlsen J., Studies of curcumin and curcuminoids: VI. Kinetics of curcumin degradation in aqueous solutions. Z. Lebensm.Unters.Forsch, (1985); 180: 402-404.

Tonnesen H. H., Masson M., and Loftsson T., "Studies of curcumin and curcuminoids. XXVII. Cyclodextrin complexation: solubility, chemical and photochemical stability, "International Journal of Pharmaceutics, (2002); 244, (1-2) 127–135.

Umbreen Saima., Ali, Shaukat. Hussain, Tanveer. and Nawaz, Rakhshanda. Dyeing properties of natural dyes extracted fron turmeric and their comparison with reactive dyeing RJTA, (2008); 12(4). 1-11

Zebib. Bachar., Moloungui, Zephirin and Noirot, Virginie., Stabilization of curcumin by Complexation with divalent cations in Glysrol/Water system, Bioinorganic Chemistry and Applications, (2010); 1-8.

Zaibunnisa. A.H, Norashikin, S., Mamot,S. and Osman, H, Stability of Curcumin in turmeric oleoresin-β-cyclodextrin inclusion complex during storage, The Malaysian Journal of Analytical Science, 13(2), (2009); 165-169.

2013, Environmental Biotechnology *Pages* ***79–86***
Editors: **D.R. Khanna, A.K. Chopra, Gagan Matta, Vikas Singh & Rakesh Bhutiani**
Published by: **BIOTECH BOOKS, NEW DELHI**

Chapter 8

Trace Element Content in Fingernails and Hair of a Non Industrialized Population

H.P. Sapkal

Department of Zoology, Shri Shivaji College, Akola, M.S.

Sampling of human beings for trace element content can be a complex and costly procedure. The use of hair and nails simplifies the process and it is for this reason that such samples are extensively used. It is observed that there exists some positive correlation between element levels in hair and nails, hypertension, and diabetes of these subjects. When studying the relationship of minerals to human health, it becomes increasingly evident that keeping a balance level of minerals in every organ, tissue and cell of the human body may be a prominent key to maintaining a healthy existence.

Keywords: *Trace elements, Hairs, Nails, Spectrophotometer.*

Introduction

Much of what is "known" about the biological trace elements is a mixture of improbable fact and plausible nonsense. Over 99 per cent of all animal matter, including the human body is constructed from just 11 elements-hydrogen, oxygen, nitrogen, sodium, potassium, chlorine, sulphur, phosphorus, magnesium, calcium, and carbon-but about half the periodic table is represented in the remaining 0-01 per cent. With greater or lesser certainty about 10 elements have been recognized as essential-copper, iron, zinc, cobalt, iodine, molybdenum, manganese, selenium, chromium, and fluorine-but even this subgroup is heterogenous. Zinc, iron, and copper are built into many enzymes, whereas cobalt is found in only one molecule and iodine too has only one function. The inessential elements include: environmental hazards (lead, mercury, and cadmium); some that are given as drugs (bromine, aluminum, and gold); and popular poisons (arsenic and antimony). The rest seem to be contaminants. What the trace elements have in common is a capacity to enthrall both laymen and scientists. Over 2000 years ago the Greeks burnt sea sponges as a prophylactic against goiter, rightly divining a magic element in the vapor; and as recently as the 1960s a locally common cardiomyopathy was traced to selenium deficiency in central China. Against this background of legend and newsworthiness the analytical advances of the past decades have created new problems as well as new knowledge. Only eight trace elements are generally accepted as being essential for health and wellbeing in higher animals through the consumption of food and beverages; these are cobalt, copper, iodine, iron, manganese, molybdenum, selenium, and zinc. Persuasive evidence has recently appeared that indicates two other trace elements, boron and chromium, may also be essential; however, general acceptance of their essentiality is still lacking. (Weiss, D *et al.*, 1972).

The objective of this study was to assess the relationships between lead, cadmium, zinc, and copper and other trace elements contents in fingernails and hair, and the level of environmental exposure in the subjects' places of residence. After hundreds of fingernails and hair analysis, Trace Elements has created a unique system of interpreting nail mineral analysis results. Each test report will provide the clinician with the most complete and comprehensive evaluation and discussion of significant mineral levels, ratios and toxic metals as tested in the fingernails and hair. Included is a listing of individual foods and food groups that the doctor can recommend to eat or avoid in accordance with food allergy indicators and individualized metabolic requirements.

Material and Methods

Sampling Procedure

Contamination of the nail can occure from air, water and nail paints. This type of contamination is removed during sample washing procedure. Nails, particularly toenails, are relatively sheltered from environmental contaminants. They are free from or less likely to have contaminants introduced through shampooing, hair treatments and medication. As for nail polish, the chemicals introduced by polishing can be largely washed out in the laboratory, and the element contents in nails are less likely to be affected. For instance, any external contamination in most nail specimens can be removed using ultrasonic cleaning protocols with both polar and nonpolar solvents. To obtain more nail masses, participants should be asked in advance not to trim their nails for a couple of weeks or longer. Nails are collected by clipping with a stainless steel clipper from the two great toes (or thumbs) and small toes (or other fingers) (Hambidge *et al.*, 1987; Miller *et al.*, 1979).

Contamination of the hair can occur from air, water, perspiration and shampoos. This type of contamination is removed during sample washing procedure. Howere,

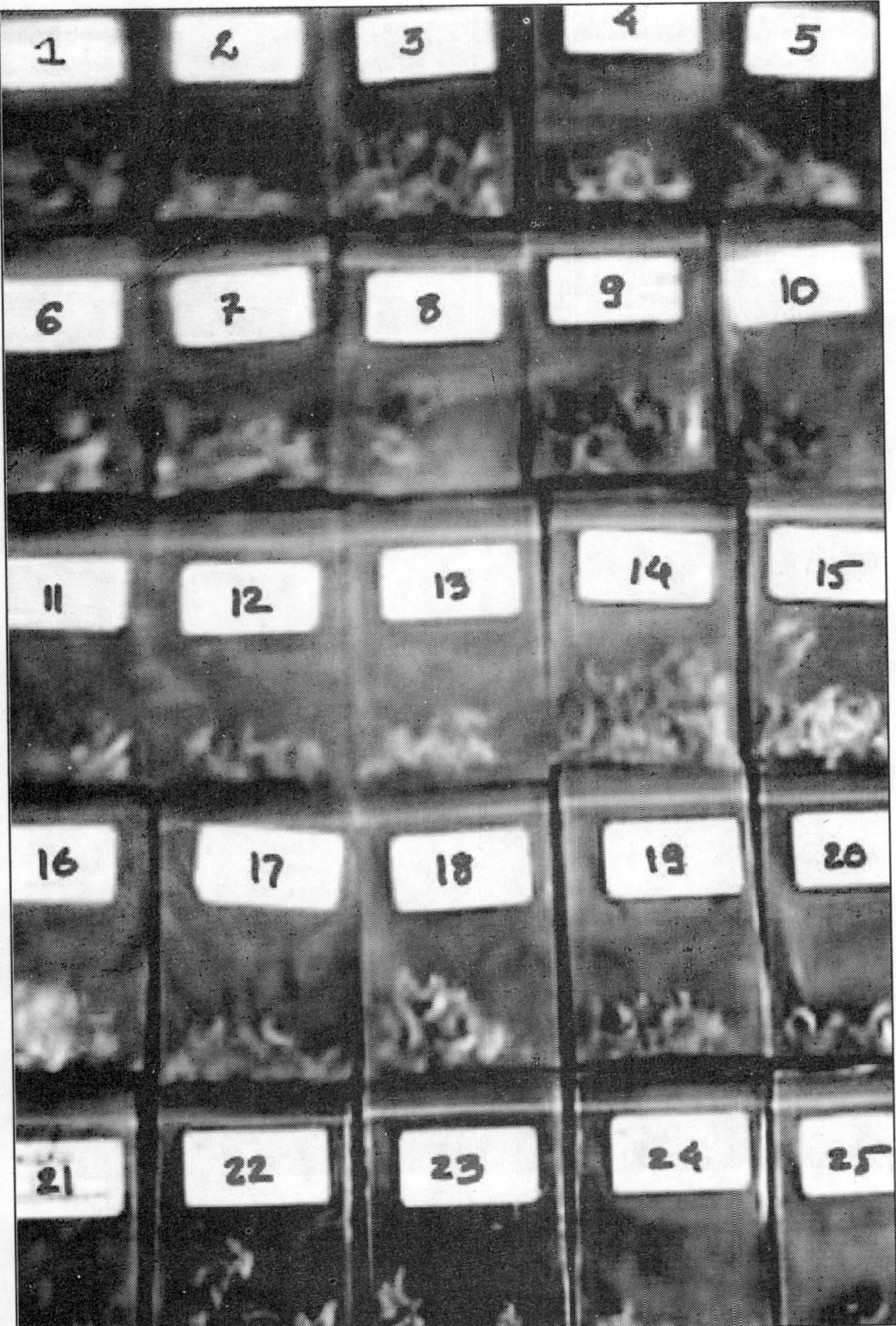

Figure 8.1: Nails Sample Collected According to Age, Sex and Locality

Figure 8.2: Hair Sample Collected According to Age, Sex and Locality

contamination from dyes, permanents or bleaching of hair cannot be removed. These cosmetic procedure permanently change the structure of the hair and therefore, only natural hair (head, pubic or beard hair) is suitable for hair minerals analysis. Underarm hair is unsuitable for analysis. As hair grows, nutrient and toxic elements

Figure 8.3: Digested Hairs

Figure 8.4: Digested Fingernails

are deposited from the blood stream into the hair follicle and hair shaft. Once a trace element has been incorporated into the hair, it remains fixed.

After drying of fingernails and hair samples, add few drops of concentrated nitric acid on the samples which is kept in vials. The procedure is used for the digestion of samples.

These ash samples now ready for atomic absorption spectrophotometer analysis (Yoshinaga *et al.*, 1993).

Observation and Result

Lead and zinc concentrations in hair and nail samples were determined by atomic absorption spectrophotometer (AAS). The mean zinc concentration in hair

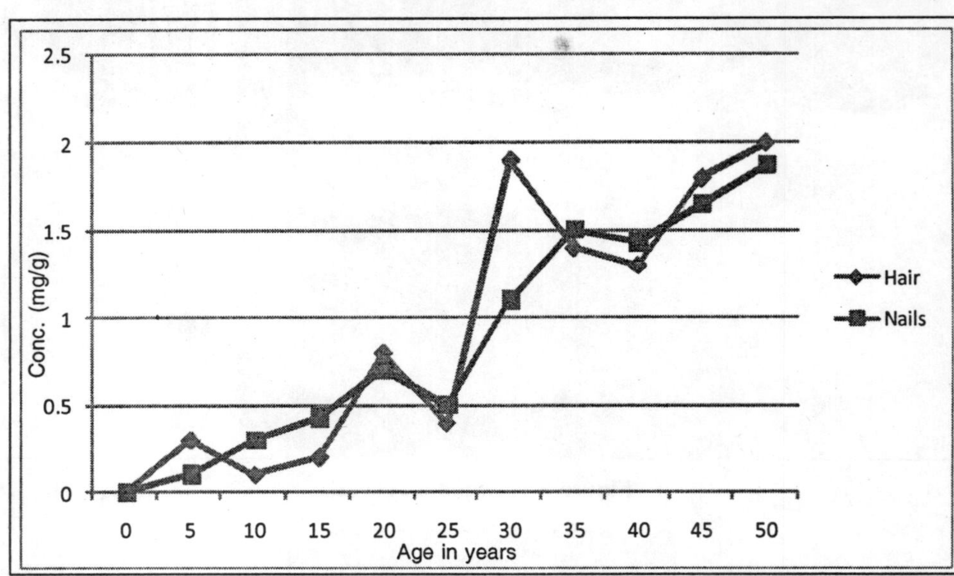

Figure 8.5: Concentration of Zinc with Respect to Age

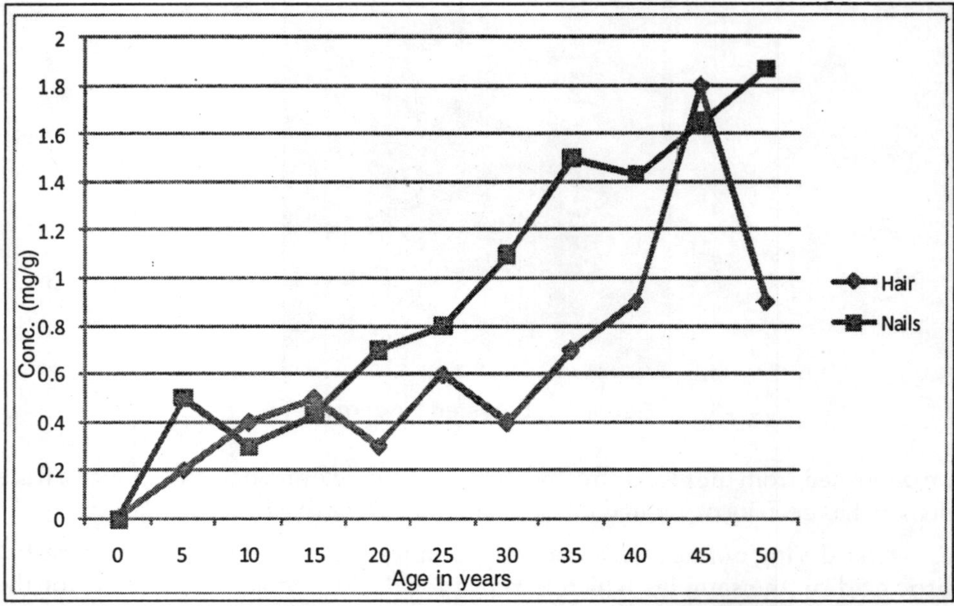

Figure 8.6: Concentration of Lead with Respect to Age

and nail were 0.743 ± 0.23 mg/g and 0.7820± 0.67 mg/g respectively while the mean lead concentrations in hair and nail were 0.334 ±0.44 and 0.460 ± 0.304 mg/g respectively. A progressive increase in zinc concentrations in hair and nails with age indicated no significant difference when their means were compared suggesting that zinc in hair and nails originate from a common source, comparing the mean lead concentrations in hair with the nails a significant difference is indicated in the 2 tissues (Strain *et al.*, 1972) . Human hair and nails are therefore recording filaments that can reflect metabolic changes of many elements over long periods of time and hence furnish a print out of post nutritional event as dietary levels of some of the essential micro-elements.

Summary

Elevated zinc in hair is an unusual finding. Occupational or hobbyist exposure to zinc through paints, metalwork, or chemical use can cause elevation. Zinc may be present in detergents and bleaches. Exposure may be related to dermatitis because zinc is a common skin irritant in allergic eczema. Some forms of zinc (*i.e.*, hexavalent) are highly toxic, but acute exposure to such compounds is rare. The high incidence of allergies among the people may in part be related to the presence of zinc in cement. Elevations of hair zinc have been reported only during the special metabolic needs of pregnancy, where the high values were present with evidence of zinc deficiency. Some Indian diets have lead-copper ratios in excess of those that produce hypercholesterolemia. Also, the mortality rate for child was correlated with the ratio of lead to copper in milk in the region. It was found that the abnormal ratio of the trace elements was due to the poor economic condition that was directly affected the diet of the people.

References

Sela H, Karpas Z, Zoriy M, Pickhardt C, Becker JS (2007): Biomonitoring of hair samples by laser ablation inductively coupled plasma mass spectrometry (LA-ICP-MS). Int J Mass Spectrom 261:199– 207

Benner BA, Levin BC (2005) :Hair and human identification. In: Tobin DJ (ed) Hair in toxicology: an important bio-monitor. RSC, Cambridge, pp 127–159

Yukawa M, Suzuki-Yasumoto M, Tanaka S (1984): The variation of trace element concentration in human hair: the trace element profile in human long hair by sectional analysis using neutron activation analysis. Sci Total Environ 38:41–54

Shamberger RJ (2002): Validity of hair mineral testing. Biol Trace Elem Res 87:1–28

Phelps RW, Clarkson TW, Kershaw TG, Wheatley B (1980): Interrelationships of blood and hair mercury concentrations in a North American population exposed to methylmercury. Arch Environ Health 35:161– 168

Weiss D, Whitten B, Leddy D (1972): Lead content of human hair (1871–1971). Science 178:69–70

Giovanoli-Jakubczak T, Berg GG (1974): Measurement of mercury in human hair. Arch Environ Health 28:139–144

Maugh TH (1978): Hair: A diagnostic tool to complement blood serum and urine. Science 202:1271– 1273

Yoshinaga J, Shibata Y, Morita M (1993): Trace elements determined along single strands of hair by inductively coupled plasma mass spectrometry. Clin Chem 39:1650–1655

Dombovári J, Papp L, Uzonyi I, Borbély-Kiss I, Elekes Z, Varga Z, Mátyus J, Kakuk G (1999): Study of cross-sectional and longitudinal distribution of some major and minor elements in the hair samples of heamodialysed patients with micro-PIXE. J Anal Atom Spectrom 14:553–557

2013, Environmental Biotechnology *Pages* **87–95**
Editors: **D.R. Khanna, A.K. Chopra, Gagan Matta, Vikas Singh & Rakesh Bhutiani**
Published by: **BIOTECH BOOKS, NEW DELHI**

Chapter 9

Target Yield Concept of Fertilizer Recommendation for Cabbage Grown on a Mollisol of Uttarakhand

Jyoti Pande and Sobaran Singh
Department of Soil Science, G.B.P.U.A &T.,
Pantnagar – 263 145, Udham Singh Nagar, Uttarakhand

The response of cabbage to the selected combinations of 4 levels of N, P, K and 3 levels of FYM with simultaneous variations in initially available soil forms of these nutrients was studied under Soil Test Crop Response (STCR) calibration in an Aquic Hapludoll of Norman E. Borlogue Crop Research Centre of G.B. Pant University of Agriculture and Technology, Pantnagar (29°N latitude and 79°29′ E longitude) in year 2008 and 2009. The head yield of cabbage and soil and plant analysis data were utilized to formulate equations each for inorganic and organic mode for fertilizer recommendation with varying yield targets at different fertility levels. The validity of the equations generated has been tested in the verification trial. The results showed that fertilizer application based on target yield gave higher yields, net benefit and B/C ratios over the farmer's practice. The results of verification trial clearly indicate the superiority of target yield over other approaches.

Keywords: *Cabbage, Soil test crop response, Yield targets, Fertilizer adjustment equations.*

Fertilizer constitutes one of the costliest inputs in present day agriculture. Greater economy in fertilizer use can be made, if fertilizers are applied on the basis of soil test. This practice ensures balanced fertilization, higher yield and more profitability. Therefore, it becomes very necessary to develop a comprehensive approach to fertilizer recommendation incorporating soil test, field experimentation and economic evaluation of results. The methodology of targeted yield approach was developed by Ramamoorthy *et al.* (1967). Targeted yield approach provides a scientific basis for balanced fertilization not only among the fertilizer nutrients but also the soil available nutrients.

Cabbage is one of the most economically important vegetable crop of genus *Brassica*. In India, it ranks third next to cauliflower and onion in area and production. India is the third largest cabbage producer in the world. In Uttarakhand the various cabbage growing belts are Almora, Chamoli, Uttarkashi and Dehradun. In Uttarakhand, cabbage covers an area of 4,434 ha with the production of 22,957 mt (Fertilizer statistics, 2007-08). Cabbage is well known for its nutritive value and health benefits. It is an excellent source of minerals such as calcium (39 mg), iron (0.8 mg), magnesium (10 mg), sodium (14.1 mg), potassium (114 mg) and phosphorus (44 mg). It has substantial amounts of β-carotene (provitamin-A), ascorbic acid, riboflavin, niacin and thiamine. Cabbage has anti cancer property, it protects against bowel cancer due to the presence of indole-3-carbinol. The leaves are used to cover ulcers and wounds. It is said to help in digestion. Despite its wide popularity, nutritive value and health benefits its productivity is still low. The main reason for this appears to be inadequate and imbalanced fertilizer application. Moreover no systematic study has been conducted so far for fertilizer requirements based on soil test values for cabbage grown in Mollisols. The present study was, therefore, undertaken with a view of evolving soil test based fertilizer recommendations for cabbage and to test their adaptability under farmer's field condition.

Materials and Methods

The experimental plot was located on a sandy loam soil and classified as Aquic Hapludoll (Deshpande *et al.*, 1971). The experiment was conducted in two phases during 2008-09 and verification trial during 2009-10. In a preparatory trial during kharif 2008, the whole plot was divided into 3 equal strips and fertility gradients were created artificially through differential application of fertilizers and growing fodder sorghum (var. NTJ-2) as a fertility gradient stabilizing crop in the preceding crop season to enable thorough interaction between the nutrients in the soil and those added through fertilizer (Figure 9.1). In the main experiment in Rabi 2008-09 with test crop cabbage, every strip was further sub-divided into 24 equal plots of 12 m^2 (4m X 3m) having 21 treated and 3 controls as per the treatment plan. The response of cabbage var. 'Pride of India' under selected treatment combinations 4 levels for each N (0, 90, 120 and 150 kg ha^{-1}), P (0, 30, 60 and 90 kg ha^{-1}), K (0, 30, 60 and 90 kg ha^{-1}) and 3 levels of FYM (0, 10 and 20 t ha^{-1}) was studied. Before fertilizer application, soil samples from individual plots at 0-15 cm depth were collected and analyzed for Alkaline $KMnO_4$ oxidizable nitrogen (Subbiah and Asija, 1956), Olsen's Phosphorus (Olsen *et al.*, 1954) and ammonium acetate extractable potassium (Hanway and

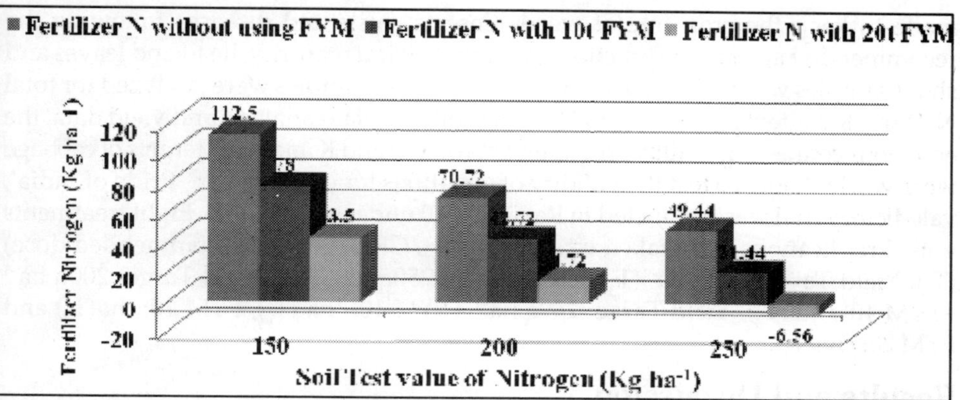

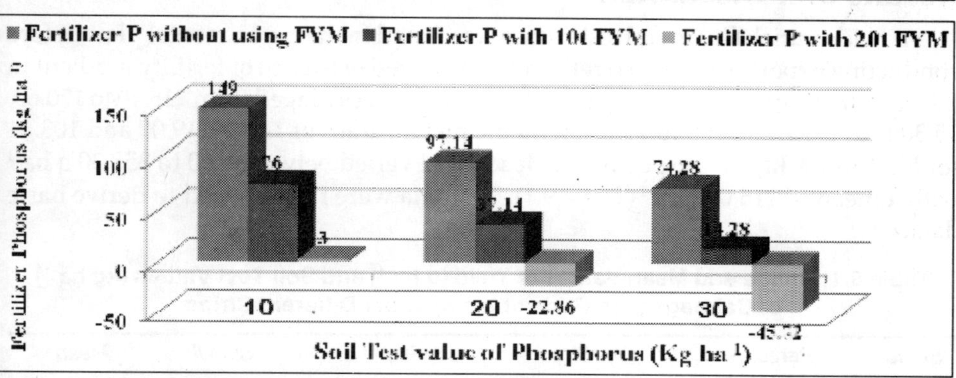

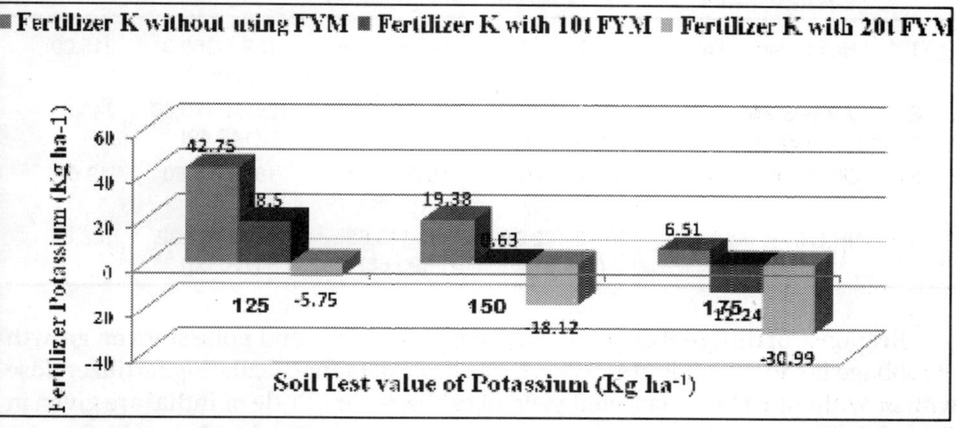

Figure 9.1: Soil Test Based N, P and K Fertilizer Doses for 200q ha^{-1} Yield Target of Cabbage

Heidal, 1952). Full dose of P and K and half dose of N were applied as basal and remaining half of nitrogen was applied in two splits one at one month after transplanting and another at two months after transplanting. Healthy disease free 4 weeks old cabbage seedlings were transplanted in the field at 50×40 cm spacing and light irrigation was given soon after transplanting to ensure better establishment of plants in soil. Irrigation was given to the crop as and when required. An attempt was made to keep the crop free of weeds, insects, pests and diseases following the recommended agronomic practices. At physiological maturity, head and leaves and shoot samples were collected and processed. These samples were analyzed for total N, P and K contents (Jackson, 1973). Using soil and plant analysis and yield data, the equations connecting fertilizer requirements of N, P and K and yield targets of cabbage were worked out. To test the validity of equations for cabbage var. 'Pride of India', valedictory trial was conducted in Rabi 2009-10 under simple RBD. Eight treatments with 3 replications were taken *viz.* control (T_1), GRD (general recommended dose) (T_2), Yield Target 200q ha^{-1}(T_3), Yield Target 250q ha^{-1} (T_4), Yield Target 200q ha^{-1} +FYM 10t/ha (T_5), Yield Target 250q ha^{-1} +FYM 10t/ha (T_6), FYM 10t/ha(T_7) and FYM 20t/ha(T_8).

Results and Discussion

A wide variability in soil test values was found in the experimental plots before conducting experiment, which reflected the desired existence of fertility gradient in the field. In the present investigation, the soil test values ranged from 115.40 to 170.60, 15.30 to 25.50 and 84.00 to 152.32 with the mean values of 143.94, 19.07 and 108.84 for N, P and K kg ha^{-1} respectively. Head yield varied between 100 to 358.30 q ha^{-1} with a mean of 215.07q ha^{-1} (Table 9.1). The data were further used to derive basic data.

Table 9.1: Range and Mean Values of Yield (q ha^{-1}) and Soil Test Values (kg ha^{-1}) of Cabbage (var. Pride of India) under Different Strips

Sl. No.	*Particular Range(Mean)*	*Strip I Range(Mean)*	*Strip II Range(Mean)*	*Strip III*	*Mean*
1.	Head yield (q ha^{-1})	100-250.6 (182.4)	143.6-298.8 (214.6)	158.5-358.3 (248.2)	215.00
2.	Alkaline $KMnO_4$-N (kg ha^{-1})	115.40-163.07 (137.64)	132.97-163.07 (146.71)	125.44-170.60 (147.49)	143.94
3.	Olsen's-P (kg ha^{-1})	15.30-19.01 (17.50)	16.23-19.93 (17.85)	18.08-25.50 (21.88)	19.07
4.	NH_4OAc-K (kg ha^{-1})	90.72-122.08 (106.21)	84-135.52 (99.68)	97.44-152.32 (120.63)	108.84

Response of different doses of nitrogen, phosphorus and potassium on growth of cabbage is shown in Figures 9.2–9.4. The basic data for calculating fertilizer dose with or without FYM for targeted yield of cabbage var. 'Pride of India' are given in Table 9.2. The perusal of these data indicate that 0.92, 0.10 and 1.03 kg of N, P and K are required for both with or without FYM to produce 1 quintal cabbage heads. The contribution of soil in supplying nitrogen, phosphorus and potassium for the crop was 59.00, 66.00 and 123.00 per cent, respectively. The values of fertilizer efficiency were 85.00, 9.00 and 126.00 for without FYM and 104.00, 11.00 and 162.00 per cent with FYM for N, P and K respectively. The values of per cent contribution from applied FYM were 59.00, 14.00 and 53.00 for N, P and K respectively. More than 100 per cent contribution of potassium from fertilizer as per cent of its applied potassium

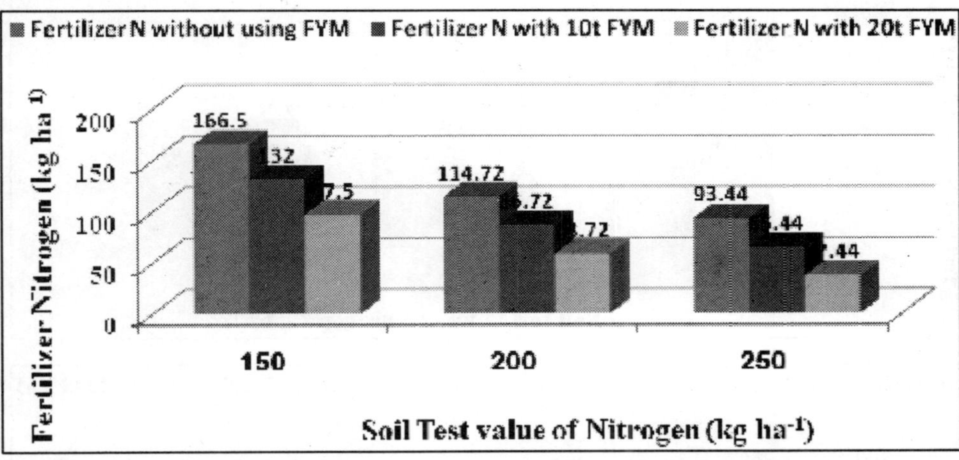

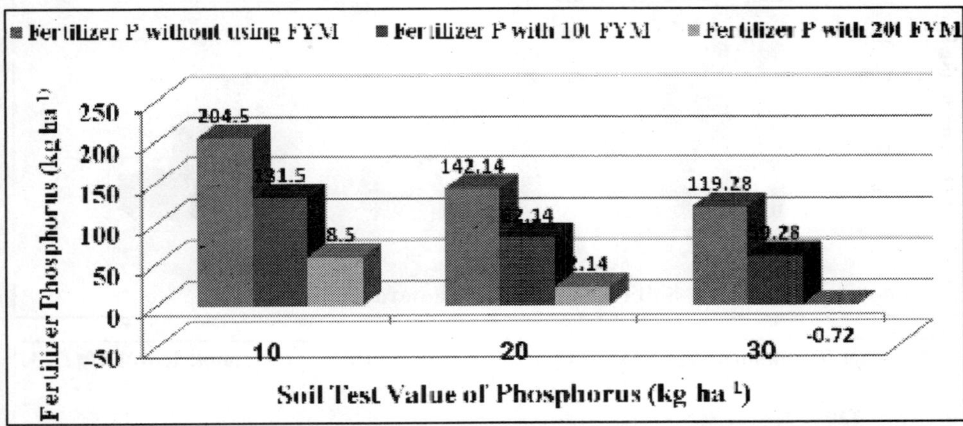

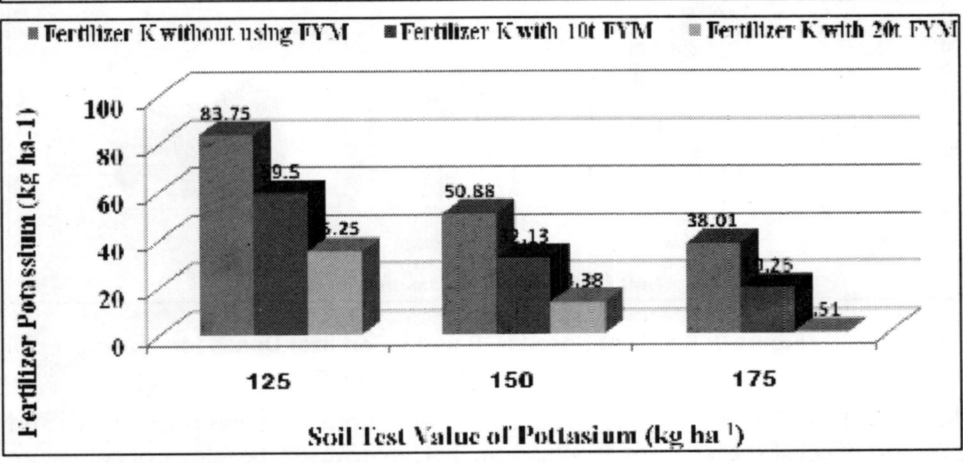

Figure 9.2: Soil Test Based N, P and K Fertilizer Doses for 250q ha^{-1} Yield Target of Cabbage

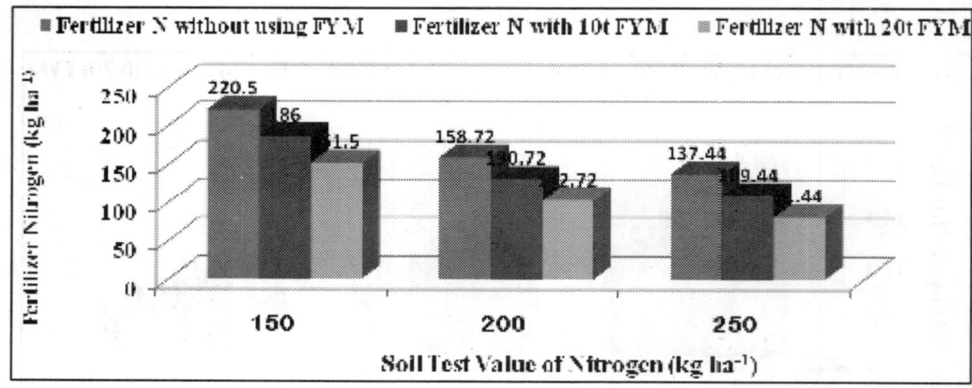

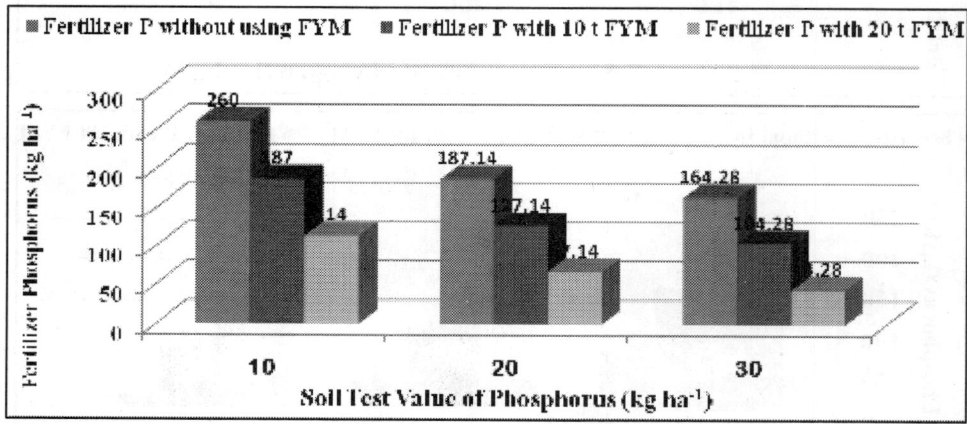

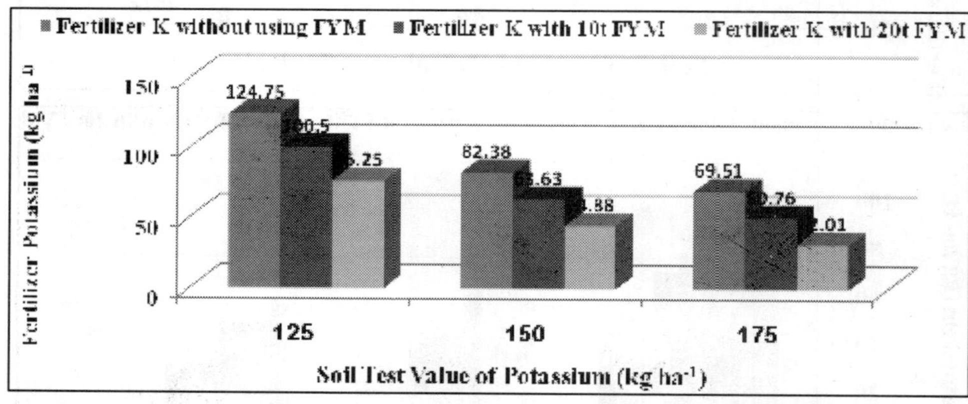

Figure 3: Soil Test Based N, P and K Fertilizer Doses for 300 q ha^{-1} Yield Target of Cabbage

(K_2O) is due to the additive effect of higher doses of nitrogen and phosphorus and 'priming' effect of starter doses of potassium that have caused the release of potassium from non-labile pool to labile pool, which resulted in increased uptake of potassium from native soil sources. Thus, here fertilizer potassium overestimated which had enhanced the efficiency of fertilizer potassium.

Final computations by using these basic data following simple fertilizer adjustment equations for targeted yield of cabbage for both with or without FYM were worked out (Table 9.2). Fertilizer prescription equations were transformed in to ready reckoners for requirement of fertilizer for three different yield targets *i.e.* 200, 250 and 300 q ha^{-1} of cabbage on soils with varying soil test values for both without FYM and with 10 t ha^{-1} and 20t ha^{-1} FYM (Figures 9.1–9.3). Fertilizer rates increased with increasing yield targets of cabbage and decrease with increasing the soil test values. It is obvious from these findings that there was net saving of fertilizers with the use of FYM for each target.

Table 9.2: Basic Data and Fertilizer Adjustment Equations for Calculating Fertilizer Dose with and without FYM for Targeted Yield of Cabbage

Particulars of Basic Data	*Without FYM*			*With FYM*		
	N	*P*	*K*	*N*	*P*	*K*
Nutrient required (kg q^{-1} of Head Yield)	0.92	0.10	1.03	0.92	0.10	1.03
Percent contribution from soil (per cent)*	59	66	123	59	66	123
Percent contribution from applied fertilizer (per cent)	85.00	9	126	104	11	162
Contribution from applied FYM nutrients (per cent)	—	—	—	59	14	53

Fertilizer Dose	*Fertilizer Adjustment Equations*	
	Without FYM	*With FYM*
Nitrogen Dose (kg ha^{-1})	FN = 1.08T- 0.69 SN	FN = 0.88T-0.56 SN-0.56 FYM-N
Phosphorus Dose (kg ha^{-1})	FP = 1.11T- 7.3SP	FP = 0.9T-6SP-1.27 FYM-P
Potassium Dose (kg ha^{-1})	FK = 0.82T-0.97SK	FK = 0.63T-0.75SK-0.33 FYM-K

* Soil test values (0-15 cm. depth); alkaline $KMnO_4$-N (kg ha^{-1}), Olsen's-P (kg ha^{-1}) and NH_4OAc-K (kg ha^{-1})

Validity of Target Yield Equations Under Verification Trial

The validity of the targeted yield equations developed for cabbage by conducting verification trial on the same location. Among the treatments, all treatments gave significantly higher yield over control (Table 9.3). Among the treatments yield target 250 q ha^{-1} (T_4) and yield target 250 q ha^{-1} +20 t FYM ha^{-1}(T_6) gave significantly higher yield as compared to general recommendation (T_2). The variation in yield obtained from the targeted yield ranged from 14.94 to 30.14 per cent. The control (T_1) and application of only FYM@ 10t ha^{-1} (T_7) and 20t ha^{-1} (T_8) were least efficient in producing head yield of cabbage. Net benefit was found to be highest with treatment yield target 250 q ha^{-1} (T_4). Fertilizer application based on targeted yield approach was found to be superior over general recommended dose (GRD). Highest benefit cost ratio was found with yield target 200 q ha^{-1} (T_3) and highest response ratio was found with yield target 250 q ha^{-1} +20 t FYM ha^{-1} (T_6). An increase in profit over farmers' practice and general recommended dose of fertilizers was observed with increasing yield

Table 9.3: Economics of Verification Trial of Cabbage

Treatments	*Fertilizer Dose N-P-K-FYM (kg ha^{-1})*	*Actual Mean Yield (kg ha^{-1})*	*Additional Yield (kg ha^{-1})*	*Value of Additional Yield (Rs.)*	*Cost of Fertilizer (Rs.)*	*Net Benefit (Rs. ha^{-1})*	*B/C Ratio*	*Response Ratio*	*Yield Deviation (per cent)*
T1 Control	0-0-0-0	145	–	–	–	–	–	–	–
T2 General Recommendation	120-80-40-0	280	135	101250	3634	97616	26.86	56.25	–
T3 Yield Target 200 Q ha^{-1}	78-3-19-0	250	105	78750	1145.85	77604.15	67.73	105	+25
T4 Yield Target 250 Q ha^{-1}	132-59-60-0	300	155	116250	3428.65	112821.35	32.90	61.75	+20
T5 Yield Target 200 Q ha^{-1} + FYM 10 t ha^{-1}	43-0-1-10	260.28	115.28	86460	8021.35	78438.65	9.78	262	+30.14
T6 Yield Target 250 Q ha^{-1} + FYM 10 t ha^{-1}	87-22-32-0	287.36	142.36	106770	9302.15	97467.85	10.48	100.96	+14.94
T7 FYM 10 t ha^{-1}	0-0-0-10	170	25	18750	7500	11250	1.5	–	
T8 FYM 20 t ha^{-1}	0-0-0-20	190	45	33750	15000	18750	1.25	–	

Cabbage Rate = Rs. 7.5/kg; FYM cost = Rs.75/Q; N, P_2O_5 and K_2O cost (Rs./kg) = 11.95, 23.75 and 7.5 Rs./kg.

targets in cabbage and other vegetable crops with or without FYM which might be due to efficiency factor tended to increase in crop yield (Kadam and Sonar, 2006, Hariprakash and Subramanian, 1994 and Anonymus, 2000).

Conclusion

The fertilizer applied on the basis of yield targets gave the highest yield and net benefit over farmer's practice and also provided higher benefit to cost ratio, indicating superiority over other methods of fertilizer application. These equations therefore, could be used for making fertilizer recommendations for targeted yields of cabbage (Pride of India) in a Mollisol of Uttarakhand. The practice of fertilizing crops on the basis of yield targets is precise, meaningful and eco-friendly and it needs to be popularized among farmers to obtain higher productivity and profitability.

References

Anonymous. 2000. A Report of the Research Work on Soils 1999-2000, Mahatma Phule Agricultural University, Rahuri.

Deshpande, S.N.; Fahrenbacher, J.B. and Ray, B.W. 1971. Mollisols of *tarai* region of Uttar Pradesh, Northern India. 2. Genesis and classification. Geoderma 6: 195-201..

FAI (2008) Fertilizer Statistics 2007-08, The Fertilizer Association of India, New Delhi.

Hanway, J.J. and Heidel, H. 1952. Soil analysis methods as used in Iowa State soil testing laboratory. *Iowa Agric.* 57: 1-31.

Hariprakash Rao M. and Subramanian T.R. 1994. Fertlizer needs of vegetable crops based on yield goal approach in Alfisols of southern India. *Journal of the Indian Society of Soil Science* 42, 565.

Jackson, M.L. 1973. *Soil Chemical Analysis.* Prentice Hall of India Pvt Ltd. New Delhi.

Kadam, B.S. and Sonar, K.R. 2006. Targeted yield approach for assessing the fertilizer requirements of onion in Vertisols. *Journal of the Indian Society of Soil Science* 54 (4), 513-515.

Olsen, S.R.; Cole, C.V.; Watanabe, F.S. and Dean, L.A. 1954. Estimation of available phosphorus in soils by extraction with sodium bicarbonate. Circ. USDA No. 939.

Ramamoorthy, B.; Narasimham, R.L. and Dinesh, R.S. 1967. Fertilizer application for specific yield targets of Sonora 64. *Ind. Farming*. 17 (5): 43-44.

Subbiah, B.V. and Asija, G.L. 1956. A rapid procedure for determination of available nitrogen in soils. *Current Science* 25, 259-260.

2013, Environmental Biotechnology *Pages 97–100*
Editors: **D.R. Khanna, A.K. Chopra, Gagan Matta, Vikas Singh & Rakesh Bhutiani**
Published by: **BIOTECH BOOKS, NEW DELHI**

Chapter 10

Adoption of *Rhizobium* Biofertilizer Increases Yield of Groundnut Pods by Tribal Farmers of Nasik District

D.K. Aher

K.R.A. Art's, Science and Commerce College, Deola, Nasik

Use of excessive chemical fertilizers causes soil unfit for cultivation. Groundnut (*Arachis hypogea* L.) is an important oil seed crop. Asia accounts for 58 per cent of the global Groundnut area and 67 per cent of the Groundnut production. Uncertain rains, pest diseases, socio-economic status of farmers, and lack of awareness and traditional knowledge of farmers about agro-practices affects yield of Groundnut. Adoption of *Rhizobium* Bio-fertilizer technology, different communication modes in Groundnut cultivation in the year 2007 and 2008 by Tribal farmers yield of Groundnut pods.

Keywords: *Rhizobium, Groundnut, Biofertilizer.*

Introduction

"Biofertilizer is a preparation containing living or latent cells of efficient strains of nitrogen fixing, phosphate solublising of cellulotic micro-organisms used to seed soil or composting area".

The micro-organisms are known as biological nitrogen fixers. They are free living *Rhizobium* bacteria, BGA *etc. Rhizobium* is symbiotic nitrogen fixing gram-negative bacterium. It can enter in symbiosis with leguminous plants and from root noodles by infecting the root hairs of legume plants *Rhizobium* is abundant in soil all of them are notable to nodulate all types of legumes.

Response of *Rhizobium* inoculation has been shown in the improvement of grain legume yield. Residual effect of *Rhizobium* culture on yield of wheat, rice or subsequent crop, etc is always higher.

Groundnut is an important oil seed crop of India occupying about 46.72 per cent of total area under major oil seeds contributing about 67.27 per cent of the total major production. The states like Andra Pradesh, Gujrath, Tamil Nadu, Karnataka, Orisa and Maharashtra contribute 92 per cent of the country production. About 77 per cent of the crops are produced in kharif season. The average yield of kharif groundnut in India are extremely low, 750 kg/ha, compared to yields at over 3000kg/ha in the developed countries. As the average yield of groundnut are much higher in rabbi (1583 kg/ha) than in kharif season (871kg/ha). The major constraints of groundnut production in India are unreliable rain fall patterns, pest and diseases, socio-economic status of farmers, failure of communication media to reach in the area, lack of awareness and traditional mind psychology of farmers about improved agro-practices. The high percentage of illiteracy amongst farmers acts as main barrier for non-adoption of improved practices. It is need to develop new practices, use of improved varieties, use of chemical and organic fertilizers, bio-fertilizers, efficient management of pest, etc. are some important practices to meet the challenge.

Residual effect of *Rhizobium* culture on yield of wheat, rice or subsequent crop, etc. is always higher.

Material and Methods

The present research paper is aimed to develop effective communication media for communicating the said technology to target group by result and method demonstrations in field (Subba Rao, 1977). The study motivates the other farmers of the locality to adopt leading to horizontal spread of technology information.

The present study was carried out in the twelve villages of two Rural Tahasil of Northern part of Nasik District of Maharashtra state during the crop season of 2007 and 2008. For the study ninety-six farmers from twelve villages *i.e.* eight from each village were selected. Primary data of Groundnut cultivation were collected by discussion, interview about their cultural practices and yield (Holsten *et al.*, 1971). In each village 8 farmers were grouped into 4 classes. Two sample farmers in each class. Field area for demonstrations was half acre for all 4 classes, which can be tabulated in Table 10.1.

The primary data related with research study was collected by giving questionnaires by schedule method, Interview method etc. for this purpose 48 question were asked in relation with research project and answers of sample farmers were recorded. The questionnaires can be grouped in to 5 classes. It is tabulated in Table 10.2.

Table 10.1: Sample Farmers with Class and Demonstration Purpose

Sl.No.	Sample Farmers	Class	Demonstration Purpose
1.	First and Second	A	Use of *Rhizobium* Biofertilizer
2.	Third and Forth	B	Use of *Rhizobium* Biofertilizer and chemical fertilizer
3.	Fifth and Sixth	C	Use of only chemical fertilizer
4.	Seventh and Eighth	D	Use of Tradition Method

Table 10.2: Characters of Questionnaires

Class	Character of Questionnaires	Questions	Total Questions	Percentage
1	Individual information of the sample farmers	1 to 24	24	50 per cent
2	Problems and fertilizers used policy of farmers	25 to 29	5	10.41 per cent
3	Primary knowledge of *Rhizobium* Biofertilizer	30 to 38	9	18.75 per cent
4	Experiments of *Rhizobium* Biofertilizer *i.e.* seed dressing	39 to 42	4	8.33 per cent
5	Awareness of *Rhizobium* Biofertilizer adoption	43 to 48	6	12.5 per cent
	Total		48	100 per cent

Initially communication was made by discussion, interview, providing Printed matter, Pamphlets about *Rhizobium* Biofertilizer, its importance and application technology (Date, R.A. 1976). After communication the sample farmers of Group 'A' and 'B' provide 250 gm Of *Rhizobium* Biofertilizer packets for half acre plot. Field

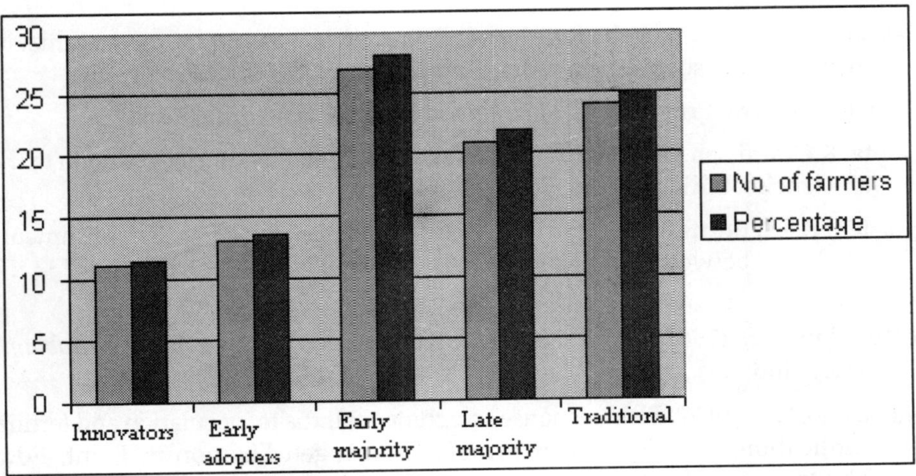

Figure 10.1: showing Attitudinal Percentage of Sample Farmers on Adoption of *Rhizobium* Biofertilizer Technology

demonstrations about application of *Rhizobium* to seeds of Groundnut, *i.e.* seed dressing at the homes of each sample farmers. They were trained to apply *Rhizobium* Biofertilizer to seeds of Groundnut before sowing in the field. Out of 96 farmers 48 farmers were an experimental group and 48 were in control group. The attitude analysis of sample farmers on Adoption of *Rhizobium* Biofertilizer technology shown in Table 10.3.

Table 10.3: Attitude Analysis of Sample Farmers on Adoption of Technology

Sl.No.	*Innovation Adoption Category*	*No. of Farmers*	*Percentage*
1	Innovators	11	11.45
2	Early adopters	13	13.54
3	Early majority	27	28.12
4	Late majority	21	21.87
5	Traditional mind	24	25.00
	Total	96	100

Results and Discussion

The data obtained from group 'A' farmers shows average increase in pod yield is 17.02 per cent/acre and group 'B' farmers is 18.83 per cent/acre. But control group 'C' farmers those used only chemical fertilizer the average increase in Groundnut pod is 5.11 per cent/acre. And control group 'D' farmers those cultivated Groundnut by traditional method before and after communication, the average increase in Groundnut pod is only 0.5 per cent/acre and which is negligible.

From above results it is concluded that the yield of Groundnut pod increases by 16 to 19 per cent due to adoption of *Rhizobium* Biofertilizer.

References

Balasundaran, V.R. and Subb Rao, N.S. (1977). A review of development of rhizoidal inoculants for soybeans in India. Fertilizer news, 22, 42-46.

Date, R.A. (1976). Principles in *Rhizobium* strain selection, pp.137-150.

Gupta, K.C. and Sen., A. (1963). Utilization of combined legumes in relation to their efficiencies. Indian J. Agric.Sci. 33, 240-243.

Holsten, R. D., Burns, R. C., Hardy, R.W.F. and Hebert, R. (1971). Establishment of symbiosis between *Rhizobium* and plant cells *in vitro*, Nature, Lond, 232, 173-176.

Subba Rao, N.S. (1967). Mechanism of infection of Legume roots by *Rhizobium*. J. Scient. Ind. Res., 26, 24-37.

Subba Rao, N.S. (1976). Field response of legumes in India to inoculation and fertilizer application, pp. 255-268. In Symbiotic Nitrogen Fixation in Plant, Ed. P. S.Nutman, Cambridge Univ. press.

2013, Environmental Biotechnology *Pages* **101–105**
Editors: **D.R. Khanna, A.K. Chopra, Gagan Matta, Vikas Singh & Rakesh Bhutiani**
Published by: **BIOTECH BOOKS, NEW DELHI**

Chapter 11

Evaluation of Eco-Safe Botanicals in Storage Against Khapra Beetle (*Trogoderma granarium* Everts) and Grain Qualities of Wheat (*Triticum astivium* cv. Lok-1)

Prachi Lambat[1], *Sanjeev Charjan*[2], *Rajesh Gadewar*[3] *and Ashish Lambat*[3]
[1]*Research Scholar, Hislop College, Nagpur, M.S.*
[2]*Assistant Professor, Dr. P.D.K.V's College of Agriculture, Nagpur, M.S.*
[3]*Assistant Professor, Sevadal Mahila Mahavidyalaya, Nagpur, M.S.*

The Khapra beetle (*Trogoderma granarium* Everts) is a serious store grain pest of wheat during storage. The use of inorganic pesticides for control of store grain pests can be dangerous to human being and cattle as well due to their residual toxicity. With view to find out eco-safe and organic seed or grain protectants, the present investigation was taken to know the effect of different botanicals on infestation percentage of Khapra beetle and seed or grain qualities of wheat during storage. It was observed that grain treated with sweet flag rhizome powder (2.5 per cent) and custard apple seed powder (2.5 per cent) were showed significantly higher 100-seed weight, germination percentage, seedling vigour, field emergence percentage and Khapra beetle adult mortality as compared to neem leaf (2.5 per cent), pongamid (2.5 per cent), tulsi (2.5 per cent), turmeric (2.5 per cent) and untreated control.

Keywords: *Germination Percentage, Seedling vigour, Field emergence percentage, Wheat, Khapra beetle.*

Introduction

Khapra beetle (*Trogoderma granarium* Everts) is a serious pest of stored grains causing considerable damage to almost all cereals in storage. It is largely responsible for damage and frequently harboring stores, mills and ware houses. *Khapra beetle* (*Trogoderma granarium Everts*) is cosmopolitan in nature attributing about 50 percent loss in seed weight during storage.

The steady rise in the use of pesticides for control of store grain pests can be dangerous to human beings and cattle s well due to their residual toxicity,. With a view to find out safe and organic seed protectant, present investigation was taken up to evaluate the organic grain protectant, in seed storage against *Khapra beetle* (*Trogoderma granarium Everts*) in wheat.

Materials and Methods

Wheat (Triticum aestivum L. Cv. HD- 2329) seeds were used in various phases of this study, produced in 2008-09. The seeds were cleaned and dried (moisture content 10.5 per cent). The wheat seeds were treated (May, 2009) with six plant product viz, Neem leaf, Sweet flag, Tulsi, custard apple, turmeric and pogamia powder each in proportion of 2.5 per cent by weight of the seeds. The experiment as conducted in glass bottle of one liter capacity with seven treatments including untreated control. Each glass bottle was then filled with 500 grams of wheat seeds. Ten pairs of 2-3 days old *Khapra beetle* (*Trogoderma granarium Everts*) were released in each glass bottle covered with muslin cloth. The set of experiment was kept in well ventilated wire mesh almirah in mesonary building having cemented walls, roof and floor under ambient temperature (24.1to 45.8°C) and relative humidity (20 to 85 per cent) from May to July, 2009. After three months the seeds from each treatment ere keenly observed and those found infested were separated out weighted to determine the infestation percentage on weight basis, 100-seed weight and germination were tested in quadruplicate with 100 seeds in each replication. The germination percentage was evaluated on the value for normal seedlings (Anon, 1985). The vigour index was workout following the method of Abdul – Baki and Anderson (1973). For field emergence test, sowing of wheat seeds was done in randomized block design, with four replication with inter and intra- row spacing of 1 feet and 6 inches respectively. Observations for field emergence were recorded daily and finally the established seedlings were counted after one month of sowing. The experimental data was statistically scrutinized as per Panse and Sukhatme (1967)

Results and Discussion

The data regarding the effect of the different organic grain protestant on population behavior (adult mortality), infestation percentage 100-seed weight, germination percentage, vigour index and field emergence percentage after three months of storage are given in Table 11.1.

Table 11.1: The Effect of Different Organic 4 Protectant on Rice Weevil Mortality, Infestation Percentage, 100–Seed Weight, Germination Percentage, Seedling Vigour and Field Emergence Percentage

Treatments	*Adult (Khapra Beetle (Trogoderma granarium Everts) Mortality per cent*	*Per cent Weight Loss due to Infestation per cent*	*100–Seed Weight (g)*	*Germination per cent*	*Seedling Vigour Index (svi)*	*Field Emergence (per cent)*
Neem leaf (2.5 per cent)	68	5.4	3.04	66	1212	53
Sweet flag (2.5 per cent)	100	0.08	3.31	94	1957	86
Tulsi (2.5 per cent)	40	7.1	2.81	62	1189	48
Custard apple (2.5)	96	2.1	3.19	90	1912	80
Turmeric (2.5 per cent)	38	7.8	2.62	60	1113	47
Pongamia (2.5 per cent)	52	06.0	2.89	63	1191	48
Untreated control	05	16.9	2.19	30	608	15
SE (+)	0.35	0.09	0.03	0.12	—	0.1
CD at 5 per cent	1.06	0.29	0.10	0.36	—	0.31

The results indicated that the variation in number of *Khapra beetle* (*Trogoderma granarium Everts*) (adult) in each treatment. The *Khapra beetle* (*Trogoderma granarium Everts*) adult mortality was significantly highest in wheat seeds treated with sweet flag (100 per cent) which is closely followed by custard apple (96 per cent) Neem leaf (68 per cent), Pongamia (52 per cent), tulsi (40 per cent) and turmeric (38 per cent). Where in significantly lower mortality was observed in untreated control (5 per cent). Saxena *et al.* (1976) Tikku *et al.*, 1978 and Khan and Borle (1995) found Acorus calamus L. oil vapour responsible for causing infecundity among the females of a number of stored grain pest. Bireda (2000) who reported that Neem leaf has got insecticidal and ovicidal effect. The significantly wheat seed weight loss was observed in untreated control followed by turmeric, tulsi, pongamia, neem leaf, custard apple and sweet flag treatment during.

Entire Period of Storage

Charjan and Tarar (1994) and Deshpande *et al.* (2010) reported that the sharp declined in infestation percentage of store grain pest in seed treated with Acorus calamus powder.

The seed quality parameters *viz.*, 100-seed weight, germination percentage and seedling vigour index was highest in seed treated with 2.5 per cent concentration of sweet flag powder followed by custard apple, neem leaf, pongamia, tulsi, turmeric and untreated control. The 100 – seed weight, germination percentage and vigour index decreased with increasing infestation of store grain pest (Howe 1972), Charjan and Tarar (1994) and Deshpande *et al.* (2010). Since the store grain pest have been eaten of major portion o the endosperm which lead to reduction in weight of the wheat seeds and intern affect the seed germination and vigour index because of lack of stored food and is in conformity with the findings of Narayanswamy)1985). Handerson and Christenser (1961) reported that pulse beetle attack the embryo and germination potential of seed reduced or totally destroyed.

The field emergence percentage of wheat seed follows the same trend of the seed quality parameters. The field emergence percentage was highest in seed treated with 2.5 per cent concentration of sweet flag powder as compared control. This might be due to the least infestation of *Khapra beetle* (*Trogoderma granarium Everts*) and higher 100 – seed weight, germinability and seedling vigour index. The results are agreement to those reported by Charjan and Tarar (1994) and Deshpande *et al.* (2010).

Conclusion

Among all the other plant products sweet flag powder (2.5 per cent) and custard apple seed powder (2.5 per cent) were found to be significantly effective against *Khapra beetle* (*Trogoderma granarium Everts*) throughout the period of investigation. These findings are in agreement with Siva Srinivasu (2001) and Deshpande *et al* (2010). Thus sweet flag and custard apple naturally occurring botanicals which are nontoxic can be sued as pre-storage seed treatment, dispensing with the use of costly and toxic chemicals, to control *Khapra beetle* (*Trogoderma granarium Everts*) damage without adversely affecting the germination of wheat seeds.

References

Abdul – Baki, A. A. and Anderson, J. D. 1973. Vigour determination in soyabean seed by multiple criteria Crop Sci. 13: 630 – 633.

Anonymous, 1985. International rules for seed testing seed sci. and Technol. 13: 29 – 513.

Biradar, B. S. 2000. Prevention of cross infestation by *Sitophillus oryzae* L. and *Rhizopertha dominica* in stored wheat. M. Sc. (Agri.) Thesis, university of Agricultural Sciences. Dharwad.

Charjan, S. K. U. and Tarar, J. L. 1994. The influence of some plant products on seed quality of lobia during storage. Ann. Plant Physiol. 8(2): 153 – 156.

Deshpande, V. K. Deshpande, H. H. and Masuthi, D. 2010. Evaluation of grain protectant in seed storage against *Sitophillus Oryzae* (L.) in sorghum.

Handerson, L. s. and Christensen C. M. 1961. Pre-harvest control of insect and fung U. S. Dept. Agri. YbK, pp. 348 – 356.

Khan. M. I. and Borle, M. N. 1985. Efficacy of some safer grain protectant against th pulse beetle, Callosobruchus Chinensis L. infesting stored Bengal gram (Cicer arietinum L.) P.K.V. Res. J. 9 (1):53 – 55

Narayanaswamy, S. 1985. Effect of pulse beetle damage on seed quality of field bea and pigeon pea. Seed Res. 13 (2): 138 – 141.

Panse, V. G. and Sukhatme, P. V. 1967. Statistical methods for agriculture workers I. C. A. R. Pub. New Delhi.

Saxena, B. P., Koul, O. and Tikku, K. 1976. Non-toxic protectant against the stored grain insect pests. Bull. Grain Technol. 14 (5): 190 – 193.

Tikku, K. Koul, O. and Sexena B. P. 1978. The influence of *Acorus calamus* L. Oil vapours on the *histo-cytological pattern* of the ovaries of *Trogoderma granarium* Evert. Bull. Grain Technol. 16 (1): 3 – 9.

2013, Environmental Biotechnology *Pages* ***107–125***
Editors: **D.R. Khanna, A.K. Chopra, Gagan Matta, Vikas Singh & Rakesh Bhutiani**
Published by: **BIOTECH BOOKS, NEW DELHI**

Chapter 12

Biotechnological Strategies for Conservation of an Endangered Forest Tree Species *Wrightia tinctoria*: Important in Toy Making Industry

Kairamkonda Madhusudhan[1], Godishala Vikram[2], Kagithoju Srikanth[1], Talari Samatha[1] and Nanna Rama Swamy[1]

[1]*Department of Biotechnology, Kakatiya University, Warangal – 506 009*
[2] *Center for Biotechnology, Jawaharlal Nehru Technological University, Hyderabad*

The species *Wrightia tinctoria* (Rox.) R.Br (Apocynaceae) is used by local artisans for making toys. Many artisans in Chennapatna, Etikoppaka and Kondapally depend upon this wood for their lively hood. Due to lack of rapid natural regeneration and over- exploitation, the species has become an endangered. The protocol for multiplication of the species through meristem tip and nodal cultures was developed. The explants, shoot –tip and nodal segments of *W. tinctoria* were cultured on MS medium supplemented with different concentrations of BAP/Kn/TDZ individually. *In vitro* rooting was established from the micro-shoots developed through shoot-tip and nodal cultures on MS medium supplemented with IBA. After, *in vitro* rooting, the plants were hardened and established. These plants were found to be normal, healthy and similar to donor plants. The protocol developed in the present studies can be used for rapid multiplication of this endangered forest tree species.

Keywords: *Wrightia tinctoria, Endangered forest tree species, Meristem tip culture, Nodal culture, In vitro rooting.*

Introduction

Forest tree improvement programs by selection and breeding techniques began in India recently. Traditional breeding methods in trees are limited by their large size and long life cycles. Clonal forestry is now gaining increasing recognition as a quicker alternative for tree improvement. *In vitro* techniques are being increasingly applied to supplement conventional methods of vegetative propagation. The benefits of this technique include high multiplication rates, generation of disease-free stocks and stress tolerant varieties and also long term storage of valuable germplasm. It has application in horticulture, agriculture and forestry and currently it is expanding worldwide. This method plays a great role in conservation and large-scale production of endangered, economically/commercially and industrially important forest tree species.

Commercially important forest tree species have been multiplied by using this technique. They are: *Eucalyptus* (Lakshmisita and Vaidyanathan, 1979; Grewal *et al.*, 1980; Gupta *et al.*, 1981); *Maytenus emerginata* (Rathore *et al.*, 1992); *Melia azaderach* (Raghuraman and Ramanujam, 1998); *Zizyphus mauritiana* (Sudershan *et al.*, 2000); *Celastrus paniculatus* (Arya *et al.*, 2002); *Balanites aegyptica, Citrus lemon, Syzygium cuminii* (Rathore *et al.*, 2004a, b); *Swietenia mahagoni* (Nagarajan *et al.*, 2006); *Terminalia alata* (Lakshman, 2006); *Givotia rottleriformis* (Rambabu, 2007).

Lack of suitable method for natural regeneration and over exploitation of *W. tinctoria*, its population drastically has been reduced and the species has listed as an endangered species. Therefore, there is a need for an alternate method to produce a large number of plants within a limited period. Hence, *in vitro* multiplication has been attempted in *Wrightia tinctoria* for conservation of the species using shoot tip/ nodal segments/axillary bud culture during the present investigations.

Methodology

Plant Material

For shoot tip culture, terminal portions (5-7 cms long) of new shoot growths were collected from two years old healthy plants growing in the Research field, Department of Botany, Kakatiya University. These shoot tips with adjacent leaves were excised and washed thoroughly under running tap water. Then treated with Bavistin (1 per cent) for 15 minutes followed by thorough washing with sterile distilled water (5 times). Surface sterilization was carried out with 0.1 per cent (w/v) $HgCl_2$ for

5 minutes followed by sterile distilled water for 5 times and dried them on sterile tissue paper. Shoot tips were trimmed aseptically consisting of 1.0-1.5 cm size with leaf primordia and meristem. Later, these were implanted on MS (Murashige and Skoog, 1962) medium supplemented with different concentrations of plant growth regulators.

For nodal culture, nonwoody branches from one and half years old plants growing in the Research field, Department of Botany, Kakatiya University, were used as explants. Shoots were excised free of leaves, trimmed into (3cm) segments containing a single node with opposite axillary buds. These explants first washed for 30 minutes in running tap water. Later treated with Bavistin (1 per cent)for 5 minutes, followed by rinsing in sterile distilled water five times. Surface sterilization was carried out with 0.1 per cent (w/v) $HgCl\text{-}_2$ for 5 minutes followed by five rinses in sterile distilled water. These sterilized explants were dried on sterile tissue paper. The nodal segments were further cut into 1.0-2.0 cm long having one node and were used as explants.

Culture Media and Culture Conditions

Shoot tip and nodal explants were cultured on MS medium containing 30 g/L with out growth regulators and also sucrose supplemented with different concentrations (0.2-5.0 mg/L) of BAP/Kn/TDZ (0.2-1.0 mg/L) individually. For further proliferation and elongation of shoots the cultures were shifted on to MS medium fortified with 0.8 mg/L BAP.

The pH of medium was adjusted to 5.8 with either 0.1N NaOH or 0.1N HCl before adding 0.8 per cent (w/v) Difco-bacto agar. The culture medium was dispensed into different culture tubes (15 ml) and also culture bottles and autoclaved at 121°C under 15 psi for 15-20 minutes. After inoculation, all the cultures were incubated under cool white-fluorescent lights at an intensity of 40-60 µ?mol $m^{-2} s^{-1}$ for 16 hrs photoperiod at 25°C. The cultures were maintained by regular subculturing for every 3 weeks on fresh media with the same concentration of plant growth regulators.

Elongation of Shoots

The microshoots developed through shoot tip and nodal cultures were inoculated on MS medium supplemented with 0.5 mg/L GA_3 +1mg/L BAP for shoot elongation.

In vitro Rooting

The micro shoots developed from nodal and shoot tip cultures were used for *in vitro* rooting. The micro-shoots were excised consisting of 3-4 cms in height and having at least 2-3 nodes. These were cultured on ½ strength MS, MSO and MS medium containing 3 per cent (w/v) sucrose supplemented 0.1-5.0 mg/L IBA.

Hardening of Plantlets

After induction of roots *in vitro*, the plantlets having atleast 5-6 roots of 5-6 cm length were washed carefully with sterile distilled water to remove traces of agar. Later these plantlets were transferred to plastic pots containing different sterilized substrates for hardening (Figure 12.5). The plastic pots were covered with plastic bags to prevent desiccation, avoid rapid changes in the environment and also to maintain the RH (85-90 per cent). The plantlets were acclimatized in the walk-in-chamber at 25°C-27°C temperature. During the hardening procedure, plastic bags were gradually perforated after 15 days and after one month they were removed. These plantlets were transferred to earthenware pots containing garden soil and maintained under shady conditions in the research field.

Data Analysis

Data were recorded at the end of 4th week and 30 replicates were maintained for each experiment. Each experiment was repeated at least twice.

Results

Shoot Tip Culture

Shoot tip explants of *W. tinctoria* were cultured on MS medium supplemented with different concentrations of BAP/Kn/TDZ (0.2 - 5.0 mg/L) as a sole growth promoter (Tables 12.1–12.3 and Plate 12.1). In the present investigation, it was observed that the number of shoots per explant and the mean length of the shoots were found to be varied with the growth regulators and their concentrations used (Figures 12.1–12.2).

Table 12.1: Effect of BAP on *In vitro* Shoot Tip Culture in *W. tinctoria*

Conc. of PGR (mg/L)	*Percentage of response*	*Average No. of shoots/ explant ± (SE)[a]*	*Mean length of shoot (cm) ± (SE)[a]*
MSO	20	1.0	1.8 ± 0.02
0.2	73	6.0 ± 0.15	4.5 ± 0.15
0.4	85	12.0 ± 0.15	5.0 ± 0.13
0.6	88	14.0 ± 0.03	6.0 ± 0.50
0.8	98	16.0 ± 0.05	6.5 ± 0.15
1.0	95	7.0 ± 0.05	4.5 ± 0.13
2.0	90	5.5 ± 0.10	3.0 ± 0.40
3.0	80	5.0 ± 0.1	3.0 ± 0.20
4.0	75	4.0 ± 0.2	3.0 ± 0.40
5.0	60	3.0 ± 0.05	3.0 ± 0.05

a: Mean ± Standard Error.

Shoot tips cultured on MSO medium without growth regulators showed the elongation of single shoot (Plate 12.1a). High percentage (95 per cent) of response and increased number of multiple shoots formation were recorded at 0.8 mg/L BAP as a sole growth regulator, within 3 weeks of culture. Of all the concentrations of BAP studied, at 0.8 mg/L was most favorable in terms of enhanced shoots formation and mean length of shoots (Plate 12.1b-c). Maximum multiples (16.0±0.03) was recorded at the concentration of 0.8 mg/L BAP (Table 12.1; Figure 12.2). At higher concentration of BAP, low percentage of response as well as the number of shoots per explant were found to be reduced (Figure 12.1).

Table 12.2: Effect of Kn on *in vitro* Shoot Tip Culture in *W. tinctoria*

Conc. of PGR (mg/L)	*Percentage of Response*	*Average No. of Shoots/ Explant ± (SE)[a]*	*Mean Length of Shoot (cm) ± (SE)[a]*
0.2	65	5.5 ± 0.50	4.0 ± 0.15
0.4	83	6.0 ± 0.05	4.5 ± 0.05
0.6	95	8.5 ± 0.30	4.5 ± 0.15
0.8	85	5.0 ± 0.05	4.0 ± 0.13
1.0	80	4.0 ± 0.15	3.5 ± 0.04
2.0	80	3.5 ± 0.13	3.0 ± 0.05
3.0	78	3.0 ± 0.04	2.5 ± 0.02
4.0	60	2.5 ± 0.01*	2.5 ± 0.05
5.0	52	2.5 ± 0.01*	2.0 ± 0.03

a: Mean ± Standard Error * with callus.

Table 12.3: Effect of TDZ on *in vitro* Shoot Tip Culture in *W. tinctoria*

Conc. of PGR (mg/L)	*Percentage of Response*	*Average No. of Shoots/ Explant ± (SE)[a]*	*Mean Length of Shoot (cm) ± (SE)[a]*
0.2	50	3.0 ± 0.15	2.5 ± 0.10
0.4	73	4.0 ± 0.05	3.5 ± 0.15
0.6	88	5.0 ± 0.10	4.5 ± 0.10
0.8	98	5.0 ± 0.10	4.0 ± 0.13
1.0	95	5.0 ± 0.06	6.0 ± 0.15
2.0	70	3.5 ± 0.15	4.0 ± 0.05
3.0	65	3.5 ± 0.10	3.5 ± 0.15
4.0	60	3.0 ± 0.02	3.0 ± 0.13
5.0	58	3.0 ± 0.10	2.5 ± 0.10

a: Mean ± Standard Error.

Shoot tip explants were cultured on MS medium supplemented with 0.2-5.0 mg/L Kn. More number of multiple shoots and high percentage (95 per cent) of response were found at 0.6 mg/L Kn with 8.5±0.30 multiple shoots per explant (Table 12.2). At higher (5.0 mg/L) levels of Kn, less number of multiple shoots formation was found (Figures 12.1–12.2). As the concentration of Kn increased from 0.2-0.6 mg/L Kn, gradual enhancement in the percentage of response and as well as adventitious shoots induction were recorded (Plates 12.1d).

Effect of TDZ on shoot tip culture was also studied in *W. tinctoria* and the results are presented in Table 3. TDZ showed less percentage of response compared to BAP and Kn (Figures 12.1–12.2). Maximum percentage (90 per cent) of response was observed at 0.8 mg/L TDZ, and high frequency of shoots development per explant (5.0±0.05) was found at the same concentration of TDZ. At higher concentration of

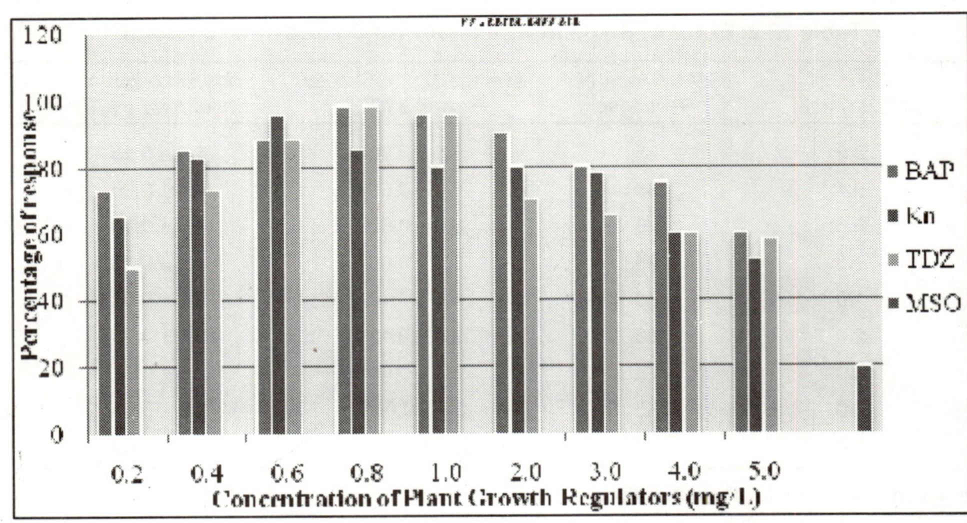

Figure 12.1: Effect of BAP/Kn/TDZ/MSO on *In vitro* Shoot Tip Culture of *W. tinctoria*

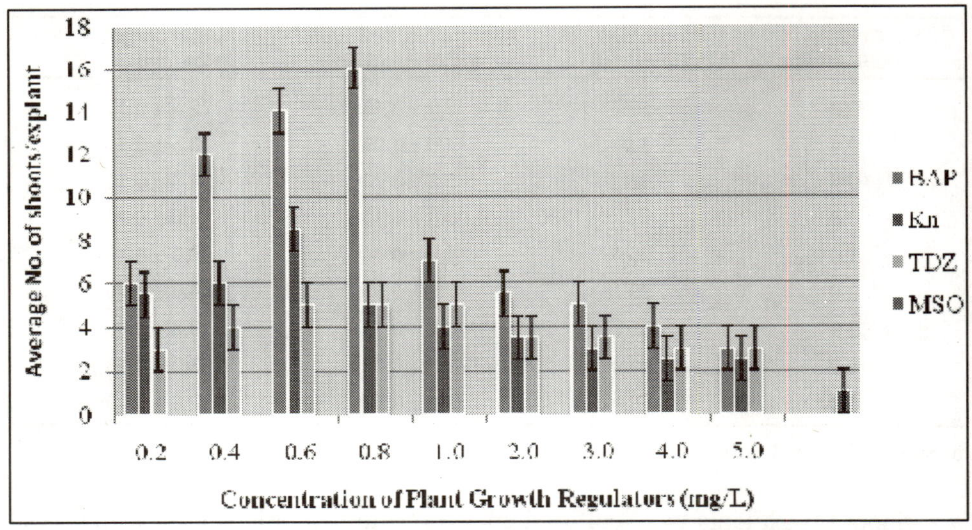

Figure 12.2: Effect of BAP/Kn/TDZ on *In vitro* Shoot Tip Culture of *W. tinctoria*

TDZ less percentage of response and also low number of shoots per explant were noted (Table 12.3).

However, multiple shoots obtained by this method did not proliferate further into additional shoots. After the formation of multiple shoots, the explants were subcultured on 0.8 mg/L BAP, produced maximum frequency number of shoots (21±0.08) per explant. It was also noticed that the formation of axillary branching at the same concentration after first subculture with the shoot elongation and multiple shoot buds proliferation too (Plate 12.1f-g).

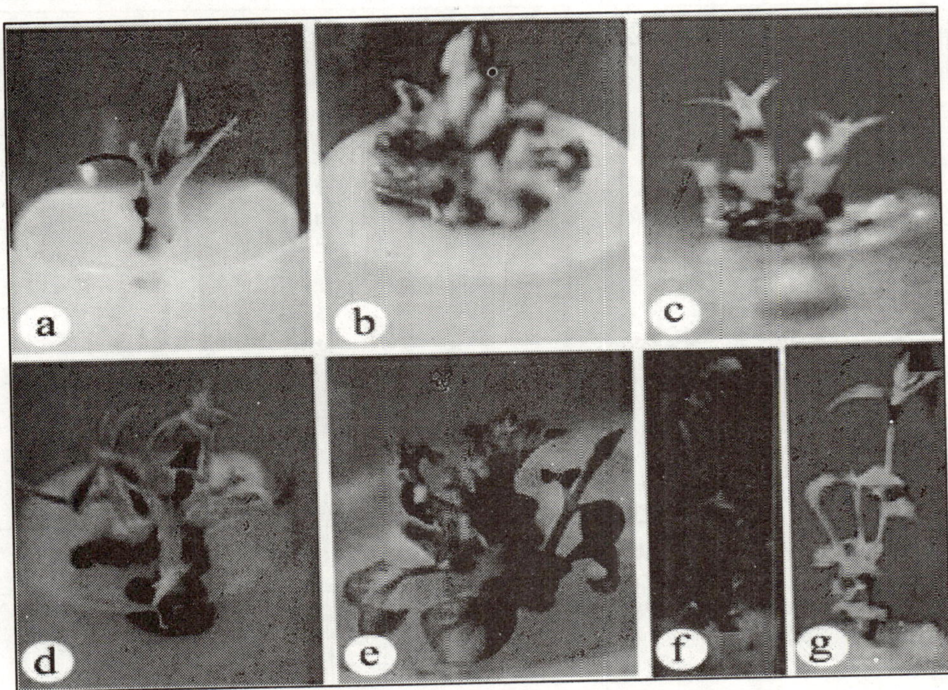

Plate 12.1: a–g: Shoot Tip Culture of *W. tinctoria*

(a) Single shoot elongation on MS basal medium (after 4 weeks of culture); (b) Multiple shoots induction on MS+ 0.8 mg/LBAP; (c) Multiple shoots formation at 0.6 mg/L BAP; (d) Multiple shoots formation at 0.8 mg/Kn; (e) 1st subculture on MS+ 0.8 mg/L BAP; (f and g) Multiple shoots with axillary branching on the same.

Nodal Culture

The initial response exhibited by the nodal explants was an enlargement and subsequent breakage of the axillary bud. The time taken for bud breaking was found to be varied depending on the type of growth regulator, concentration and combination of growth regulators used. New shoots were developed after two weeks of culture and attained height of 3-4 cms after 3 weeks.

Nodal segments of *W. tinctoria* were cultured on MS medium without plant growth regulators (MSO), MS medium supplemented with different concentrations of BAP/Kn/TDZ (Tables 12.4–12. 6). MSO medium didn't induce the bud breakage, whereas the nodal explants cultured on all the concentrations of growth regulators supported the induction and proliferation of axillary bud break in *W. tinctoria* (Plates 12.2 and 12.3).

Nodal segments of *W. tinctoria* were cultured on MS medium augmented with different concentrations of BAP. The results are presented in Table 12.4 and shown in Plates 12.2–12.3 and Figures 12.4–12.5.

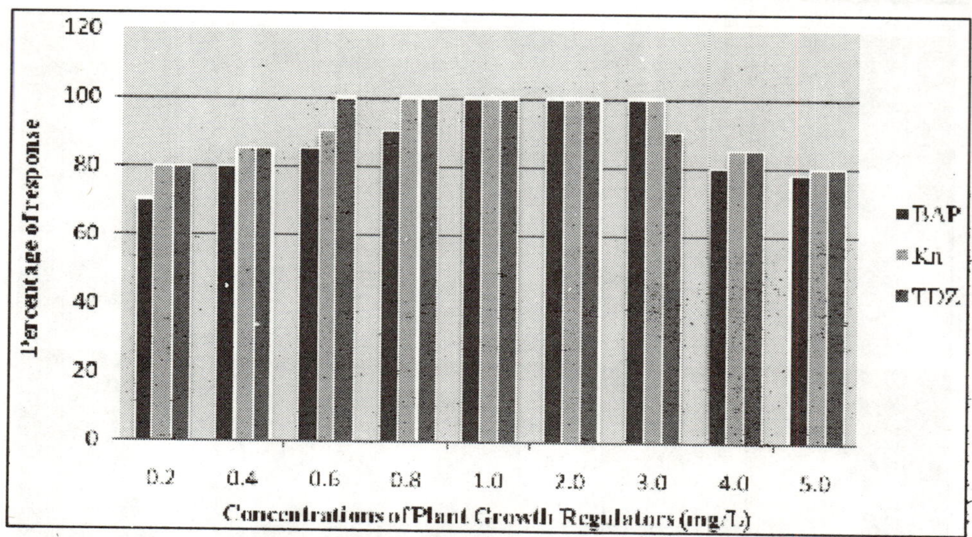

Figure 12.3: Effect of BAP/Kn/TDZ on *In vitro* Nodal Culture of *W. tinctoria*

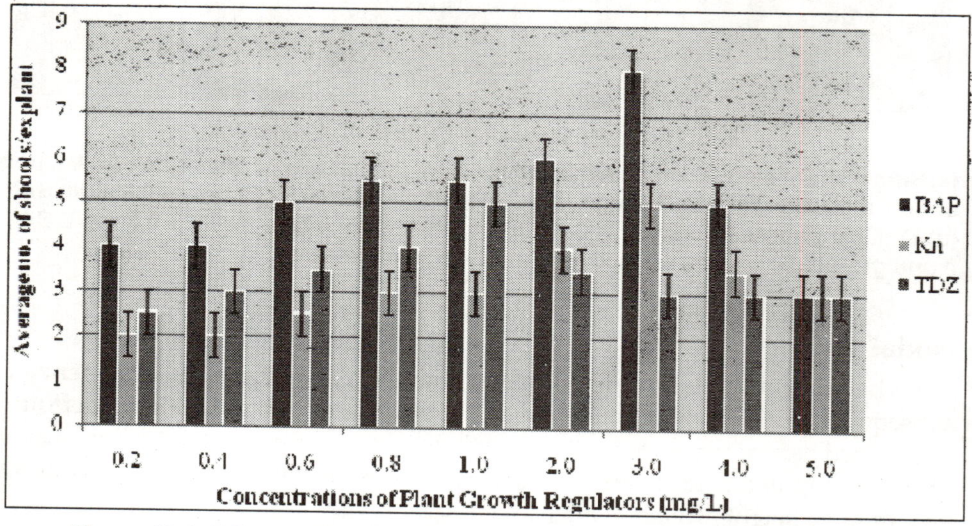

Figure 12.4: Effect of BAP/Kn/TDZ on *In vitro* Nodal Culture of *W. tinctoria*

As the concentration of BAP increased upto 3.0 mg/L, the percentage of response was also increased and absolute percentage of response was observed at 1.0-3.0 mg/L BAP. At high concentrations of BAP the response from nodal explants were decreased (Table 12.4). Callusing was observed at 5.0 mg/L BAP along with the multiple shoots formation. Maximum percentage of response and more number of multiple shoots formation per explant was found at 3.0 mg/L BAP followed by 2.0 mg/L BAP.

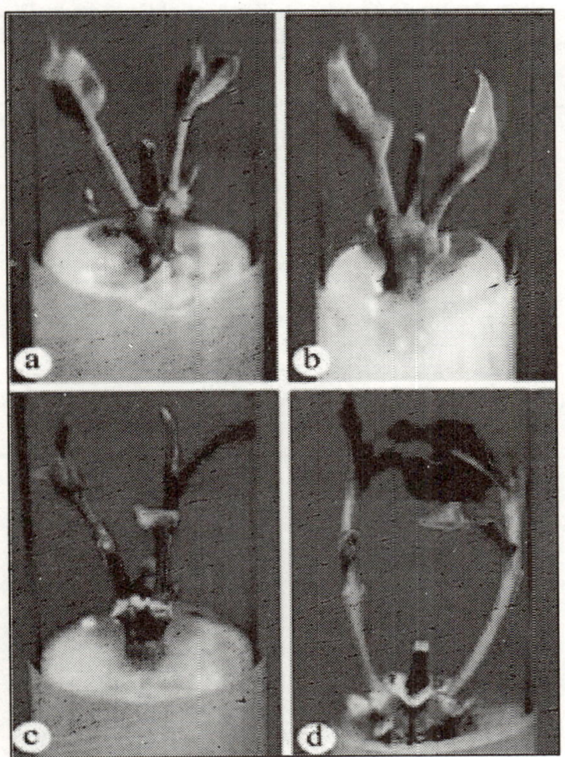

Plate 12.2: a–d: Nodal Culture on MS + BAP in *W. tinctoria*

(a) Axillary node proliferation and elongation at 0.2 mg/L BAP (Note the axillary branching); (b) Shoots elongation on the same; (c) Multiple shoots formation at 0.4 mg/L BAP; (d) Further elongation at the same with axillary branching.

Table 12.4: Effect of BAP on *in vitro* Nodal Culture in *W. tinctoria*

Conc. of PGR (mg/L)	*Percentage of Response*	*Average No. of Shoots/ Explant ± (SE)*[a]	*Mean Length of Shoot (cm) ± (SE)*[a]
0.2	70	4.0 ± 0.15	2.0 ± 0.15
0.4	80	4.0 ± 0.13	2.0 ± 0.10
0.6	85	5.0 ± 0.01	2.5 ± 0.50
0.8	90	5.5 ± 0.01	2.5 ± 0.12
1.0	100	5.5 ± 0.02	2.5 ± 0.03
2.0	100	6.0 ± 0.01	2.7 ± 0.10
3.0	100	8.0 ± 0.10	3.0 ± 0.10
4.0	80	5.0 ± 0.10	2.2 ± 0.10
5.0	78	3.0 ± 0.03*	3.2 ± 0.03

a: Mean ± Standard Error; *: Callusing at base.

Table 12.5: Effect of Kn on *in vitro* Nodal Culture in *W. tinctoria*

Conc. of PGR (mg/L)	*Percentage of Response*	*Average No. of Shoots/ Explant ± (SE)[a]*	*Mean Length of Shoot (cm) ± (SE)[a]*
0.2	80	2.0 ± 0.10	1.7 ± 0.30
0.4	85	2.0 ± 0.03	2.5 ± 0.01
0.6	90	2.5 ± 0.02	2.7 ± 0.03
0.8	100	3.0 ± 0.03	1.5 ± 0.10
1.0	100	3.0 ± 0.50	1.5 ± 0.04
2.0	100	4.0 ± 0.13	2.0 ± 0.01
3.0	100	5.0 ± 0.10	2.2 ± 0.01
4.0	85	3.5 ± 0.10*	2.1 ± 0.03
5.0	80	3.0 ± 0.01*	2.0 ± 0.04

a: Mean ± Standard Error; *: Callusing at base.

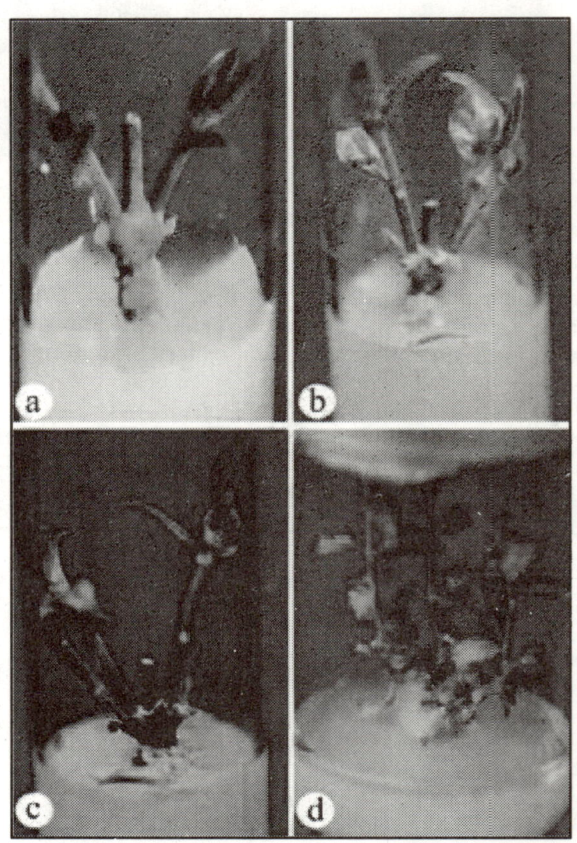

Plate 12.3: a–d: Nodal Culture on MS+ Kn/TDZ/BAP of *W. tinctoria*

(a) Shoots proliferation at 0.2 mg/L Kn; (b) Further proliferation on the same; (c) Multiple shoots at 1.0 mg/L TDZ; d) Multiple shoots at 3.0 mg/L BAP.

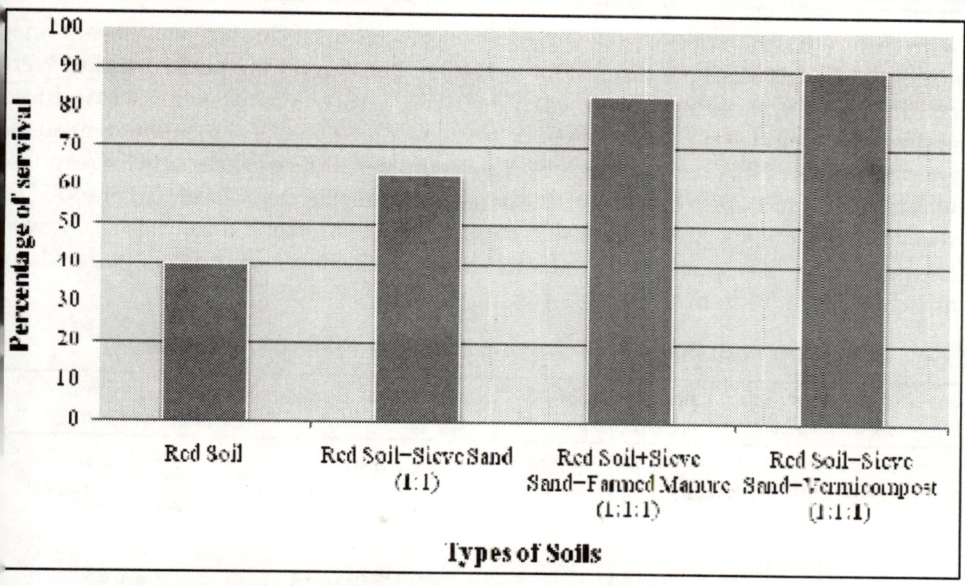

Figure 12.5: Effect of Different Soil on Hardening of *In vitro* Rooted Plantlets in *W. tinctoria*

Nodal segments cultured on MS medium fortified with different concentrations of Kn, showed varied results (Table 12.5). Absolute percentage of response was observed at 0.8-3.0 mg/L Kn. Whereas maximum number of multiple shoots was observed at 3.0 mg/L Kn. Less percentage (80 per cent) of response was noted at low and high concentrations of Kn (Figures 12.4–1.25). It was also found that at 4.0-5.0 mg/L Kn, callus was developed along with the formation of multiple shoots.

The results on nodal segments cultured on MS medium with different concentrations of TDZ, are presented in Table 12.6. Maximum percentage of response (100 per cent) was observed with the development of 5.0± 0.5 average number of shoots per explant at 1.0 mg/L TDZ (Figures 12.4–12.5). Low percentage of response with less number of shoots per explant was recorded at low and high concentrations of TDZ. Callusing was also observed along with the formation of multiple shoots at 4.0-5.0 mg/L TDZ.

Shoot Elongation

The micro-shoots developed from shoot tip and nodal cultures were cultured on MS medium augmented with 0.5mg/L GA_3+1.0 mg/L BAP for shoot elongation (Plate 12.5). The micro-shoots start to elongate after one week of inoculation. These shoots were elongated without further bud proliferation and callus formation at basal region of the shoot. The elongated shoots were also normal and healthy. These shoots were used for *in vitro* rooting in the present investigations.

In vitro Rooting and Plantlet Establishment

Micro-shoots that developed from nodal and shoot tip cultures on MS+BAP/Kn/TDZ were transferred on to ½ strength MS, MSO and MS medium supplemented with different concentrations of IBA (Table 12.7). Roots induction was absent on ½ strength MS and MSO medium without auxins. The regenerated micro-shoots were cultured on MS medium fortified with 0.1-5.0 mg/L IBA. *In vitro* rooting was induced at the basal region of micro-shoots of *W. tinctoria* after 2 weeks of incubation in all the concentrations of IBA used (Plate 12.5). Maximum percentage (95 per cent) of response was recorded at 0.4 mg/L IBA with induction of more number (18.0±0.01) of roots per shoot (Plate 12.5a-c). At low and high concentrations of IBA (1.0-5.0 mg/L),showed less percentage of response and moderate rhizogenesis. Roots were developed without any callus formation in all the concentrations of IBA tested.

Table 12.6: Effect of TDZ on *in vitro* Nodal Culture in *W. tinctoria*

Conc. of PGR (mg/L)	Percentage of Response	Average No. of Shoots/ Explant ± (SE)[a]	Mean Length of Shoot (cm) ± (SE)[a]
0.2	80	2.5 ± 0.10	2.7 ± 0.43
0.4	85	3.0 ± 0.03	2.1 ± 0.56
0.6	100	3.5 ± 0.02	1.7 ± 0.76
0.8	100	4.0 ± 0.03	1.7 ± 0.85
1.0	100	5.0 ± 0.50	1.5 ± 0.48
2.0	100	3.5 ± 0.13	2.0 ± 0.79
3.0	90	3.0 ± 0.10*	2.2 ± 0.84
4.0	85	3.0 ± 0.10*	1.1 ± 0.97
5.0	80	3.0 ± 0.01*	2.5 ± 0.34

a: Mean ± Standard Error; *: With callus.

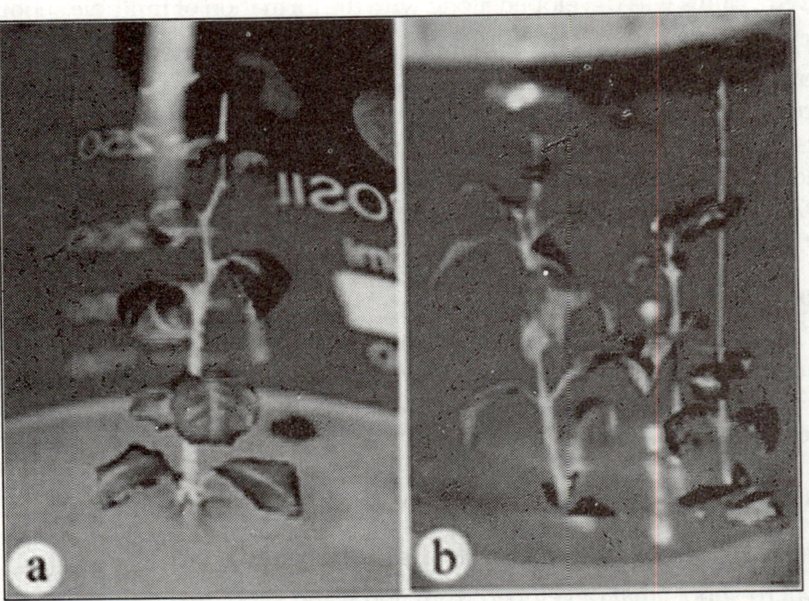

Plate 12.4: a–b: Elongation of Microshoots Developed from Shoot Tip and Nodal Cultures on MS + 0.5 mg/L GA_3 + 1.0 mg/L BAP

(a and b) Micro-shoots developed from shoot tip.

Table 12.7: Effect of IBA on *in vitro* Rooting in *W. tinctoria*

Conc. of PGR (mg/L)	*Percentage of Response*	*Average No. of Shoots/ Explant ± (SE)[a]*	*Mean Length of Shoot (cm) ± (SE)[a]*
0.1	60	6.0 ± 0.02	2.0 ± 0.05
0.2	65	8.0 ± 0.10	3.0 ± 0.01
0.3	75	10.0 ± 0.07	4.0 ± 0.50
0.4	95	18.0 ± 0.01	4.5 ± 0.01
0.5	88	8.0 ± 0.13	3.5 ± 0.05
1.0	85	8.0 ± 0.09*	2.0 ± 0.01
2.0	78	6.0 ± 0.07*	2.2 ± 0.10
3.0	60	7.0 ± 0.13*	2.0 ± 0.12
4.0	65	6.0 ± 0.91*	1.5 ± 0.09
5.0	50	5.0 ± 0.05	1.2±0.07

a: Mean ± Standard Error; *: Callusing+ roots.

In the present investigations, the *in vitro* regenerated plants of *W. tinctoria* hardened to transfer them from *lab-to-land*. The *in vitro* rooted plantlets from shoot tip and nodal cultures were hardened on four different types of soil mixtures *viz.*, red soil, red soil+sieved sand (1:1), red soil + sieved sand + farmyard manure (1:1:1) and red soil + sieved sand + vermicompost (1:1:1) (Figure 12.5). Out of four potting mixtures used for hardening procedure, the survival percentage was maximum (90 per cent) in red soil + sieved sand + vermicompost compared to all other substrates used. New leaves were formed after 12 days after transplantation in the same soil mix. Low percentage of survival (40 per cent) was recorded in only red soil and the new leaf appeared after 22 days. Remaining substrates showed 63-90 percentage of plantlet survival and new leaves appeared after 12-15 days respectively (Figure 12.5). These plants were shifted to earthenware pots containing garden soil and were maintained

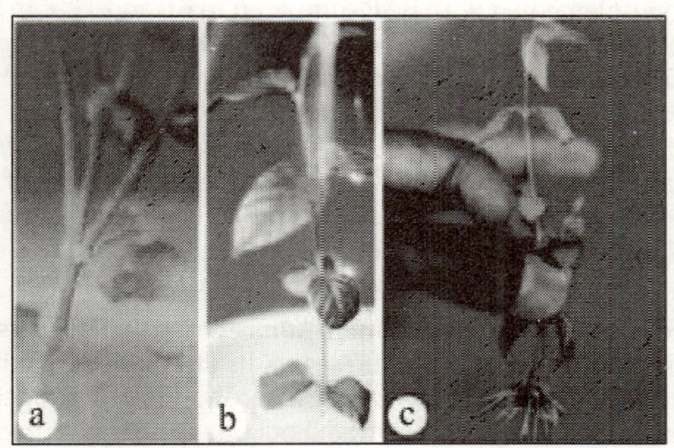

Plate 12.5: a–c: *In vitro* rooting in *W. tinctoria*
(a–c) Rooting on MS + 0.2, 0.3 and 0.4 mg/L IBA respectively.

Plate 12.6: a–c: Acclimatization/Hardening of *in vitro* Regenerated Plants of *W. tinctoria*

(a) Plantlet regenerated through shoot tip culture; (b) The same plantlets after 4 weeks of hardening; (c) Acclimatized plants under shady place in the Research field.

under shady conditions in the research field. The acclimatized plants were normal, healthy and showing similar morphological features as that of donor plants (Plate 12.6).

Discussion

In the present study the apical dominance of the meristem in shoot tips continued to persist there by enhancing the scope for rapid multiplication. BAP had shown superiority over Kn followed by TDZ in the present investigations. Anitha and Pullaiah (2001, 2002) have also reported the maximum frequency of adventitious shoots on MS medium supplemented with BAP compared to Kn and TDZ in *Sterculia foetida*. Amin and Jaiswal (1993) and Tyagi and Kothari (2001) have also reported the same findings in *Artacarpus heterophyllus* and *Capparis deciduas* respectively. Combination of two cytokinins was found to be better than those of single cytokinin in red sandalwood (Lakshmi Sita *et al.*, 1992) and *Sapindus trifoliatus* (Appa Rao, 2004).Nagarajan *et al*. (2006) have also reported the requirement of both auxin and cytokinin for induction of multiple shoots in *Eucalyptus tereticornis*, and *Swietenia mahagoni and* Jack fruit in contrast to our present findings. Whereas single cytokinin is sufficient for developing multiple shoots in *W. tinctoria*.

In the present investigation maximum number of multiple shoots was developed in the presence of BAP alone. From these results it is evident that the endogenous hormones are variously distributed and therein, addition of exogenous hormones causes multiple shoots formation from the shoot tip explants in *W. tinctoria.* Maximum frequency number of shoots per explant was developed after 1st subculture at 0.8 mg/L BAP in *W. tinctoria*.

The size of the explant is also important for induction of multiples from shoot tip. In general larger the explants, the better chance of survival. The stages of development of the explant also have influence on the regeneration from the shoot tip. The shoot tip collected from the tip of a shoot showed higher percentage of response and giving multiples compared to basal region of shoot in *W. tinctoria*. Thus, multiple shoots were induced from the shoot tips cultured on MS medium augmented with BAP/Kn/TDZ used. This protocol can be used for rapid multiplication of an endangered forest tree species *W. tinctoria*.

Axillary buds breakage and proliferation was observed in all the concentrations and combinations of growth regulators used *W. tinctoria* except on MS basal medium. Early bud breakage (8 days) and shoot emergence was observed at 3.0 mg/L BAP/Kn alone. Whereas the axillary bud breakage taken for 10days on TDZ. Maximum percentage of response and high frequency number of multiple shoots formation were found at the same concentration of BAP in comparison to Kn. BAP showed superiority in inducing more number of multiple shoots per explant in comparison to other concentrations of BAP/Kn/TDZ used in *W. tinctoria*.

Similarly Purohit and Kukda (2004) have reported the multiple shoots formation from nodal explants of forest tree *Wrightia tinctoria* on MS medium with 2.0 mg/L BAP. The season of explant collection greatly influenced the establishment of *W. tomentosa* cultures *in vitro* (Purohit *et al.*, 2004) as observed in *W.tinctoria*. Effect of season on bud sprouting was also noted in many tree species *viz.*, *Tectona grandis* (Gupta *et al.*, 1980); *Guava* (Jaiswal and Amir, 1987); *Tecomella undulata* (Rathore *et al.*, 1991); *Prosopis cineraria* (Shekhawat *et al.*, 1993) and *W. tinctoria* (Purohit and Kukda, 2004). Tiwari *et al.* (2002) have found significant results on BAP in combination with IAA with regard to shoot multiplication from nodal segments of *Lagerstromia parviflora*.

Anitha and Pullaiah (2002) found that BAP was the best cytokinin in enhancing the shoot buds proliferation in *Sterculia foetida* than to Kn and TDZ. Sunnichan *et al.* (1998) have recorded the highest frequency number of shoots from axillary buds on MS + 6.62 µm BAP than to Kn in *Sterculia urens*. At higher concentration of BAP and Kn reduced markedly the number of shoots as found in the present investigation. These finding are similar to our present observations that BAP showed superiority over Kn in *W. tinctoria*. Whereas Tiwari *et al.* (2004) have reported the multiple shoots induction from nodal cultures on MS/B_5 medium supplemented with 0.2 mg/L IBA in an endangered forest tree *Pterocarpus marsupium*.

The size of the nodal explant was also found to play an important role in initiation, proliferation and elongation of the shoots. Sharon and D'souza (2000) and D'souza and Sharan (2001) and Rama Swamy *et al.* (2004) have reported that the smaller (0.5 cm) explants could initiate more multiples than that of longer nodal explants (1.5 cm). Similar findings were also noted in *Sapindus trifoliatus* (Appa Rao, 2004), *Terminalia alata* (Lakshman, 2006) and *Givotia rottleriformis* (Rambabu, 2007).

Nodal segment orientation at the time of culture also influences on the proliferation of shoots. Vertical orientation of *Wrightia tinctoria* explants was found better than horizontal orientation in terms of number of proliferated shoots (Purohit *et al.*, 2004).

In vitro rooting was achieved in the present investigations from the shoots developed after shoot tip and nodal cultures. It was quite difficult to standardize the protocol for rooting from *in vitro* developed microshoots in *W. tinctoria*. In comparison to all other concentrations of auxins used in *W. tinctoria*, MS medium supplemented with 0.4 mg/L IBA showed the best hormonal concentration for induction of profuse rhizogenesis. Even the best rooting efficiency was also recorded in the same concentration of IBA with lengthy and healthy roots (Plate 12.Va-c). Profuse rhizogenesis without callusing was observed in all the concentrations of IBA used. Similar efficiency of NAA compared to IAA/IBA was also observed in *Gymnema sylvestre* by Komalavalli and Rao (2000) in inducing *in vitro* rooting. Sunnichan *et al.* (1998) have reported that the induction of profuse rhizogenesis at 9.82 µM IBA from shoots developed from nodal cultures in *Sterculia urens*. They have also recorded that in any concentration of IAA used, rooting was not observed.

The ability of *in vitro* shoots to rooting is affected by several factors including difference between genotypes. The level of tissue maturity (Bonga, 1982) and physiological characteristics attributable to seasonal changes (Lakshman, 2006). Requirement of auxin to induce rooting and the subsequent interaction between applied auxin and the tissue at the base has been the subject of much research. Shoot characteristics such as size and shoot culture origin can also lead to variable rooting response (Rama Swamy, 2006).

Purohit and Kukda (2004) have also optimized the *in vitro* rooting in *W. tinctoria*. A pulse treatment with IBA for excised shoots improved rooting and minimized the callusing. Maximum root induction percentage (60 per cent) with more number of roots per shoot (2.8±0.81) was found at 100 mg/L IBA for 10 min when shoots cultured on ½ MS semi- solid medium. Whereas in the present investigations on MS solidified medium at 1.0 mg/L NAA produced 23 roots per shoot in *W. tinctoria*. Joshi *et al.* (2009) have developed the protocol for *in vitro* rooting in *W. tomentosa*. Pulse treatment of IBA (200mg/L) for 10 min was found effective for root induction (4-24 roots). The physical state of rooting media strongly influenced *in vitro* rooting response of IBA pulse treated shoots. Liquid medium was found to be significantly superior with high number of roots (4-60 roots) produced per shoot as compared to semi solid medium in *W. tomentosa*.

Thus, the protocol developed for *in vitro* rooting in the present investigations *is a break-through*. Because, it is quite difficult for inducing *in vitro* rooting in tree species like *W. tinctoria*. The reproducible protocol developed for *in vitro* rooting during the present investigations can be used for *in vitro* rapid multiplication and conservation of an endangered forest tree species which has importance in toy making industry.

Hardening/Acclimatization plays an important role in tissue culture technology for establishment of plantlets developed *in vitro*, for *Lab-to-Land Program*. In conservation and multiplication of especially endangered, medicinal and economically important forest tree species, this technique plays a vital role for an establishment of plantlets. Recently, we are successful in transfering the *in vitro* regenerated plants from forest tree species in to field conditions. They are: *Terminalia alata* (Lakshman, 2006), *Givotia rottleriformis* (Rambabu *et al.*, 2006). Acclimatization of *in vitro* rooted plantlets has been established in *W. tinctoria* for the first time. The potted mixtures containing red soil + sieved sand + vermicompost (1:1:1) showed 90 per cent survival for *in vitro* rooted plants.

Hardening is a critical step prior to transplantation of plants to the soil. The *in vitro* plantlets survive in 100 per cent relative humidity and they also dependent on the medium for supply of sugar and other nutrients. Plants are, therefore, allowed to grow on rooting media for about one month after root initiation. During this phase the nutrients in the culture go on gradually depleting and plants become sturdy and easy to acclimatize in walk-in-chamber (Ahuja, 1993a, b).

Thus, the hardening/acclimatization process includes the physiological changes which are important for the adaptation of tissue cultured plants to *ex vitro* conditions is a continuous process requiring more extended time than generally expected.

From the foregoing, it is evident that the species can be multiplied by using nodal cultures and mericlone technology to regenerate *true-to-type* of plants. The protocol developed during the present investigations can be used for *in vitro* rapid multiplication of an endangered forest tree *W. tinctoria* an important in toy making industry.

Acknowledgements

We thank the University Grants Commission, New Delhi for providing the financial assistance in the form of Major Research Project (2007-2010) F.No.32-363/2006(SR).

References

Ahuja MR (1993a) Biotechnology and clonal forestry. In: *Clonal Forestry 1.* (eds.) Ahuja MR and Libby WJ, pp.135-144.

Ahuja MR (1993b) *Micropropagation of woody plants*, Vol.41, (ed.) Ahuja MR, Kluwer Academic Publishers, The Netherlands, pp. 3-9.

Amin MN and Jaiswal VS (1993) *In vitro* response of apical bud explants from mature trees of Jack fruit (*Artacarpus heterophyllus*). *Plant Cell Tiss. and Org. Cult.*, 33: 59-65.

Anitha S and Pullaiah T (2001) *In vitro* propagation of *Sterculia foetida* Linn. (Sterculiaceae). *Plant Cell Biotech. Mol. Biol.*, 2: 139-144.

Anitha S and Pullaih T (2002) Shoot regeneration from hypocotyls and shoot tip explants of *Sterculia foetida* L. derived seedlings. *Taiwania*, 47: 62-69.

Appa Rao K (2004) *In vitro micropropagation of Sapindus trifoliatus* L. M.Phil, Thesis, Kakatiya University, Warangal.

Arya V, Singh RP and Shekhawat NS (2002) A micropropagation protocol for mass multiplication and off-site conservation of *Celastrus paniculatus* – A vulnerable medicinal plant of India. *J. Sustainable Forestry.*, 14: 107-120.

D'Souza MC and Sharon M (2001) *In vitro* clonal propagation of *Annatto* (*Bixa orellana* L.). *In vitro Cell Dev. Biol. Plant.*, 37: 168-172.

Grewal S, Ahuja A and Atal CK (1980) *In vitro* proliferation of shoot apices of *Eucalyptus citroidera* Hock. *Indian J. Exp., Biol.*, 18: 775-777.

Gupta PK, Nadgir AL, Mascarenhas AF and Jagannathan V (1980) Tissue culture of forest trees. Clonal multiplication of *Tectona grandis* (teak) by tissue culture. *Plant Sci. Lett.*, 17: 259-268.

Gupta PK, Mascarenhas AF and Hagannathan V (1981) Tissue culture of forest trees-clonal propagation of mature trees of *Eucalyptus citrioidora* Hook, by tissue culture. *Plant Sci., Lett.*, 20: 195-201.

Jaiswal VS and Amir MN (1987) *In vitro* propagation of guava from shoot cultures of mature trees. *J. Plant Physiol.*, 130: 7-12.

Joshi P,Trivedi R and Purohit SD (2009) Micropropagation of *Wrightia tomentosa:* Effect of gelling agents, carbon source and vessel type. *Indian J. Biotechnol.*, 8: 115-120.

Komalavalli N and Rao MV (2000) *In vitro* micropropagation of *Gymnema sylvestre*–a multipurpose medicinal plant. *Plant Cell Tiss. and Org. Cult.*, 61: 97-105.

Lakshmi Sita G and Vaidyanathan CS (1979) Rapid multiplication of *Eucalyptus* by multiple shoot production. *Curr. Sci.*, 48: 350-351.

Lakshmi Sita G, Sreenatha KS and Sujatha S (1992) Plantlets production from shoot tip cultures of red sanders (*Pterocarpus santalinus* L.). *Curr. Sci.*, 62: 532-535.

Lakshman A (2006) *Tissue culture studies in forest tree Terminalia alata an important for Tassar Silk Industry*. Ph.D. Thesis, Kakatiya University, Warangal.

Murashige T, Skoog F. 1962. A revised medium for rapid growth and bioassays with tobacco tissue cultures. *Physiologia Plantarum.* **15:** 473-497.

Nagarajan SM, Rajashekharan S, Kandasamy S and Jaychandra R (2006) *In vitro* studies on *Swietenia mahagoni* (L.) Jacq. *Asian J. Microbiol. Biotech. Environ. Sci.*, 8: 473-474.

Purohit SD and Kukda G (2004) Micropropagation of an adult tree – *Wrightia tinctoria. Indian J. Biotech.*, 3: 216-220.

Purohit SD, Joshi P, Tau K and Nagori R (2004) Development of high efficiency micropropagation protocol for an adult tree – *Wrightia tomentosa.* In: *Plant Biotechnology and Molecular Markers,* (eds.) P.S. Srivastava, Alka Narula and Sheela Srivastava, Anamaya Publishers, India, pp. 217-227.

Raghuraman G and Ramanujam MP (1998) Micropropagation of *Melia azaderach* (L.). *J. Swamy Bot. Cl.*, 15: 1-5.

Rama Swamy N, Ugandhar T, Praveen M, Lakshman A, Rambabu M and Venkataiah P (2004) *In vitro* propagation of medicinally important *Solanum surattense. Phytomorphology.*, 54: 281-289.

Rama Swamy N (2006) *Biotechnological applications for improvement of Solanum surattense* – A medicinal plant. Daya Publishing house, New Delhi.

Rambabu M, Ujjwala D, Praveen M, Upender M, Ugandhar T and Rama Swamy N (2006) *In vitro* zygotic embryo culture of an endangered forest tree *Givotia rottleriformis* and factors affecting its germination and seedling growth. *In vitro Cell Devt. and Biol. Plant*, 42: 418-421.

Rambabu M (2007) *Development of protocols for micropropagation of an endangered forest tree species Givotia rottleriformis Griff.* Ph.D thesis, Kakatiya University, Warangal.

Rathore TS, Singh RP and Shekhawat NS (1991) Clonal propagation of desert teak (*Tecomella undulata*) through tissue culture. *Plant Science*, 79: 217-222.

Rathore TS, Deora NS and Shekhawat NS (1992) Cloning of *Maytenus emerginata* (wild) Ding Hou – a tree of the Indian Desert, through tissue culture. *Plant Cell Rep.*, 11: 449-451.

Rathore JS, Vinod Rathore, Shekhawat NS, Singh RP, Liler G, Mahendra Phulwaria and Dagla HR (2004a) Micropropagation of woody plants. In: *Plant Molecular Biology and Molecular Markers.*(eds) P.S. Srivastava, Alka Narula and Sheela Srivastava, Anamaya Publishers., New Delhi, pp.195-205.

Rathore JS, Shekhawat NS, Singh RP, Rathore JS and Dagla HR (2004b) Cloning of adult trees of Jamun (*Syzygium cuminii*). *Indian J. Biotech.*, 3: 241-245.

Sharon M and D'Souza M (2000) *In vitro* clonal propagation of annatto (*Bixa orellana* L.). *Curr. Sci.*, 7: 1532-1535.

Shekhawat NS, Rathore TS, Singh RP, Deora N and Rao SR (1993) Factors affecting *in vitro* clonal propagation of *Prosopis cineraria*. *Plant Growth Reg.*, 12: 273-280.

Sudershan C, Aboel MN and Hussain J (2000) *In vitro* propagation of *Zizyphus mauritiana* cultivar umrdn by shoot tip and nodal multiplication. *Curr. Sci.*, 80: 290-292.

Sunnichan VG, Shivanna KR and Mohan Ram HY (1998) Micropropagation of gum karaya (*Sterculia urens*) by adventitious shoot formation and somatic embryogenesis. *Plant Cell Rep.*, 17: 951-956.

Tiwari SK, Kashyap MK, Ujjaini MM and Agarwal AP (2002) *In vitro* propagation *Lagerstromia parviflora* Roxb. from adult tree. *Indian J. Expt. Biol.*, 40: 212-215.

Tiwari S, Shah P and Singh K (2004) *In vitro* propagation of *Pterocarpus marsupium* Roxb. an endangered medicinal tree. *Indian J. Biotech.*, 3: 323-464.

Tyagi P and Kothari SL (2001) Continuous shoot production for micropropagation of *Capparis decidua*. *J. Indian Bot. Soc.*, 80: 5-8.

2013, Environmental Biotechnology *Pages* **127–133**
Editors: **D.R. Khanna, A.K. Chopra, Gagan Matta, Vikas Singh & Rakesh Bhutiani**
Published by: **BIOTECH BOOKS, NEW DELHI**

Chapter 13

The Immunomodulatory Effect of Immunex DS and H-Treat on Serum Aspartate Transaminase (AST) and Alanine Transaminase (ALT) in *Labeo rohita* (Hamilton) during Aeromoniasis

V. Viveka Vardhani and V.V. Padmavathi
Department of Zoology and Aquaculture,
Acharya Nagarjuna University, Nagarjuna Nagar, A.P.

In aquaculture practices infectious diseases are mainly controlled by antibiotics or chemotherapeutics but their use may create problems with drug resisting bacteria, toxicity, and accumulation both in fish and environment. On the contrary, natural immunostimulants have various beneficial activities like antistress, appetizer and antimicrobial. Moreover, they are cheaper and safer, non-toxic, biodegradable and biocompatible. In the present investigation, an attempt has been made to determine the immunomodulatory effect of two types of immunostimulants -Immunex DS (IDS) and H-Treat (manufactured from PVS Laboratories, Vijayawada) on the activity of serum AST and ALT in *L. rohita* during aeromoniasis Marked changes were observed in all treated fish (IDS/H-Treat) compared to controls and fish which received only infection.

Keywords: Immnostimulants–Immunex DS and H-Treat, Aeromonas liquefaciens, Labeo rohita.

Introduction

Immunostimulants are the alternatives to vaccination and antibiotics. Though the antibiotics and other chemicals are useful in aquaculture operations for treatment of various diseases, they cannot be suggested due to their residual and other adverse effects. Immunotherapy is an approach that has been actively investigated in recent years as a method for disease prevention to improve fish health and reduce the losses by diseases and other stressors. It does not involve recognition of a specific antigen but causes an overall immune response that hastens recognition of foreign proteins (Campos *et al.*, 1993, Scombes, 1994 and Sardello *et al.*, 1997). Efficacy of dietary doses of *Withania somnifera* (L-Dunal) on immunological parameters and disease resistance to *A. hydrophila* were assessed in *L. rohita* (Hamilton) fingerlings (Arun Sharma *et al.*, 2010). The enhancement of immune system and resistance to *A. hydrophila* in common carp (*Cyprinus carpio*) supplemented with plant extracts were assessed (Mohamad and Abasali, 2010). Recent approach of use of immunostimulants in fish feed will be of immense help to farmers to combat aflatoxin - induced immune suppression and other stressful situations in fish culture operations to enhance production. Since use of immunostimulants in Indian carps is an active area of research, an attemt was made in the present study to evaluate the effect of two types of immunostimulants, Immunex DS (Commercially available) and H-Treat (Herbal) on the level of serum Aspartate Transaminase (AST) and Alanine Transaminase (ALT) in *Labeo rohita* during aeromoniasis. Significant decrease of AST and ALT was found in fish treated with Immunex DS and H-Treat compared to controls and fish infected with *A. liquefaciens.* The AST and ALT levels in fish treated with H-treat were comparatively lower than in fish treated with Immunex DS.

Materials and Methods

Experimental Fish

Experiments were performed on the common freshwater carp, *Labeo rohita* (Indian major carp) which is extensively cultured in India and is valued as an important food fish.

Procurement and Maintenance of Fish

Healthy fish with an average weight of 50-60 g were obtained from Jalipudi fish farm, Jalipudi Mandal, West Godavari District, Andhra Pradesh (India) and kept in the laboratory's recirculation units until used. Fish were brought to the laboratory and maintained in cement tanks. Both experimental and control fish were acclimated to the laboratory conditions for about 4-5 days before they were used for experimentation. Dechlorinated groundwater was used during acclimation and experimental period. The water in acclimation tanks were frequently oxygenated with electrical aerators.

Bacterial Strain and Cultivation

Aeromanas liquefaciens strain, MTCC 2654 (Virulent Strain) was obtained from MTCC, Chandigarh, India. From this parent culture, sub cultures of *A. liquefaciens* were prepared and doses were made under aseptic conditions.

Antigen Dose

Various doses like 10^{-1}, 10^{-2}, 10^{-3}, 10^{-4}, 10^{-5}, 10^{-6}, 10^{-7}, 10^{-8} and 10^{-9} CFU/Fish were injected to the fish to induce aeromoniasis and determined 10^{-4} dose as LD_{50}. So,10^{-5} was selected for experimentation as optimum dose.

Route of Infection

Aeromonas liquefaciens bacterial suspension was injected to the fish, *Labeo rohita* intramuscularly near the anal region.

Immunostimulants

Immunex DS and H-Treat: manufactured from PVS Laboratories Ltd., Vijayawada, Krishna District, A.P, India were used in the present study. Test doses of immunostimulant were selected as per the recommended dosage given by PVS Lab, *i.e.* 5 g per kg of pellets.

Culture Method of *Aeromonas liquefaciens*

Culture of *Aeromonas liquefaciens* was done following the method of Pelczar *et al.* (1993).

Blood samples were collected from day 1 to 5 and tested for serum AST and ALT according to the protocol presented by IFCC (Biosystems, S.A. Costa Brava 30, Barcelona, Spain) recommended method.

Blood was collected by caudal cut method and serum was separated following standard procedures

Healthy fishes were divided into six groups and maintained in separate tanks each with twenty fish as detailed below:

- ☆ Group A (fed with Immunex DS @ 40 mg/100 g of feed)
- ☆ Group B (infected with *A.liquefaciens* @ 10^{-5} CFU/Fish)
- ☆ Group C (untreated and uninfected controls)
- ☆ Group a (fed with H-Treat @ 40 mg/100 g of feed)
- ☆ Group b (infected with *A.liquefaciens* @ 10^{-5} CFU/Fish)
- ☆ Group c (untreated and uninfected controls)

Results

Results showing in Tables 13.1–13.2 and Figures 13.1–13.2.

Discussion

The state of the liver and other organs are considered as important in determining the pathogenic effect of any infection. Therefore, it is focused on the activity of serum AST and ALT. In two groups of fish fed with Immunex Ds (group A) and H-Treat (group a) supplemented diet, the level of serum AST and ALT decreased when compared with controls. These data agree with those of Metwally (2009) studied the effects of garlic(*A. sativum*) on antioxidant system in *Tilapia nilotica* and observed significant growth performance and reduced mortality, significant increase of

Table 13.1: Aspartate Transaminase (AST) (IU/L) and Alanine Transaminase (ALT) (IU/L) in the Serum of Control (Group C, Untreated and Uninfected), Treated (Group A, Treated with Immunex DS @ 40 mg/100g of Feed) and Infected (Group B, Infected with *A. liquefaciens* @ 10^{-5} CFU/Fish) Fish, *L. rohita* at Different Days of Experiment. Values are expressed in mean derived from five observations.

Days of Necropsy	*Experimental Groups*				*Control Groups*	
	Treated with Immunex DS (Group A)		*Infected with A. liquefacies 10^{-5} CFU/Fish (Group B)*		*Untreated and Uninfected (Group C)*	
	AST	*ALT*	*AST*	*ALT*	*AST*	*ALT*
1	69	90	244	472	65	110
2	65	89	245	475	65	110
3	60	86	251	476	64	110
4	55	84	249	481	65	109
5	51	84	256	481	65	110

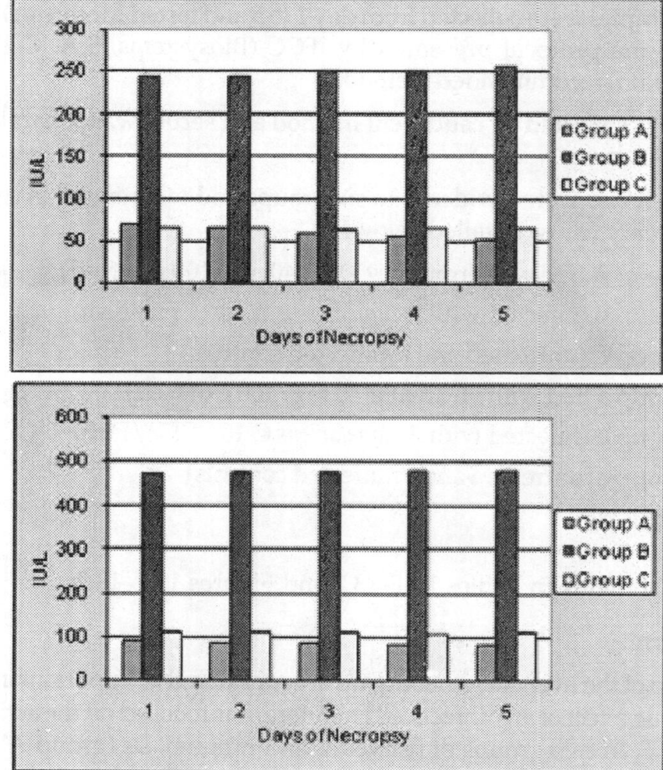

Figure 13.1: Aspartate Transaminase (AST) (IU/L) and Alanine Transaminase (ALT) (IU/L) in the Serum of Control (Group C, Untreated and Uninfected), Treated (Group A, Treated with Immunex DS @ 40 mg/100 g of Feed) and Infected (Group B, Infected with *A. liquefaciens* @ 10^{-5} CFU/Fish) Fish, *L. rohita* at Different Days of Experiment

Table 13.2: Aspartate Transaminase (AST) (IU/L) and Alanine Transaminase (ALT) (IU/L) in the Serum of Control (Group c, Untreated and Uninfected), Treated (Group a, Treated with H-Treat @ 40 mg/100 g of Feed) and Infected (Group b, Infected with *A. liquefaciens* @ 10^{-5} CFU/Fish) Fish, *L. rohita* at Different Days of Experiment. Values are expressed in mean derived from five observations.

Days of Necropsy	*Experimental Groups*				*Control Groups*	
	Treated with Immunex DS (Group A)		*Infected with A. liquefacies 10^{-5} CFU/Fish (Group B)*		*Untreated and Uninfected (Group C)*	
	AST	*ALT*	*AST*	*ALT*	*AST*	*ALT*
1	20	54	242	470	64	110
2	24	52	247	475	66	111
3	22	50	249	478	67	110
4	22	52	252	480	67	109
5	20	52	255	484	65	108

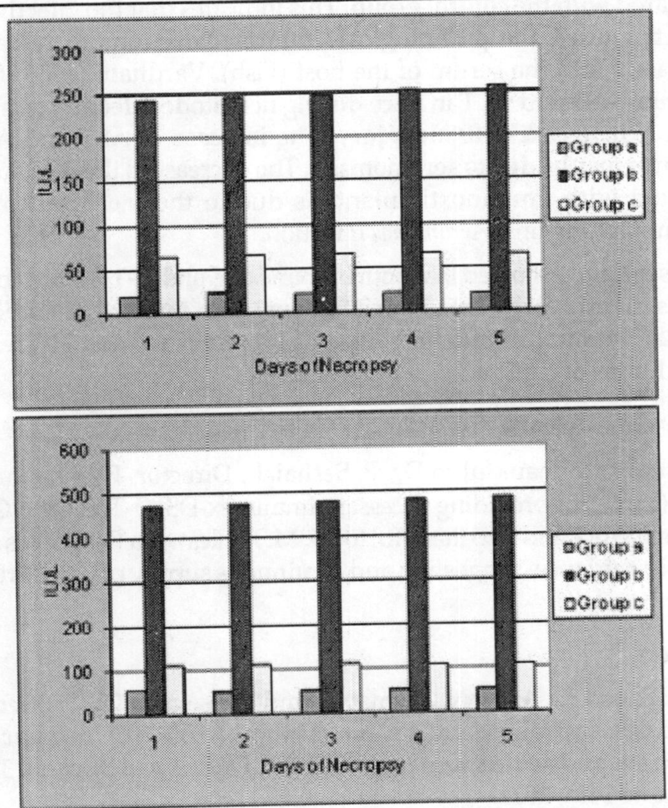

Figure 13.2: Aspartate Transaminase (AST) (IU/L) and Alanine Transaminase (ALT) (IU/L) in the Serum of Control (Group c, Untreated and Uninfected), Treated (Group a, Treated with H-Treat @ 40 mg/100g of Feed) and Infected (Group b, Infected with *A. liquefaciens* @ 10^{-5} CFU/Fish) Fish, *L. rohita* at Different Days of Experiment

glutathione peroxidase, SOD, catalase (CAT), significant decrease of aspartate aminotransferase (AST) and alamine aminotransferase (ALT) in test animals. Akhtar *et al.* (2010) assessed the effect of pyridoxine on stress mitigation and immunomodulation in *L. rohita* and found significant decrease in aspartate aminotransaminase (AST), alanine aminotransaminase (ALT), lactate dehydrogenase, malate dehydrogenase, super oxide dismutase and catalase levels and significant increase in acetyl cholinesterase levels in pyridoxine fed groups. Erythrocyte count, haemoglobin content and total serum protein, albumin, globulin, nitroblue tetrazolium and lysozyme activity were significantly elevated during pyridoxine treatment in test fish. The results indicated that dietary supplement of pyriodoxine reduced endosulfan induced stress and triggered immune response in rohu fingerlings. Ahmad *et al.* (2011) assessed the efficacy of dietary nucleotides on growth, haemotological parameters, serum proteins and serum enzymes during acute stress conditions in rainbow trout (*O. mykiss*) and found increased percentage of body weight and decreased levels of serum enzymes such as ALP, AST, LDH and ALT and suggested that dietary nucleotide administration promotes growth and enhances resistance against handling and crowding stress. The activity of serum AST, ALT showed significant increase in fish (groups B and b) infected with 10^{-5} CFU/Fish when compared with the control group. This indicates that the infective doses of *A. liquefaciens* triggered the pathological/immune reactions to release enzymes (Transaminases) into the serum of the host (Fish). Vardhani (1986) found higher levels of serum AST and ALT in mice during nematode infection due to disturbed physiological changes like tissue injury. The increase of AST and ALT levels in infected groups may be due to aeromoniasis. The decrease in the AST and ALT levels in fish treated with immunostimulants is due to the increased protection by immunostimulants against *Aeromonas* infection.

The present study showed that both Immunex DS and H-Treat are able to enhance the immune system of carp even under infectious stress; but further studies are needed to find out the immunostimulation effect of Immunex DS and H-Treat in rohu in various conditions of culture.

Acknowledgements

The authors are thankful to Dr. P. Seshaiah, Director, PVS Laboratories Ltd., Vijayawada (India) for providing necessary Immunex-DS, H-Treat and Ciprofloxacin for the research work and also thankful to Sri. M. Malleswara Rao for his spontaneous help, valuable guidance, suggestion and continuous supply of fish during the entire course of study.

References

Ahmad T-K., Saeed K., Amin N., Nemat M. and Hossein P. (2011) Effects of dietary nucleotides supplementation on rainbow trout (*Oncorhynchus mykiss*) performance and acute stress response. *Fish Physiol. and Biochem.* 37. Published online, June 14th, 2011.

Akhtar M.S., Kumar Paul A., Prasad Sahu N., Alexander C., Sanjay Kumar G., Arup Kumar Ch., Ashish kumar J. and Govindrajan M. (2010). Stress mitigating and immunomodulatory effect of dietary pyridoxine in *Labeo rohita* (Hamilton) fingerlings. Aqua. Res. 41(7): 991-1002

Arun Sharma., Ashutosh D. D., Ritesh Kumar T.S., Chanu T.I. and Arabinda D. (2010) Effect of *Withania somnifera* (L. Dunal) root as a feed additive on immunological parameters and disease resistance to *Aeromonas hydrophila* in *Labeo rohita* (Hamilton) fingerlings. Fish and Shellfish Immunol. 29(3): 508-512.

Campos M., Godson D., Hughes H., Babiuk L. and Sordillo L. (1993) The role of biological response modifiers in disease control. J. Dairy Sci. 76: 2407-2417

Metwally M.A.A. (2009) Effect of Garlic (*Allium sativum*) on some antioxidant activities in *Tilapia nilotica* (*Oreochromis niloticus*). World. J. Fish and Marine. Sci. 1(1): 56-64.

Mohamad S. and Abasali H. (2010) Effect of plant extracts supplemented diets on immunity and resistance to *Aeromonas hydrophila* in common carp *Cyprinus carpio.* Agricultural Journal, 5(2): 119-127.

Pelczar M.J.Jr.,Chan E.C.S. and Noel R.K. (1993) Characterization of microorganisms. Microbiology concepts and Applications.5thEd.Tata Mcgraw. Hill Publishing co.Ltd. PP. 81-83 (Robert Koch, 1883).

Scombes C.J. (1994) Enhamncement of fish phagocytic activity. Fish and Shellfish Immunol. 4: 421-436

Sordello L.M., Shafer W.K. and De rosa D. (1997) Immunobiology of the mammary gland. J. Dairy Sci. 80: 1851-1865

Vardhani V.V. (1986) Serum levels of Aspartate Transaminase, Alanine Transaminase and worm burden in mice infected with *Ancylostoma caninum* larvae. Folia Parasitologica. 33: 163-167.

2013, Environmental Biotechnology *Pages 135–146*
Editors: **D.R. Khanna, A.K. Chopra, Gagan Matta, Vikas Singh & Rakesh Bhutiani**
Published by: **BIOTECH BOOKS, NEW DELHI**

Chapter 14

Morphophysiological Characterization of Pea (*Pisum sativum* L.) Genotypes for Drought Tolerance

Usha Rana and Kavita Kumari*
Assistant Professor and Research Scholar
Department of Biology and Environmental Sciences
College of Basic Sciences CSKHPKV, Palampur – 176 062

The present study was conducted to assess drought tolerance amongst four leading garden pea genotypes (Arkel, Azad Pea 1, Palampriya and Lincoln) based on certain Morphophysiological parameters *viz.*, plant dry weight, root length, No. of leaves, leaf area,leaf area index, relative water content, Drought susceptibility index and harvest index (per cent) The experiment was designed by imposing water stress at vegetative stage T_1(30- 45 DAS) flower initiation stage T_2 (60 -75 DAS) and pod filling stage T_3 (90 -100 DAS).All the treatment affected the performance of the pea genotypes. Palampriya and Azad Pea 1 exhibited significantly more plant dry weight, root length, No. of leaves, leaf area,leaf area index, relative water content, harvest index and low Drought susceptibility index than Arkel and Lincoln under all the treatments. Similar trend was noticed for all the Morphophysiological parameters under different treatments, however, Palampriya and Azad Pea 1 showed highest increase in T_3 treatment stage in comparison to other treatments in all the genotypes. Based on different

* Corresponding Author: E-mail: rana.usha@rediffmail.com

morphophysiological responses of all the four pea genotypes under water stress it has been concluded that Palampriya proved to be the water stress tolerant and Azad pea-1 was moderately water stress tolerant pea genotype where as Lincoln was susceptible and Arkel was moderately water stress susceptible genotypes. Water stress at vegetative stage was found to be more sensitive than flower initiation and pod filling stage.

Keywords: *Drought tolerance, Pea, DSI, Harvest Index, Water stress.*

Introduction

Pea (*Pisum sativum* L.) is one of the most important *Rabi* vegetable crop grown throughout the world. Commercially, it is grown as winter season crop in Northern Indian plains and summer crop in the high hills. Pea is grown as major cash crop in Himachal Pradesh under rainfed conditions. More than 80 percent of the cultivated area in the state is rainfed as the crop is cultivated largely on conserved moisture in the soil profiles from the preceding monsoon rains which gradually decreases with time. Due to erratic and untimely distribution of rainfall, runoff losses due to undulating topography and low water holding capacity of the soils further accentuate the problem of drought in the hills; that is why average productivity level of pea in the state (1,5MT/ha) is much below the national average productivity (8.2MT/ha Anonymous 2009). Drought management is therefore a major challenge to rainfed grain and legume crop production as growth and yield of crop depends on the interaction between environmental factors with numerous physiological

processes of the plant (Sanap *et al.*, 2004). Sinha (1985) defined drought as "inadequacy of water availability including precipitation during the life cycle of crop to restrict expression of its full genetic yield potential" hence water deficit at any stage, especially critical stages of the crop, cause severe damage to the yield of the crop. Water stress during grain filling reduces the thousand grain weight and grain yield than the well watered control (Hochman, 1982). It also reduces number of days to flowering and harvest index compared with the irrigated crop (Yadav *et al.*, 1989). Grain yield decreases with increasing water stress from 0.65 kg/plant at 10 per cent soil moisture depletion in the 1 to 30 cm to 0.32 kg at 25 per cent depletion in the 30 cm depth. Reduction in grain yield due to significant reduction in pods/plant, grains/pod and grain weight under water stress conditions. The lack of nitrogen under stressed condition caused a 77 per cent reduction in grain yield compared with that of high nitrogen treated irrigated plants (Satchithananthamet *al.*, 2000). Water deficit subsequent to flowering can induce the abscission of flower buds and small pods, and when it occur later towards maturity, promote pre mature leaf loss and reduced seed size. Susceptibility to moisture stress is a major constraint limiting productivity of pulses, thereby leaving the best option for crop production, yield improvement and yield stability under water deficit conditions is to develop drought tolerant crop varieties. It is reported that the tolerant genotypes very often maintain better water relation for osmotic adjustment and higher photosynthetic activity than susceptible ones (Raptan *et al.*, 2001). Keeping this in view, the present study was undertaken to analyse the magnitude of changes due to water stress in four genotypes differing in drought tolerance in order to understand drought tolerance mechanism of this crop by using the parameter total leaf area, no. of leaves, leaf area index, dry matter accumulation, relative water content and drought susceptibility index and harvest index. The present study describe evaluation of genotypes on the basis of inherent stress tolerant characters as tolerant genotypes can be named as most potential genotypes and the study will be useful for the plant breeder to incorporate these characters while selecting genotypes for crosses.

Materials and Methods

The seeds of four pea genotypes namely Arkel, Azad Pea-1, Palampriya and Lincoln were procured from Seed Technology unit of CSKHPKV, Palampur and sowing was done in 30 cm diameter pots filled with soil, vermicompost and sand in the 3:2:1 ratio along with the addition of NPK in the ratio 12:32:16 during winter season. Water stress was created at three stages of plant development *i.e.* vegetative stage (30-45 DAS), flower initiation stage (60-75 DAS) and pod filling stage (90-100 DAS) by withholding water till the appearance of slight wilting symptoms. Soil moisture content was taken to be15 per cent for unstressed plants and 5 to 8 per cent in different stressed genotypes. (Susceptible genotypes contain more moisture content in the soil whereas tolerant use for growth and development so contain only 5 per cent in the soil) Measured quantity of water was supplied whenever needed in stressed plants and unstressed plants were watered regularly. Tagging of the crop was done for three replication, staking to give support to crop plants and fungicide (Hexaconazole) was sprayed whenever necessary and regular hand weeding was done

Morphophysiological Parameters

Number of Leaves

The leaf number was taken on counting the leaves of whole plant. Counting was done at 30days interval from stress imposition.

Root-Shoot Dry Weight/Plant Dry Weight

Plant dry weight was taken by weighing dried root and shoot. Dry weight was taken after 30 days interval from stress imposition.

Length of Roots

Length of roots of plant was taken by uprooting the plant and measure with the help of meter rod. Observations were recorded at 30 days of interval from stress imposition. For analyzing the root length, the roots were washed gently under the running water to avoid any type of damage and to recover complete root set.

Relative Water Content (RWC)

Relative water content were measured from top leaves (Merah, 2001) method. First take fresh weight (FW) at excision, after taking fresh weight, leaves were left to rehydrate in distilled water for 24 hours at 4° C in darkness to obtain the weight at full turgour ((TW). The leaf dry weight (DW) was taken after 48 hours at 80° C. The relative water content were calculated by the equation:

$$RWC = \frac{FW - DW}{TW - DW} \times 100$$

Observation were taken at 30 days interval after stress imposition.

Leaf Area

Leaf area of all the leaves of the plant was measured at 30 days of interval from stress imposition with the help of leaf area meter (CI-203 Leaf area meter. CID)

Leaf Area Index (LAI)

Leaf area index was measured at 30 days interval from stress imposition with plant canopy analyzer (LAI-2000, Licor).

Harvest Index (HI)

Harvest index is called the coefficient of effectiveness of formation of economic part of total yield. It characterizes the movement of dry matter to the economic part of plant. The Harvest index was calculated by formula given by Donald and Hamblin (1976).

$$HI = \frac{\text{Economic yield}}{\text{Biological yield}} \times 100$$

Drought Susceptibility Index (DSI)

Drought susceptibility index was calculated by Fischer and Maurer (1978) method by applying formula.

$$S = \frac{(1 - Ys/Yp)}{D}$$

Ys: Mean grain yield of genotype under stress

Yp: Mean grain yield of genotype under irrigated environment

D: Stress intensity

$$\text{Stress Intensity (D)} = \frac{1 - \text{Mean Ys of all genotypes}}{\text{Mean Yp of all genotypes}}$$

Results

Water stress imposed at vegetative (T_1) stage (Table 14.1) showed that numbers of leaves were increased towards maturity the genotype Azad pea-1 showed maximum number of leaves followed by Palampriya. The numbers of leaves of all the genotypes were less as compared with unstressed plants. Azad pea-1 showed less and Arkel showed maximum reduction in comparison to unstressed conditions. Water stress at pod filling (T_3) stage also indicate that the Number of leaves in the genotype Azad pea-1 was significantly maximum than the other three genotypes. Palampriya and Azad Pea-1 were at par with each other and significantly lower number of leaves was observed in the genotypes Arkel and Lincoln than the Azad pea-1 under this stage.

Table 14.1: Effect of Water Stress on Plant Dry Weight (g/plant), Root Length (cm), No. of Leaves, Leaf Area (cm²/plant), Leaf Area Index, Relative Water Content (per cent), DSI and Harvest Index (per cent) in Different Genotypes of Pea at Vegetative Stage

Pea Genotypes	Plant Dry Weight (g/plant)		Root Length (cm)		No. of Leaves		Leaf Area (cm²/plant)		Leaf Area Index		Relative Water Content (per cent)		Drought Susceptibility Index	Harvest Index (Per cent)	
	S	US	S	US	S	US	S	US	S	US	S	US	T_1	S	US
Arkel	0.57	1.43	7.4	11.9	35	41	73.1	141.6	0.4 8	0.78	65.5	79.8	3.126	54.2	82.0
Azad Pea-1	1.33	1.74	13.1	9.53	42	44	89.1	110.9	0.97	1.12	76.9	81.5	1.133	67.4	83.2
Palampriya	0.97	1.09	7.93	7.83	47	49	95.2	116.7	0.97	0.99	82.8	87.5	0.540	80.6	94.6
Lincoln	0.64	0.93	7.03	9.96	39	51	64.4	105.9	0.44	1.32	59.1	67.0	3.303	45.1	81.8
CD(5 per cent)	0.292	0.392	1.72	NS	NS	3.606	1.86	5.55	0.236	0.282	1.21	1.57	0.039	5.49	2.68

S: Stressed Plants; US: Unstreassed Plants; T_1: Treatment (stress at vegetative stage).

The plants under this treatment level showed higher number of leaves than the first two treatments but all the genotypes had reduced number of leaves than unstressed plants and in this stage tolerant genotypes showed almost same no. of leaves than their unstressed plants. Length of roots increased in the genotype Azad pea-1 at all stages of stress induction and Palampriya showed more root length only at initial time intervals. Lincoln and Arkel were at par to each other at all stages of stress imposition except at maturity where Lincoln try to adjust it self by growing its root (Table 14.3).

Azad pea-1 has maximum length of root among all the genotypes at flower initiation (T_2) stage (Table 14.2). Azad Pea-1 was significantly more than Arkel and Lincoln which were at par to each other and Palampriya followed the Azad pea-1. Palampriya showed maximum increase in root length till 30 days from stress imposition. At maturity Azad pea-1 showed less and Lincoln showed maximum reduction in comparison to unstressed conditions.

Plant dry weight increased progressively with time intervals in all the genotypes. Azad pea-1 and Arkel showed best result in accumulating plant dry weight in comparison to unstressed conditions. Arkel initially accumulate more dry weight but Azad pea-1 accumulates towards maturity. Azad pea-1 and Palampriya were almost in same pattern of accumulating dry weight where as maximum reduction shown by genotype Lincoln.

As in T_1 plant dry weight increased in T_2 from stress imposition. Azad pea-1 accumulates maximum plant dry weight and significantly more than Arkel and Lincoln but at par with Palampriya. Palampriya and Azad pea-1 showed initially increase in dry weight but after that showed sudden decline in dry weight but maximum decline in plant dry weight was seen in Arkel and Lincoln. Arkel is moderately susceptible where as Lincoln is highly susceptible genotype in this stage.

Leaf area reduced in all the genotypes in stressed than unstressed plants. Palampriya showed maximum leaf area among all other genotypes at all stages of stress imposition. As compare to control Azad pea-1 had shown maximum reduction in leaf area at 30 days of interval and less reduction in leaf area at maturity. The maximum reduction in leaf area observed in Arkel and minimum in Palampriya was seen in comparison to unstressed condition

In Water stress at flower initiation (T_2) stage (Table 14.2) Azad pea-1 and Palampriya were shown best result in leaf area. Azad pea-1 and Palampriya had maximum leaf area than Arkel and Lincoln. Palampriya and Azad pea-1 were at par with each other at 30 days from stress imposition and were resistant genotypes but Arkel and Lincoln were susceptible one.

Leaf area of the plants in which stress imposed at pod filling stage (Table 14.3) was more than the plants in which the stress was imposed at vegetative and flower initiation stage. Leaf area increased upto 30 days interval of time in all the genotypes under stressed conditions and then decreased towards maturity. Palampriya was very much tolerant genotype followed by Azad pea-1 at this stage and Arkel was moderately susceptible and Lincoln was susceptible genotype but Lincoln was also trying to show more leaf area in pod filling stage under water stressed condition.

Table 14.2: Effect of Water Stress on Plant Dry Weight (g/plant), Root Length (cm), No. of Leaves Leaf Area (cm²/plant), Leaf Area Index, Relative Water Content (per cent), DSI and Harvest Index (per cent) in Different Genotypes of Pea at Flower Initiation Stage

Pea Genotypes	*Plant Dry Weight (g/plant)*		*Root Length (cm)*		*No. of Leaves*		*Leaf Area (cm²/plant)*		*Leaf Area Index*		*Relative Water Content (per cent)*		*Drought Susceptibility Index*	*Harvest Index (Per cent)*	
	S	*US*	*S*	*US*	*S*	*US*	*S*	*US*	*S*	*US*	*S*	*US*	T_2	*S*	*US*
Arkel	1.62	1.76	8.06	15.7	35	41	107.4	145.2	0.68	1.98	68.7	72.6	0.726	67.5	82.0
Azad Pea-1	1.64	1.92	19.7	20.6	42	44	166.7	176.7	0.75	0.81	86.4	92.1	0.530	70.4	83.2
Palampriya	1.54	1.80	15.3	18.4	46	49	166.8	196.1	0.94	0.95	93.3	94.5	0.263	82.1	94.6
Lincoln	0.93	2.26	8.23	13.7	39	51	93.2	144.0	0.72	1.66	61.9	74.3	0.796	61.6	81.8
CD(5 per cent)	0.33	0.37	1.77	3.62	NS	3.60	6.12	4.60	0.94	0.21	1.78	1.02	0.036	6.17	2.68

S: Stressed Plants; US: Unstreassed Plants; T_2: Treatment (stress at initiation stage).

Table 14.3: Effect of Water Stress on Plant Dry Weight (g/plant), Root Length (cm), No. of Leaves Leaf Area (cm²/plant), Leaf Area Index, Relative Water Content (per cent), DSI and Harvest Index (per cent) in Different Genotypes of Pea at Pod Filling Stage

Pea Genotypes	*Plant Dry Weight (g/plant)*		*Root Length (cm)*		*No. of Leaves*		*Leaf Area (cm²/plant)*		*Leaf Area Index*		*Relative Water Content (per cent)*		*Drought Susceptibility Index*	*Harvest Index (Per cent)*	
	S	*US*	*S*	*US*	*S*	*US*	*S*	*US*	*S*	*US*	*S*	*US*	T_3	*S*	*US*
Arkel	1.19	2.07	16.2	19.8	59	68	171.6	174.7	0.68	0.87	55.4	84.5	0.396	75.2	82.0
Azad Pea-1	1.55	2.02	18.1	24.4	89	90	188.4	189.4	0.73	0.84	82.9	83.9	0.350	78.4	83.2
Palampriya	1.16	1.39	17.4	25.7	75	77	188.5	189.3	0.82	0.98	88.6	92.1	0.130	90.0	94.6
Lincoln	1.67	3.02	11.9	22.7	68	78	183.6	225.2	0.61	0.85	48.7	80.4	0.180	73.7	81.8
CD(5 per cent)	0.18	0.19	2.12	1.86	2.82	3.30	5.33	6.94	0.236	0.34	0.62	1.99	0.044	3.28	2.68

S: Stressed Plants; US: Unstreassed Plants; T_3: Treatment (stress at pod filling stage).

Leaf area index first increased and then decreased towards maturity in all the genotypes under stressed condition. Palampriya showed maximum leaf area index and significantly more than Arkel and Lincoln but as compared to unstressed conditions Azad pea-1 showed best results in all the observations of stress imposition at vegetative stage. Lincoln was drought susceptible as it showed maximum reduction in leaf area index in comparison to all other genotypes at every time interval from stress imposition and minimum in the genotype Palampriya. Azad pea-1 and Palampriya were at par with each.

Palampriya increased its leaf area index at T_2 (flower initiation) stage at 30 days from stress imposition than other genotypes. Next to genotype Palampriya, Azad pea-1 maintained higher leaf area index. Palampriya was followed by Azad pea-1 and significantly more than Arkel and Lincoln and also showed minimum reduction in leaf area index at maturity. This proves that Palampriya was more tolerant than Azad pea-1 at flower initiation stage.

Similarly, in Water stress at pod filling (T_3) stage (Table 14.3) leaf area index of all the genotypes was reduced as compared to unstressed plants. Palampriya significantly showed more leaf area index than all other genotypes. In this stage all the genotypes were also showed less decline than flower initiation stage which showed that at pod filling stage, maximum development of plant parts has already taken. There was less reduction in leaf area index found.

Relative water content reduced in all the genotypes under stressed than unstressed plants. Palampriya showed maximum relative water content and significantly more than other genotypes and followed by Azad pea-1 which was showing that these two genotypes maintain its relative water content even under stressed condition. Arkel and Lincoln showed almost similar responses but significantly lower than Palampriya and Azad pea-1 all time intervals of this stage but, in this stage Arkel is more susceptible than Lincoln as Lincoln showed more RWC than Arkel compare to non stress treatment where as Azad pea-1 and Palampriya were drought tolerant. At flower initiation (T_2) stage Relative water content reduced in the all genotypes under stressed than unstressed plants. Relative water content was maximum in the genotype Palampriya followed by Azad pea-1. Palampriya proves to be more tolerant than Azadpea-1.

At flower initiation (T_2) (Table 14.2) stage Palampriya had maximum relative water content similarly as T_1 and T_2 stages. Azad pea-1 showed very less decline in relative water content under stress as compare to unstressed plants followed by Palampriya. Arkel and Lincoln showing same trend and they were water stress susceptible genotypes. Arkel and Lincoln were shown significantly less relative water content than Palampriya. These observations indicate that pod filling stage was very sensitive to relative water content in susceptible varieties. The genotype Azad pea-1 showed minimum reduction in RWC genotype in comparison to unstressed conditions proved Azad pea-1 be the water stress tolerant genotype. Overall it is concluded that Number of leaves was maximum in Azad pea-1 compared to unstressed condition followed by Palampriya and significantly more than Arkel and Lincoln. Root length of Azad pea-1 and Palampriya showed increase in stress condition than unstressed plants but Arkel and Lincoln showed minimum root length and more decline which proves that Azad pea-1 and Palampriya were tolerant and Arkel and Lincoln were water stress susceptible genotypes. Plant dry weight accumulated more in water stress tolerant genotypes Palampriya and Azad pea-1 than water stress susceptible genotypes Arkel and Lincoln.

Leaf area, leaf area index and relative water content were maximum in the genotype Palampriya followed by Azad pea-1 and less in Arkel and Lincoln. Drought susceptibility index and soil moisture content were showed almost similar results that water stress tolerant genotypes Palampriya and Azad pea-1 having less value of DSI than drought susceptible Arkel and Lincoln Drought susceptibility index was in decreasing trend.

Opposite to DSI, all the pea genotypes showed increasing trend in harvest index from vegetative to pod filling stage. This shows that vegetative stage is very sensitive to water stress in all the genotypes. Maximum HI was observed in Palampriya followed by Azad pea-1 and significantly more than Arkel and Lincoln.

Discussions

Number of leaves were reduced in all pea genotypes under water stress as compared to unstressed plants. Azad pea-1 was the best genotype among all by showing maximum leaves and very less reduction in leaf number than unstressed plants followed by Palampriya and significantly more than Arkel and Lincoln. Here, Azad pea-1 and Palampriya were water stress tolerant and Arkel and Lincoln were water stress susceptible genotypes. Similar observation was reported by Lat *et al.* (1990). Ferreira *et al.* (1992) and Yang *et al.* (2003) also showed that increasing water stress reduced number of leaves, flowers and reproductive efficiency. The reduction in the leaves under water deficit conditions may be due to enhancement in the senescence because of increase in the ABA level under stress and retaining more leaves under stressed conditions was a stress tolerant traits.

Root length of the genotypes Palampriya and Azad pea-1 varied in all the time intervals after stress imposition of different stages. Genotypes Arkel and Lincoln showed less root length than plants under unstressed conditions. Azad pea-1 and Palampriya were at par with each other and showed significantly more root length than Arkel and Lincoln. So, Azad pea-1 and Palampriya were water stress resistant as showed by maximum increase in root length than unstressed plants. Lang-youzhang, (2003) suggested that root length, root width, root penetration ability and elongation rate of roots in water stress resistance cultivars were significantly higher than those in water stress susceptible cultivars which corroborated our findings. Taylor and Nguyen (1987) also showed similar observations that modifying root length densities with depth reduce water stress as it enable the plant to get water from deeper part of the soil. Rane *et al.* (2002) stated that there exist a positive relation with root length, shoot weight and tolerant ability of cultivar.

Plant dry weight of all the genotypes showed reduction in dry weight under stress conditions at all the three stages *i.e.* Vegetative, flower initiation and pod filling in comparison to unstressed conditions. However, water stress tolerant genotypes accumulate more dry weight as compared to susceptible plants. Similar results were observed by Ashraf *et al.* (1998) and Gupta *et al.* (2001) that increase in dry weight in better performed genotypes may be due to better absorption and translocation of nutrient from soil to the stem.

Leaf area of all the genotypes under all the levels of stress treatments reduced in comparison to the genotypes grown under irrigated conditions. Palampriya showed maximum leaf area and less reduction than unstressed plants followed by Azad pea-1 and significantly higher than Arkel and Lincoln. Reduction in leaf area increased in susceptible but maintained in resistant cultivars. The finding was supported by Renu *et al.* (2000) who gave that despite of similar difference in leaf water potential between irrigated and water stressed plants at both stages, the effects on leaf area increased from vegetative to pod filling stage. Upreti *et al.* (2000) also reported that water stress treatment resulted in considerable decline in leaf area. Arkel (susceptible cultivar) was affected most with respect to these characters.

Leaf area index was found more in irrigated plants than the genotypes grown under water stress. Palampriya was followed by Azad pea -1 which showed maximum leaf area index and less reduction as compared to unstressed and was significantly more than Arkel and Lincoln. The reduction in LAI in all the genotypes was minimum at the pod filling stage showing that at pod filling stage maximum development of plant parts has already taken place. Imtiyaz *et al.* (1990) also showed that leaf area index and number of leaves are decreased under water stressed conditions. The decrease in leaf area index is due to the reduction in leaf expansion under water stress. Vasantha *et al.* (2005) reported that under stress, the leaf area index and yield were strongly correlated, indicating the usefulness of these parameters in identifying drought tolerant genotypes.

Relative water content was maintained maximum by Palampriya in all stages of stress treatments and significantly more than Azad pea-1, Arkel and Lincoln. Although, the relative water content in all the genotypes were less than the unstressed plants at all the three stages of stress treatment. Naidu *et al.* (2001) who reported that relative water content and leaf area per plant decrease in all the genotypes of chickpea under water stressed situations.

The lowest drought susceptibility index (DSI) value showed by the drought tolerant genotypes while the which were having highest drought susceptibility index were drought susceptible. DSI showed decreased trend in every stage from vegetative to pod filling stage. The genotype Palampriya showed significantly less value of DSI than Azad pea-1, Arkel and Lincoln. Kumari (1992) reported that tolerant varieties had lowest value of DSI. In chickpea genotypes RSG 44, RSG-143-1 and ICC-4958 showed tolerance to moisture stress due to better imbibition, seedling growth, lesser membrane injury, better osmotic adjustment and water use efficiency resulting in lower DSI and higher seed yield under stress (Sexena *et al.*, 1996).

Harvest index of all genotypes showed increase trend from vegetative to pod filling stage. There was very less reduction in yield in Palampriya than Azad pea-1, Arkel and Lincoln compared to unstressed conditions. Increase in Harvest index is due to higher grain yield and the genotypes translocated maximum food material from leaves towards grain. Wu *et al.* (2001) reported that the HI of cultivars under stressed conditions is low and is directly correlated to the leaf area, leaf growth which were generally inhibited under stressed environment.

References

Anonymous, 2009.Area and production of vegetables in Himachal Pradesh,Directorate of Agriculture(H.P.), Shimla-5.

Ashraf, M.Y., Ala, S.A. and Bhatti, A.S. 1998. Nutritional imbalance in wheat (*Triticum aestivum* L.) genotypes grown at soil water stress. Acta Physiologiae Plantarum 20(3): 307-310.

Donald, C.M and Hamblin, J. 1976. The biological yield and harvest index of cereals as agronomic and plant breeding criteria. Advances in Agronomy. 28:361-905

Ferreira, C.G.R., Santos, I.F.dos, Tavora, F.J.F., Silva, J.V.-dos, Santos-I.F., Da – Silva, J.V. 1992. Effect of water deficit on groundnut (*Arachis hypogaea* L.) cultivars. Physiological Reactions and Yields oleagineux – Paris 47(8-9): 523-530.

Fischer, R.A. and Maurer, R. 1978. Drought resistance in spring wheat cultivars. Grain yield responses. Australian Journal of Agricultural Research 29: 897-912.

Gupta, N.K., Sunita Gupta and Arvind Kumar. 2001. Effect of water stress on physiological attributes and their relationship with growth and yield of wheat cultivars at different stages. Journal of Agronomy and Crop science 186(1): 56-62.

Hochman, Z. 1982. Effect of water stress with phasic development on yield of wheat grown in a semi arid environment. Field crop Research 5(1): 55-67.

Imtiyaz, M., Dwyer, S., Kumar, A., Salokhe, V.M. and Llangatileke, S.G. 1990. Effect of moisture stress on wheat. Proceeding of the international agricultural engineering conference and exhibition, Bangtok, Thailand. Soil and Water Engineering 909-918.

Kumari, V. 1992. Genetic variability for some morphophysiological character associated with drought tolerance. Department of Plant Breeding and Genetics, CSK HPKV Plamapur.

Lang – Youzhang, Hu – Jian, Yang – Jianchang, Zhang – Zujian, Zhu – Qingsen. 2003. Morphological and Anatomic traits in drought resistance rice roots. Journal of Youzhang – University, - Agriculture and Life Sciences, Edition 24(4): 58-61.

Lat, K.R.,Moralia, R.N. and Kumar, A. 1990. Physiology of drought tolerance in wheat (*Triticum aestivum* L.) growth and yield. Comparative Physiology and Ecology 15(4): 147-158.

Naidu, K.R., Seethambaram, Y. and Dass, V.S.R. 2001. Leaf and nitrate proteinase and nitrate reductase activities in relation to grain protein and grain yield in four species of Aamaranth. Proceeding of India Academy of Science 91(5): 433-441.

Rane, J., Rao, N. and Nagrajan, S. 2002. Association between early vigour and root traits in wheat (*Tricum aestivum*) under moisture stress. Indian Journal of Agricultural Sciences 72(8): 474-476.

Raptan,P.K.,Hamid,A.,Khaliq,Q.A.,Solaiman,A.R.M.,Ahmed,J.U.and Karim,M.A.(2001) Salinity tolerance in blackgram and mungbean II. Mineral ions accumulation in different plant parts. Kor. J. crop Sci. 46:387-394.

Renu Khanna – Chopra, Reddy, P.V. and Sinha, S.K. 2000. Interaction of drought stress induced senescence with monocarpic senescence in cowpea (*Vigna unguiculata*). Water relations, growth and senescence. Journal of plant Biology 2000, 27(2): 171-178, 124, 145, 157.

Sanap, M.M, Munge, H.B and Deshmukh, D.V. 2004. Growth and Yield variation in chickpea for drought tolerance. Journal of Maharashtra Agriculture University 29(2): 143-145.

Satchithananatham, S. and Bandara, D.C. 2000. Physiological responses of maize (*Zea mays* L.) to the interactive effects of nitrogen fertilizer and water regimes. Tropical Agricultural Research 2000 12:22-32.

Sexena, H.K, Yadav, R.S. and Mathur, R.K. 1996. Effect of moisture stress on metabolic activity and grain yield in wheat (*Triticum aestivum* L.) genotypes. Indian J. Physiology 1:303-306.

Sinha, S.K. 1985. Approaches for incorporating drought and salinity resistance in crop plants. In: V.L. Chopra and R.S. Paroda (eds.), Drought Resistance in Crop Plants. Physiological and biochemical analysis.pp 121-137. Oxford and IBH Pulishing Co. Pvt. Ltd. New Delhi.

Upreti, K.K., Murti, G.S.R. and Bhatt, R.N. 2000. Responses of pea cultivars to water stress: changes in morphophysiological characters endogenous hormones and yield. Vegetable Science 27(1): 57-61.

Vasantha, S., Alarmelu, S., Hemáprapha, Gad and Shanthi, R.M. 2005. Evaluation of promising sugarcane genotypes for drought. Sugar tech. 7(2-3): 82-83.

Wu, H.Q.,Duran,A.W. and Yang, C.F. 2001. Physiological and morphological responses of winter wheat to soil moisture. Acta Agriculture Boseali Sinica 15(1): 92-96.

Yadav, R.D.S., Singh, S.N. and Singh, S.B. 1989. Morphological limitations of dwarf wheat productivity under dry land. Narendradeva Journal of Agricultural Research 4(1): 96-97.

Yang, J.C., Zhang, J.H., Wang, Z.Q. and Liee, L.T. 2003. Involvements of abscissas acid and cytokinins in the senescence and remobilization of carbon reserves in wheat subjected to water stress during grain filling period. Plant cell and environment 26(10): 1621-1631.

2013, Environmental Biotechnology *Pages 147–152*
Editors: **D.R. Khanna, A.K. Chopra, Gagan Matta, Vikas Singh & Rakesh Bhutiani**
Published by: **BIOTECH BOOKS, NEW DELHI**

Chapter 15

Influence of Early and Late Mounting on Economic Parameters in Autumn Rearing of PM × CSR$_2$ Larvae of Silkworm, *Bombyx mori* L.

Amardev Singh[1] *and Shamin Ahmed Bandey*[2]
[1]*Contractual Lecturer, Department of Sericulture, Govt. Degree College, Poonch*
[2]*Assistant Professor, Department of Zoology, Govt. Degree College, Poonch*

Effect of early and late mounting larvae on some economic traits such as single cocoon weight, single shell weight, shell per cent, cocooning per cent, defective and good cocoon per cent of multi × bi silkworm hybrid (PM × CSR$_2$) is presented in this paper. All the parameters studied under this experiment showed better performance in control batches when the silkworms were mounted on recommended day of mounting *i.e,* 7th day after maturation.

Keywords: Silkworm, Mounting, Early mounting, Late mounting.

Introduction

Mounting operation is one of the times bound and labour intensive activities in silkworm rearing. After feeding on mulberry leaves for several days, silkworms of the

5th instar stop feeding and begin to spin cocoon. To spin cocoons, mature silkworms need mountages (cocoon frames) as supports. The process of moving mature larva onto the cocoons frame is called mounting. Mounting process in silkworm rearing is the most labour intensive operation to be simplified. Mounting should not be delayed when larvae mature as it results in loss of silk besides production of poor quality cocoons ([10]). The rearing of silkworm in the state of J and K is a supporting occupation for small and marginal sericulturists to earn livelihood unlike in tropical areas this enterprise is practiced as major activity by the rural masses. The sericulturists in the state generally do not have their own mulberry garden, but are dependent on wild trees growing on road sides, forests, etc. Quite often they are forced to abandon their rearing even on penultimate day due of seriposition due to non-availability of leaf ([4]). Although availability of quality mulberry leaf commensurate with the quantum of seed distributed among silkworm rearers is of paramount importance yet the rearers generally over look this fact and rear more worms and are ultimately encountered with undesirable situations like shortage of leaf, rearing space, mountage, labour etc. Keeping in view, an attempt was made to ascertain the Influence of early and late mounting on economic parameters in autumn rearing of PM × CSR_2 larvae of silkworm, *Bombyx mori* L.

Materials and Methods

The experiment was carried out at Department of Sericulture, Govt. Degree College, Poonch (J&K). A popular multi × bi hybrid (PM × CSR_2) reared as per the method advocated by ([11]). Under favourable condition this hybrid completes its larval period in seven days and starts to spin cocoons. The duration of 7 days in 5th age was taken as bench mark (Control).Larvae mounted on 6th day (24h before recommended time of 7 days) formed treatment first (T1). Similarly, larvae mounted on 7th day (20 h after maturation recommended time for 7 days) formed treatment two (T2). Mounting of worms before recommended day *i.e.*, 7th day was considered as early mounting whereas, mounting after recommended 7th day was considered as late mounting. In both the treatments, larvae were handpicked and kept separately in plastic collapsible mountages for spinning. Total four replications in each treatment including control were maintained. Each replication comprised of 250 larvae. This experiment was conducted in autumn season and the data pertaining to some important cocoons parameters were collected and presented on average performance in the Figures 15.1–15.3).

Results and Discussion

In time mounting of matured larvae have a vital influence on the quality of cocoons. The farmers are often losing their crops owing to poor knowledge about the actual time for mounting the matured larvae ([3]).It is clear from mean performance of the experiment data that early mounting and late mounting have negative influence on the economic character as compared to the actual mounting of matured larvae The important economic parameters such as cocoon weight, shell weight, shell per cent, cocooning per cent, defective and good cocoon per cent were studied under this experiment are depicted in Figures 15.1–15.3, and discussed below.

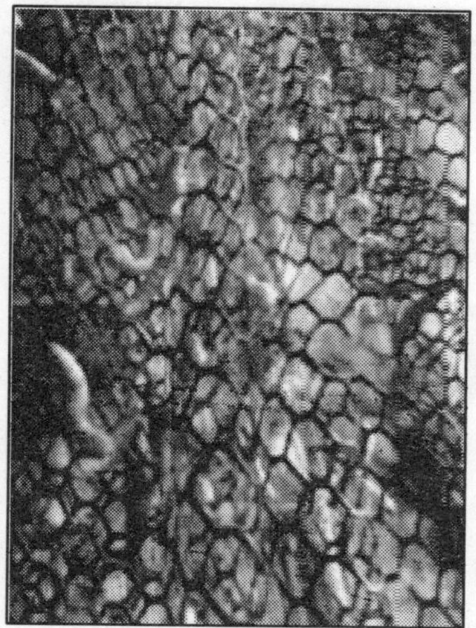

Plate 15.1: Early Mounted Larvae

Plate 15.2: Late Mounted Larvae

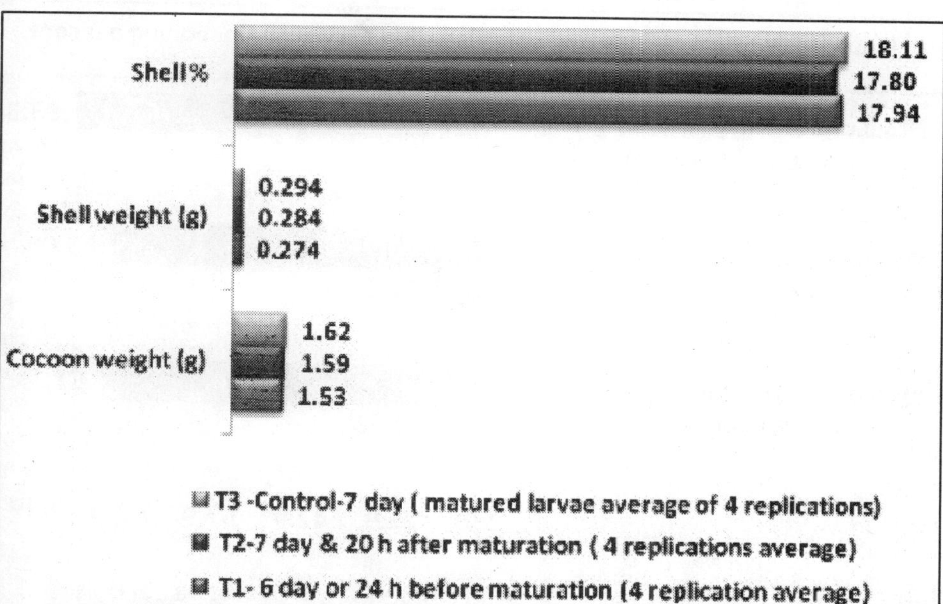

Figure 15.1: Effect of Early and Late Mounting Larvae on Cocoon Weight, Shell Weight and Shell per cent

Single Cocoon Weight

Cocoon weight is one of the important traits for the rearers point of view because the cocoons are sold by the weight only. The data relevant to this parameter showed that lowest cocoon weight was recorded in early T1 and late mounted larvae T2 (1.53 and 1.59g), but highest single cocoon weight was noticed when worms were mounted on 7th day *i.e,* control T3 (1.62g). The single cocoon weight is directly correlated to feed

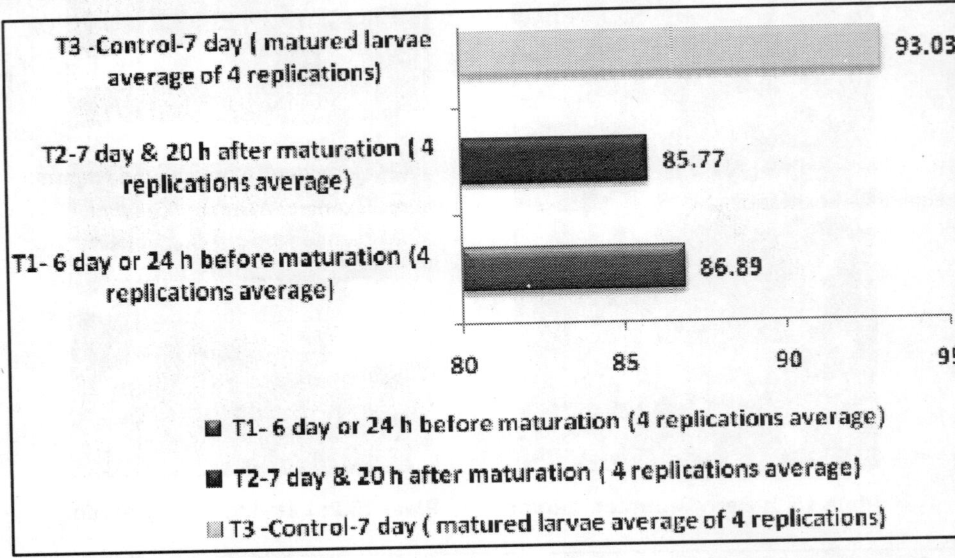

Figure 15.2: Effect of Early and Late Mounting Larvae on Cocooning per cent

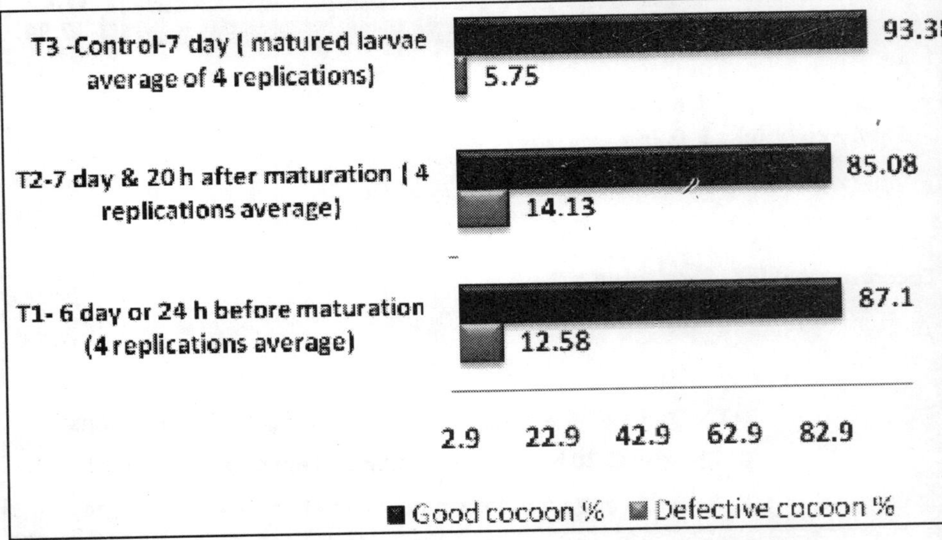

Figure 15.3: Effect of Early and Late Mounting Larvae on Defective and Good Cocoon per cent

quantum ingested during 5th instar. (1) reported that larvae mounted after taking feeds (5th day) spun lighter cocoons. (5) also reported that single cocoon weight directly correlated to feed quantum ingested during 5th instar. The present investigati also confirmed with this statement as the highest cocoon weight was found in contr when the worms were mounted at the actual time of mounting.

Single Shell Weight

The data with regard to single shell weight depicted that highest value was recorded in control T3 (0.294 g) followed by late mounted larvae T2 (0.284g) and early mounted larvae T1 (0.274g).Koul (6) reported that during last 2 instars 2-3 per cent of food consumed is converted into cocoon shell weight. Single shell weight was also affected by the quantum of mulberry leaved fed up to the end of 5th instar or up to full maturation of the silkworms ([8], [13], [14], [9], [1]). The present study also supports the observations made earlier.

Cocooning Per cent

Highest cocooning was recorded in control T3 (93.03 per cent 0 followed by early mounted larvae T1 (86.89) and late mounted larvae T3 (85.77 per cent) (Figure 15.2). ([2]) stated that the obligatory period of feeding is the first 4 days in silkworm *B.mori* L. He also reported that when larvae were starved after at least a 5 day feeding period, all moulted into pupae. ([7]) reported that starvation effects worms and cocoon characters significantly, but surviving worms spun cocoons although low in weight and skin content. The results of the present finding also showed that, when the worms are mounted at different hours after 6th day in 5th instar (before 24 h maturation and on 7th day and 20 h after maturation) these formed cocoons. This observation is supported by studies of ([15], [1]). ([12]) suggested that mounting of silkworm larvae at proper time gives good results, however the silkworm mounted on 4th day of 5th instar for seriposition also spin cocoons.

Defective and Good Cocoon Per cent

The lowest defective cocoon percentage was found in control T3 (5.75 per cent) followed by late mounted larvae T3 (14.13 per cent) and early mounted larvae T1 (12.58) (Figure 15.3). The data pertaining to good cocoon percentage depicted that highest value was recorded in control T1 (93.38 per cent) followed by early mounted larvae T1 (87.10 per cent) and late mounting larvae T2 (85.08 per cent) (Figure 15.3).

Acknowledgment

The authors express their sincere gratitude to the Deputy Director, State Sericulture Development Department, Poonch (J and K) for supplying of Dfls for this experiment.

References

1. Bora, B., Mahadevmurthy, T.S., Shivakumar, G.R. and Magdum, S. B. 1994. Effect of feed quantum during 5th instar of silkworms, *Bombyx mori* L., on economic parameters. *Uttar Pardesh J. Zool.*, 14(2):17-124.
2. Calvez, B. 1981. Progress of development programme during the last larval instar of *Bombyx mori* l. relationship with food intake ecdysteriods and juvenile hormone. *J.Insect Physiol.*, 27(4): 233-239.
3. Chandrakanth, K. S., G.K.Shrinivasa Babu, S.B.Dandin, V.B.Mathur and T.S.Mahadevmurthy, 2004. Development of improved mountages.*Indian Silk*, (43): 7-11.

4. Hague Rufaie, S.Z., Singh, T.P., Raja, T.A., Ganie, N.A., Malik, G.N. and Afifa Kamili, S. 2009. Impact of mounting larvae of different maturations on economic characters of silkworm (Bombyx mori L.) Green farming. *An International Journal of Agriculture, Horticulture and Allied Science,* 2(11); 797-800.

5. Ito, T. 1967. Nutritional requirement of silkworm, *Bombyx mori* L. *Proc.Jap.Acad.,* 43:47-67.

6. Koul, A. 1989. Relationship among leaf consumption, body weight and silk production in *Bombyx mori* L. *Agric.Sci Digest.,* 9(4): 208-210.

7. Koul, A., Ram, K. and Singh Darshan 1998. Impact of starvation shock on worm and cocoon character in silkworm (*Bombyx mori* L.) *J. Seri.,* 6 (1&2): 41-44.

8. Krishnaswamy, S. 1978. *New technology of silkworm rearing.* Bull. No.2. Central Sericultural Research and Training institute, Mysore, India, p.1-24.

9. Naito, W., Kurashima, H. and Kanematatsu, A. 1987. Studies on the production method of the silkworm cocoon for new use (1) effect of supplied leaf on the size of the cocoon filament. *Research. Bull. Aichi-Ken.,* 19:336-340.

10. Rajan, R.K., Kuribayashi, S., Meenal, R., Singh, G.B. and Himantharaj, M.T. 2000. Study on the use of saw dust to accelerate mounting of silkworm and its effect on cocoon quality. *Indian J. Seric.,* 39 (1): 72-73.

11. Rajan, R.K. and Himantharaj, M.T. 2005. *Silkworm Rearing Technology,* published by Central Silk Board, Bangalore.

12. Singh, G.B. and Kamble, C.K.1994. *Effect of mounting on different days.* Annual Report, C.S.R. and T.I., Mysore. p.95.

13. Sudo, M. and Okajima, S.Y.T. 1981. The relation between the leaf quality at different leaf order on silkworm growth on cocoon quality. *J.Seric. Sci. Jpn.,* 50:306-310.

14. Sumioka, H., Kuroda, S. and Yoshitaka, N. 1982. Relationship among food ingestion, food digestion and body weight gain in the silkworm larvae, *Bombyx mori* L. under the restricted feeding by index. *J. Seric. Sci. Jpn.,* 51:51-56.

15. Takano and Arai, N. 1978. Studies on food value on the basis of feeding and cocoon productivity in the silkworm *Bombyx mori* L. 1: The amount of food intake and productivity. J.*Seric. Science of Japan.,* 47(2):183-189.

2013, Environmental Biotechnology *Pages 153–162*
Editors: **D.R. Khanna, A.K. Chopra, Gagan Matta, Vikas Singh & Rakesh Bhutiani**
Published by: **BIOTECH BOOKS, NEW DELHI**

Chapter 16

Classical Approach to the Nucleation in Presence of Ions and Electric Field in Water Vapour Condensation

Shivani Avasthi and P.P. Pathak
Department of Physics, Gurukula Kangri University, Haridwar

Theory of drop growth in presence of external electric field and ions has been applied to water vapour condensation. In presence of electric field, Helmholtz free energy of formation of water molecule about a critical nucleus is found to be much less than that in absence of the electric field. Behavior of Helmholtz free energy is shown in figure. We see that for vapour just saturated with respect to water, the energy of the nucleus formation rapidly increases with size as r^2_{wc}. The observations suggests that nucleation rate is extremely sensitive to supersaturation ratio, since the term in the exponent varies as $S^{-2}_{v,w}$. Since in presence of ions and electric field a small value of electric field is comparable to very high supersaturation ratio to get a nucleus of given size under similar conditions of temperature.

Introduction

It is well known fact that there exists an electric field in the atmosphere which is supposed to influence many atmospheric processes. Thunderclouds are known to have strong electric fields. During a lightning discharge, the maximum electric field (10 e.s.u.) is produced near the channel. The electric field so generated may affect the rate of condensation of water vapour and ice nucleation. The magnitude of fair weather electric field near the surface of the earth, average over the whole globe, is about 130 V/m. the value in clear atmosphere in countryside is about 60V/m and in highly polluted industrial regions it is as high as 360V/m (Wallace, 1977). The growth of particles in a cloud with the electrical development was used by Levin and Ziv(1974) to explain the raingush. In an electrified cloud a sudden growth of precipitation is often observed in association with lightning. When the lightning discharge occurs, a large fraction of charge is neutralized and electric field is decreased considerably. Thus, those electrical forces which are responsible to hold the precipitation particles together are weakened drastically and the droplets fall down suddenly in the form of rain, resulting dramatic increase in precipitation rate. Thus the raingush is observed (Sharma *et al.*, 1997).

When vapour condenses in to liquid droplets or snow crystals, the first stage for this phase change is the formation of nuclei. If the size of the nucleus is smaller than the critical size it is unstable and does not grow to form droplet. When the size of the nucleus is greater than the critical size, it becomes stable and goes in to the growth stage. The critical size is the boundary between the nucleation stage and the growth stage.

Vapours and liquids are known to cool down to their dew point or freezing point without any phase change. They remain in metastable state. A slight increase in energy of the system can initiate the phase change. The initiation of formation of liquid water by condensation of water vapour is known as nucleation. Nucleation is basically a competition between the growth and evaporation of molecular clusters. For a stable cluster to be formed an initial energy barrier must be overcome. This barrier is due to the surface tension of the cluster (Enghoff and Svensmark, 2008).

Wilson (1897) showed that even in the absence of soluble and insoluble dust particles, acting as condensation centre, the condensation does take place. He also expressed the possibility of the condensation on ions. Mason (1971) has reported the condensation of water vapour on ions. Varshneya (1969) observed the path of ionizing radiation through super cooled liquid, thereby showing that the ice nucleation was possible on ions. Later on he developed a theory of homogeneous condensation and ice nucleation on ions(Varshneya, 1971). Bigg and Miles(1963) showed experimentally that the ice nuclei concentration is sufficiently higher (~ 10^5) in upper and lower atmosphere as compared to the ground. The concentration of cloud ice particles have been observed in excess than the concentration of ice nuclei at cloud top temperature (Koenig, 1963; Brahm, 1964; Mossop, 1970, 1972; Hobbs *et al.*, 1974). The above results can be explained if we consider the effect of ions on condensation and ice nucleation.

Wang *et al.* (1978) studied about the collection efficiency at atmospheric aerosols or rain drops. The collection of charged aerosols by a conducting sphere in an external electric field (Wang 1983) and enhancement in the collection efficiency of various shaped aerosol particles by hydrometeors under thunderstorm conditions as compared with fair weather electric field (Wang 1984) throws light on the importance of the electric fields. Jalaluddin *et al.* (1962), Parmar and Jalaluddin (1973) reported that the electric field influences the nucleation process. Varshneya (1969, 1971) and

Pruppacher (1973) found that when charged ions and aerosol particles come in contact with the neutral drops, the nucleation process is enhanced.

Loeb (1963) and Smith *et al.* (1971) discussed the disruptive effect of electric field and suggested that the nucleation on the fragments of a drop is responsible for the observed enhancement in nucleation. Later, Abbas and Latham (1969) gave a possible mechanism for this enhancement. Their results were totally inconsistent with Pruppacher's theory(1963), but did agree with those of Loeb(1963). Murino (1979) studied theoretically the effect of the electric field on the condensation of water vapour and concluded that under similar conditions of temperature and supersaturation the bigger size of the drops can be achieved in a given time than that obtained in absence of electric field. He considered the polarization of water vapour molecules in the electric field of the central dipole alone.

A model of global atmospheric circuit (Sapkota *et al.*, 1990) has been developed incorporating the pollution due to aerosol particles of anthropogenic and volcanic origin, which shows that ionization is caused by the coronal discharge due to intense electric field beneath thunderclouds in atmospheric layer close to the earth surface. The scavenging of aerosol particles by hydrometeors under external electric field has been discussed by Wang (1983, 1984). Although, there is a fair weather electric field in a fine day, the field is much larger in thunderstorm situation. Under thunderstorm conditions the electric field effects on scavenging are greatly enhanced.

Thus the electric field plays an important role in cloud formation. This role could be effectively verified in absence of saturation or supersaturation. Murino (1979) has shown that action of a constant and uniform electric field accelerated the condensation of water vapour by a factor depending upon the intensity of electric field. It was shown that a droplet acquires a particular size under very low supersaturation, under an electric field, which would, otherwise, requires very high supersaturation.

In the present paper, we examined, theoretically, the effect of ions and external electric field on water vapour condensation to estimate the Helmholtz free energy of formation of nucleus, nucleation rate and the radius of critically sized nuclei. Comparison has been made with that in ion induced and electric field-free case.

Theoretical Consideration of Homogeneous Nucleation Process

In Absence of Ions and Electric Field

Our theoretical understanding of the stability of phases has its origins in the classical work of Gibbs (1948) on the metastable and unstable states. Consider for simplicity a mean electric field system with a critical point, such as a water vapour transition (Gunton 1999). Near the ground, ions are produced mainly by ionizing radiations of radioactive substances such as Thoron and Radon. Generally cosmic rays (Arnold 2006) Produce ionization in atmosphere.

Dufour and Defay (1963), and Abraham (1968) have shown that the Helmholtz free energy is the proper thermodynamic potential and Gibb's function is the only approximation of Helmholtz's function. Practically the differences turn out to be negligible at constant temperature. However, it is not a constant pressure process. We may assume that the required phase change occurs at constant temperature. Thus, the energy for the formation of nucleus is given by

$$\Delta F = 4\eth\acute{o}_{W/V}\, r^2_w - 4\,\eth\Delta F_{vol} r^3_w/3 \quad (1)$$

where,

r_w is the radius of the spherical water nucleus, $\acute{o}_{W/V}$ the surface free energy per unit area of water nucleus in water vapour, ΔF_{vol}, the volume free energy of condensate per unit volume per mole.

Here, $\Delta F_{vol} = \rho_w RT \ln S_{v,w}/M_w$

where,

ρ_w is the density of water and $\ln S_{v,w}$, the supersaturation ratio.

Corresponding to the maximum free energy for the critical size of nucleus, we have

$$\partial/\partial r_w(\Delta F) = 0 \quad (2)$$

From (1) and (2), the critical radius for the uncharged case is obtained to be

$$r_{wc} -= 2\,\sigma_{W/V}/\Delta F_{vol}$$

$$r_{wc} = 2M_w\,\sigma_{W/V}/\rho_w RT \ln S_{v,w} \quad (3)$$

Therefore, critical free energy for uncharged droplet can be written as

$$\Delta F_c = 4\pi\sigma_{W/V}\, r^2_{wc}/3 \quad (4)$$

When homogeneous nucleation of water from the vapour occurs, the formation of droplets of the clouds takes place. So, we are interested to find the rate at which the droplet appear in the system as a function of saturation ratio. This rate is known as nucleation rate (J) and is measured as the number of droplets appearing per unit volume and per unit time. Rate of nucleation as given by Pruppacher and Klett (1978)

$$J = \alpha_C/\rho_w[2N^3 M_w\,\sigma_{W/V}/\pi]^{1/2}\,[e_{sat.w}/RT]^2\, S_{v,w} \exp[-\Delta F_c/KT] \quad (5)$$

where,

N is the Avogadro number and α_C, is the condensation coefficient.

Table 16.1: Values of r_{wc}, ΔF_c, in J as the Function of Temp. and Supersaturation Ratio (In absence of ions and electric field) at T= 273 K

$S_{v,w}$	r_{wc} ($\times 10^{-8}$ cm.)	ΔF_c ($\times 10^{-8}$ergs)	lnJ
0.6	–	–	–
0.8	–	–	–
1.0	infinite	–	–
1.2	62.61	118.27	–3097.70
1.5	28.51	23.91	–592.83
2.0	16.47	8.18	–175.02

Contd...

Table 16.1–Contd...

$S_{v,w}$	r_{wc} ($\times 10^{-8}$ cm.)	ΔF_c ($\times 10^{-8}$ergs)	lnJ
2.5	12.46	4.68	–81.89
3.0	10.39	3.26	–44.02
3.5	9.11	2.50	–23.69
4.0	8.23	2.04	–11.34
4.5	7.59	1.74	–3.26
5.0	7.09	1.52	2.68

In Presence of Ions

In presence of ions the water vapour molecules from cluster with ion as the centre. The condensation takes place at low supersaturations because the necessary energy for germ formation is attained at smaller radi idue to addition of electrostatic potential energy. Helmholtz energy expression (Eq. 1) is modified by Singh, Rai and Varshneya (1985) as

$$\Delta F' = 4\pi\sigma_{W/V}\, r'^2_w - 4\,\pi\Delta F'_{vol} r'^3_w/3 + 3(z\,e)^2/5\,r'_w \quad (6)$$

Where, the primes indicate the values in the presence of ions. The last term in Eq. (6) represents the electrostatic energy of the cluster in presence of an ion of charge (ze), z being the charge number, e, the electronic charge.

In this case the critical radius can be obtained by setting

$$\partial/\partial\, r_{w'} (\Delta F') = 0 \quad (7)$$

The critical supersaturation ratio in the present case can be obtained by

$$\ln S'_{v,w} = 2M_w\, \sigma_{W/V}/\rho_w\, R\,T.\, r'_{wc} - 3\, M_W\, Z^2 e^2/20\pi\, \rho_w\, RT.\, r'^4_{wc} \quad (8)$$

Nucleation rate for charged case is given by

$$J' = \alpha_C/\rho_w [2N^3\, M_w\, \sigma_{W/V}/\pi]^{1/2}\, [e_{sat.w}/RT]^2\, S'_{v,w} \exp\,[-\Delta F'_c/KT] \quad (9)$$

From Eqs. (8) and (9), it is clear that the nucleation in presence of ions is faster than otherwise.

Table 16.2: Values of r'_{wc}, $\Delta F'_c$, in J' as the Function of Temp. and Supersaturation Ratio (in presence of ions) at T= 273 K

$S_{v,w}$	r_{wc} ($\times 10^{-8}$ cm.)	ΔF_c ($\times 10^{-8}$ergs)	lnJ
0.6	4.02	5.08	–93.93
0.8	4.16	4.98	–90.99
1.0	4.28	4.89	–88.34
1.2	4.40	4.82	–86.34
1.5	4.52	4.71	–83.19
2.0	4.91	4.57	–79.19
2.5	5.06	4.43	–75.25

Contd...

Table 16.2–Contd...

$S_{v,w}$	r_{wc} ($\times 10^{-8}$ cm.)	ΔF_c ($\times 10^{-8}$ergs)	lnJ
3.0	5.50	4.29	–71.36
3.5	6.27	4.13	–66.98
4.0	–	–	–
4.5	–	–	–
5.0	–	–	–

In Presence of Ions and Electric Field

Eisenberg and Kauzmann (1969), and Hobbs (1974) reported that water is strongly polarizable medium with electric dipole moment 1.83 10^{-18} e.s.u. cm. Singh *et al.* (1985) studied the effect of electric field on relaxation time in nucleation process of water vapour condensation and ice nucleation. They found that for homogeneous nucleation in presence of electric field of 5 e.s.u., the relaxation time decreases by 43.98 per cent.

Application of electric field induces an electric dipole moment on the embryo of water as well as surrounding water vapour molecules. The moment induced on the embryo is

$$P = E\, r^3_w \qquad (10)$$

where,

r_w is the radius of the water embryo and E, the moment P_1 induced on water vapour molecule is given by

$$P_1 = \alpha E_1 \qquad (11)$$

where,

α is the polarizability of the medium. From the kinetic theory of gases (Engel, 1955), it is assumed that in a non-uniform electric field, the drift velocity is given by the transformation of potential energy lost in one mean free path in to kinetic energy. Change in potential energy of a molecule in a transition through a distance λ on to the surface of the embryo of radius r''_w is given by $9\alpha\lambda E^2/2r''_w$.

Therefore Eq.(6) is modified as

$$\Delta F'' = 4\pi\sigma_{W/V}\, r''^2_w - 4\pi\Delta F''_{vol} r''^3_w/3 + 3(z\,e)^2/5\,r''_w - 9\,\alpha\,\lambda E^2/2\,r''_w \qquad (12)$$

where,

$$\Delta F''_{vol} = \pi_w\, RT \ln S''_{v,w}/M_w$$

The critical supersaturation ratio in this case can be written as

$$\ln S'_{v,w} = 2M_w\sigma_{W/V}/\rho_w\, RT.r'_{wc} - 3M_W Z^2e^2/20\pi\rho_W RT.r'^4_{wc} + 9\,\alpha\,\lambda M_W E^2/8\pi\rho_w\, RT\, r''^4_{wc}$$

Further, Nucleation rate is given by

$$J'' = \alpha_C/\rho_w[2N^3 M_w\, \sigma_{W/V}/\pi]^{1/2}\, [e_{sat.w}/RT]^2\, S''_{v,w} \exp[-\Delta F''_c/KT] \qquad (13)$$

From these Eqs. It is clear that in presence of the electric field nucleation accelerates to higher values

Table 16.3: Values of r''_{wc}, $\Delta F''_c$, in J" as the Function of Temp. and Supersaturation Ratio and Electric Field (in presence of ions and electric field) at T= 273 K

$S_{v,w}$	r_{wc} ($\times 10^{-8}$ cm.)	ΔF_c ($\times 10^{-8}$ergs)	lnJ
0.6	3.97	4.94	–90.22
0.8	4.09	4.84	–87.28
1.0	4.19	4.76	–84.93
1.2	4.29	4.69	–82.89
1.5	4.45	4.59	–80.01
2.0	4.70	4.44	–75.75
2.5	4.98	4.31	–72.07
3.0	5.26	4.18	–68.44
3.5	–	–	–
4.0	–	–	–
4.5	–	–	–
5.0	–	–	–

Results and Discussion

In the present calculations the constant values used are:

$\rho_w = 1$ gm. cm^{-3}, R = 8.317×10^7 erg mole^{-1}K^{-1}, $\sigma_{W/V} = 72$ erg. Cm^{-2}, $\alpha_C = 1.0$, N = 6.023×10^{23}, k = 1.38×10^{-16}erg.K^{-1}, e = 4.80325×10^{-10} e.s.u., α = polarizability of the medium = 5×10^{-19}, $\lambda = 10^{-4}$ cm., E = 1.5×10^3V/m, $e_{sat.w}$= saturation vapour pressure over water (by Pruppacher and Klett (1978)

$e_{sat.w} = a_0 + T(a_1 + T(a_2 + T(a_3 + T(a_4 + T(a_5 + a_6)))))$, here T(°C)

$a_0 = 6.107799961$, $a_1 = 4.436518521\times10^{-1}$, $a_2 = 1.428945805\times10^{-2}$

$a_3 = 2.650648471\times10^{-4}$, $a_4 = 3.031240396\times10^{-6}$, $a_5 = 2.034080948\times10^{-8}$

$a_6 = 6.136820929\times10^{-11}$

In presence of ions, condensation takes place at low supersaturation compared to in absence of ions. While in presence of ions and electric field condensation takes place at very low supersaturation because of that the necessary energy for germ formation is attained at lower radii due to the addition of electrostatic energy $[3(z\,e)^2/5\,r''_w]$ and also transformation of potential energy in to kinetic energy $[9\alpha\,\lambda E^2/2\,r''_w]$.

For a given temperature in absence of ion and electric field the value of supersaturation varies approximately as $r^{-1}_{wc,}$ while in presence of ions and electric field it varies as r''^{-4}_{wc} for smaller radii and r''^{-1}_{wc} for larger radii.

At T = 273K, $S_{v,w} = 2.5$

$r_{wc} = 12.46$ A°, $r'_{wc} = 5.06$ A°, $r''_{wc} = 4.98$ A° and

At $S_{v,w} = 1.0$

r_{wc} = infinite, $r'_{wc} = 4.28$ A°, $r''_{wc} = 4.19$ A°

it means that at $S_{v,w} < 1$, uncharged water drops cannot exist and in presence of ions and electric field radius of water drops are smaller than in only presence of ions. Hence condensation takes place at smaller radii in presence of ions electric field both.

At T = 273K, $S_{v,w} = 2.5$

$\Delta F = 4.68\ \ 10^{-12}$ ergs, $\Delta F' = 4.43\ \ 10^{-12}$ ergs, $\Delta F'' = 4.31\ \ 10^{-12}$ ergs and

$\ln J = -81.89$, $\ln J' = -75.25$, $\ln J'' = -72.07$

Hence free energy reduces in presence of ions and electric field both. *i.e.* rate of nucleation is increased as compared to ion free and ion induced case. Thus we conclude that ions and electric field affect very much the process of condensation and nucleation. Ions supply more energy to the nucleus so that energy necessary and sufficient for phase change is achieved at an early stage at embryo formation and hence there is a remarkable change in nucleation rate in presence of ions and electric field both.

References

Abbas, M.A., and Latham, J., The electro freezing of supercooled water drops, J. Meteorol. Soc. Japan; 47, 67-74,1969.

Abraham, F.F.,A re-examination of homogeneous nucleation theory: Thermodynamic aspects, J Atmos. Sci.; 25,47-53,1968

Arnold, F, Atmospheric aerosol and cloud condensation nuclei formation: a possible influence of cosmic rays?, Space Sci. Rev.; 125,169-186, 2006.

Bigg, E.K., Miles, G.T., Stratospheric ice nucleus measurements from balloons, Tellus; 15,162-166, 1963.

Brahm, R.R., Jr., What is the role of ice in summer rain showers ?, J. Atmos. Sci.; 21,640 645,1964.

Dufour, L.,and Defay R., Thermodynamics of Clouds, Academic Press, New york,1963.

Eisenberg,D., and Kauzmann, W., The structure and properties of water, Oxford Univ. Press, Oxford,1969.

Engel, A. Von., Ionized gas, Geoffrey Cumberlege, Oxford, 1955.

Enghoff, M.B., and Svensmark H., The role of atmospheric ions in aerosol nucleation-a review, J. Atmos. Chem. Phys.; 8,7477-7508, 2008.

Gibbs, J.W., Collected Works, Vol. 1(Yale Uni. Press, New Haven,Connecticut,1948); 105-115,252-258.

Gunton, J.D., Homogeneous Nucleation, J. Statis. Phys; 95,903-921, 1999.

Hobbs, P.V., Ice physics, Oxford Uni. Press, Oxford, 1974.

Hobbs, P.V., Radke, L.F., Weiss, R.R., Atkinson, D.G., Locatelli, J.D., Biswas, K.R., Turner, F.M., and Robertson, C.E., Res. Report No.8.Dec.1974,Cloud physics group, Deptt. Atmos. Sci., Uni. Of Washington, 1974.

Jalaluddin, A.K., and Sinha, D.B., Effect on an electric field on the superheat of liquids, Nuovo Cim(Suppl.), 26(Band X); 234-237,1962.

Kittel, C., Introduction to solid state physics, John Wiley and sons, Newyork; 401-409,1977.

Koening, L.R., The glaciating behavior of small cumulonimbus clouds, J. Atmos. Sci.; 20,29-47,1963.

Levin,Z., and Ziv A.,The electrification of thunderclouds and the raingush, J. Geophys.Res.; 79,2699-2704,1974.

Loeb, L.B.,A tentative explanation of the electric field effect on the freezing of supercooled water drops, J. Geophys. Res.; 68,4475-4476,1963.

Mason, B.J., The physics of clouds,2nd ed., Oxford Univ. Press,London,1971.

Mossop, S.C., Concentration of ice crystals in clouds, Bull. Am. Mateorol.Soc.; 51,474-479,1970.

Mossop, S.C., The role of ice nucleus measurements in studies of ice particle formation in natural clouds, J. de Rech. Atmos.; 6,377-389,1972.

Murino,G.,Influence of electric fields on condensation of water vapours, S Afr. Tydskr. Fis.; 113-115,1979.

Parmar, D.S., and Jalaluddin, A.K.,Determination of the limit of absolute thermodynamic stability of liquid using external electric field as perturbation, Phys. Lett.; 42 A, 497-498,1973.

Pruppacher, H.R., The effect of an external electric field on the supercooling of water drops, J. Geophys. Res; 68,4463-4474,1963.

Pruppacher, H.R., Electrofreezing of supercooled water, Pure Appl. Geophys; 104,623-633,1973.

Pruppacher, H.R., and Klett, J.D., Microphysics of Clouds and Precipitation,. Reidal Publishing Co., Dordrecht(Holland,1978).

Sharma, P.K., Pathak, P.P., and Rai, J., Raingush phenomenon in thunderclouds, J. Natural and physical Sci., 1997.

Sapkota, B.K., and Varshneya, N.C., J. Atmos. &Terr. Phys.(1990).

Singh, N., Rai,J. and Varshneya N.C., The effect of ions on cloud condensation and ice nucleation, Ann Geophys; 3,343-350, 1985.

Smith, M.H., Griffiths, R.F., and Latham, J., The freezing of raindrops falling through strong electric fields, Quart. J. Roy. Meteorol. Soc.,97,495- 505,1971.

Varshneya N.C., Theory of radiation detection through supercooled liquid, nuclear instruments. and methods; 92,147-150,1971

Wallace, J.M., and Hobbs, P.V., Atmospheric science and introductory survey, Academic Press, Newyork, 1977.

Wang,P.K.,Collection of aerosol particles by a conducting sphere in an external electric field continuum regime approximation, J. Colloid Interface Sci.; 94,301-318, 1983

Wang, P.K., The influence of Atmospheric electricity on the precipitation Seventh international conference on atmospheric electricity; june 3- 8,Albany, Newyork, 1984

Wang,P.K.,Grover S.N., and Pruppacher, H.R., On the effect of electric charges on the scavenging of aerosol particles by clouds and small raindrops,J.Atmos.Sci; 35,1735-1743,1978.

Wilson,C.T.R.,Condensation of water vapour in the presence of dust free air and other gases, Phil. TYrans. Roy. Soc.; A-189,265-274,1897.

2013, Environmental Biotechnology
Pages 163–170
Editors: D.R. Khanna, A.K. Chopra, Gagan Matta, Vikas Singh & Rakesh Bhutiani
Published by: BIOTECH BOOKS, NEW DELHI

Chapter 17

Synthesis and Characterizations of Sr_2CeO_4, Nano Phosphor

Raghoba Nagpure[1], Naresh Sarkar[2], Shashikant Rokade[3] and S.R. Thakare[3]

[1]*Department of Physics, Sevadal Mahila Mahavidyalaya, Nagpur – 440 009, M.S.*

[2]*Head, Department of Library, Sevadal Mahila Mahavidyalaya, Nagpur – 440 009, M.S.*

[3]*Department of Nano Technology, Shivaji Science College, Nagpur, M.S.*

A new material of Sr_2CeO_4 has been synthesized by vapor condensation in addition of modified solid state method. The solid material is evaporating by heating and it rapidly condenses. This technique is very useful to produce composite materials by mixing two or more evaporation sources. The particle size can be controlled by the evaporation rate and condensation gas pressure. The first step is use to synthesis of nano particle and second step involves self assembly on suitable substrate. (Solid-state method at 200°C, 400°C, 600°C for 4 hours) The XRD pattern reveals the grain size of particle. The material is confirming from the morphology study of material gives from the SEM. In the present paper we report the excitation and emission of Sr_2CeO_4 phosphor, which gives the emission with good intensity at 389nm (600°C). [$2SrCO_3 + CeO_2 \rightarrow Sr_2CeO_4 + 2CO_2\uparrow$]

Keywords: *Photoluminescence, Vapor condensation, In addition of solid state method, XRD phosphor.*

Introduction

In 1998, Danielson and co-workers reported unusual luminescence of inorganic oxide compound Sr_2CeO_4 using combinatorial technique. Sr_2CeO_4 exhibits photoluminescence due to the charge transfer mechanism, Sr_2CeO_4 consists of infinite edge-shearing CeO_6 octahedral chain separated by 'Sr' atom [2]. The luminescence originates from a ligand to metal Ce^{4+} charge transfer [1], since Sr_2CeO_4 was found as a novel and promising blue luminescence material by vapor condensation method. The Sr_2CeO_4 phosphor has been widely studied because of its importance in realization of new generation of optoelectronic and displaying devices [3]. The conventional solid state reaction process has disadvantages in controlling the morphology and maintaining the uniformity in composition of phosphor particle. Therefore new process to overcome these disadvantages are under investigation *i.e.* 'vapor condensation method' which was applied before solid state. Hence the research for oxide base phosphors has been increasing due to their applications in many field such as cathode ray tube, LED, field emission display, the oxide-based phosphors attract the researchers attention due to the advantage of their good stability upon excitation by electron beam.

Materials and Experimental Methods

The sample was prepared in the present work for the synthesis of the Sr_2CeO_4 were namely Strontium Carbonate ($SrCO_3$) (merk 99.9 per cent purity), Cerium Oxide CeO_2, (merk 99.9 per cent purity). These materials were taken in stoichometric proportions *i.e.* Sr, Ce, as 2:1, $SrCO_3$ and CeO_2 are weighed in molecular stoichometry and the complete solution were taken inside the round bottom flux and covered it with a Lebig condenser with continuous flow of water and heated the solution at (80-100°C) for 12 hrs such that the creating vapor becomes condensed. After a long time, liquid solution has been changed to powder form inside the flux. This powder was taken in alumina crucible, after closing the cover the crucible was located in a furnace and heated to the different temperature at 200°C, 400°C and 600°C for four hours then cooled down naturally and again grained thoroughly to get fine powder, the final product was soft cream colour powder.

Results and Discussion

Figure shows XRD patterns of the Sr_2CeO_4 heating at different temperature of 200°C, 400°C, 600°C, gives rise the crystalline nature with increasing temperature like X-ray diffraction pattern, with 100 per cent peaks at 26.98, 27.32 and 28.77 (2è) position respectively at relative intensity 397.39 (200°C), 378.28 (400°C), 766.78 (600°C).has been compare with JCPDS {file 00-0221422} The diameter of Sr_2CeO_4 was found to be 107.67 nm, 174.95 nm and 111.45 nm with respective temperature of 200°C, 400°C and 600°C which is calculated from the full width at half-maximum of the peak using Debye Scherrer's equation. The XRD study confirms the structure of the system as orthorhombic. The XRDA3.1 software has been used to calculate the lattice parameters and found to be a = 6.07Å, b = 10.32Å and c = 3.62Å

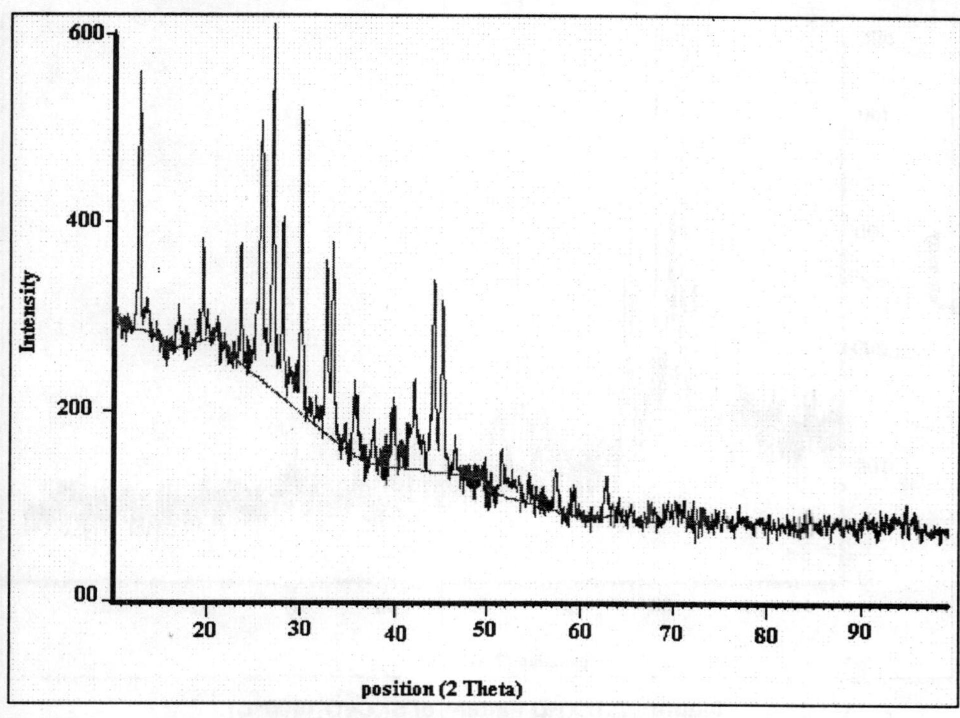

Figure 17.1: XRD Pattern of Sr_2CeO_4 (200°C)

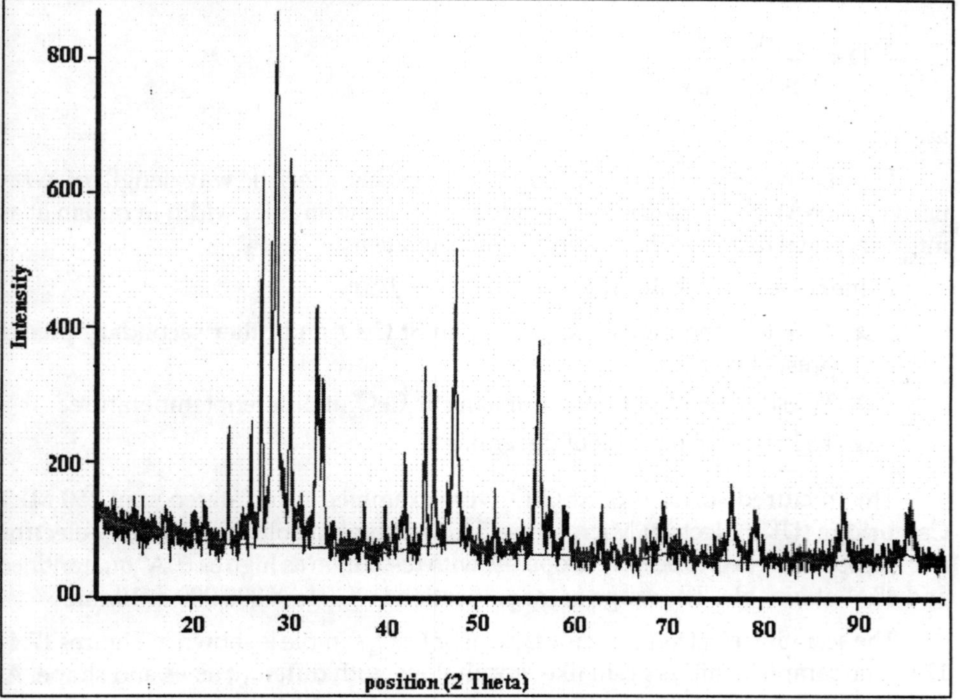

Figure 17.2: XRD Pattern of Sr_2CeO_4 (400°C)

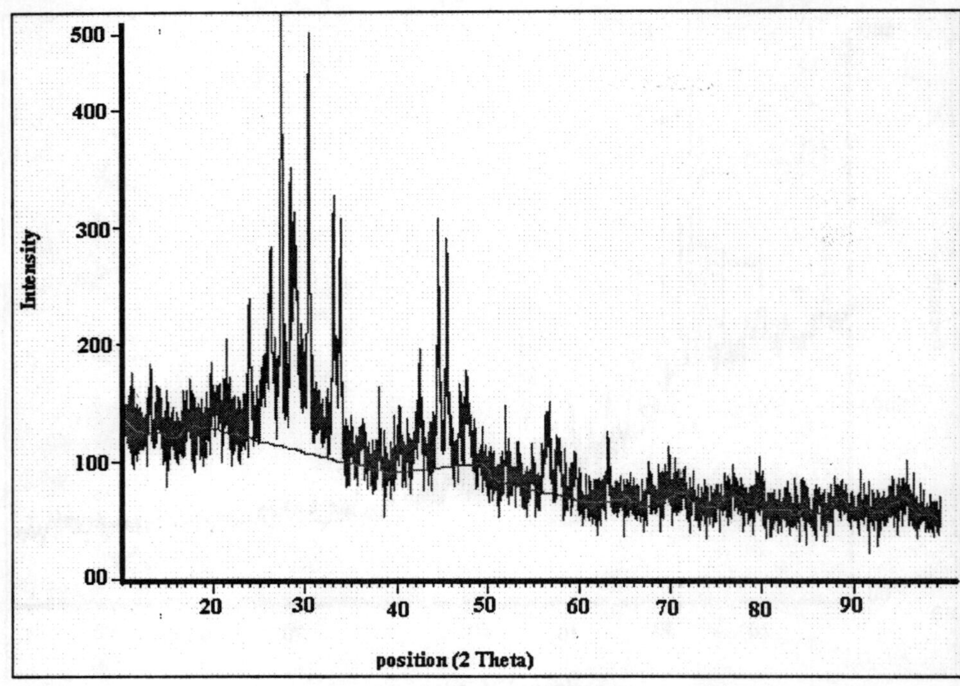

Figure 17.3: XRD Pattern of Sr_2CeO_4 (600°C)

$$D = \frac{0.9\lambda}{B_{2\theta} \cos \theta_{max}}$$

where,

D is the average crystal size in nm, λ is the characteristic wavelength of X-ray used (1.5406Å), è is the diffraction angle and $B_{2è}$ is the angular width in radian at an intensity equal to half of the maximum peak intensity.

In the present work the SEM study is carried out:

- ☆ To know the morphology of nano Sr_2CeO_4 and other secondary phases present if any
- ☆ To estimates the grain size of nano Sr_2CeO_4 at different temperature.
- ☆ To know the packing of the grains.

The fractured surfaces of Sr_2CeO_4 were examined using stereo scan 250 Mk3, Cambridge (UK), Electron Microscope. This is a high resolution scanning electron microscope designed to be easy to operate with resolution as high as 60A° quarantined and 45A° attainable. The range of magnification is X20 to X 800,000 at 10mm.

The scanning electronic micrograph of Sr_2CeO_4 sample is shown in Figures 17.4–17.6. The sample exhibits grain like morphology with different sizes and shape. At

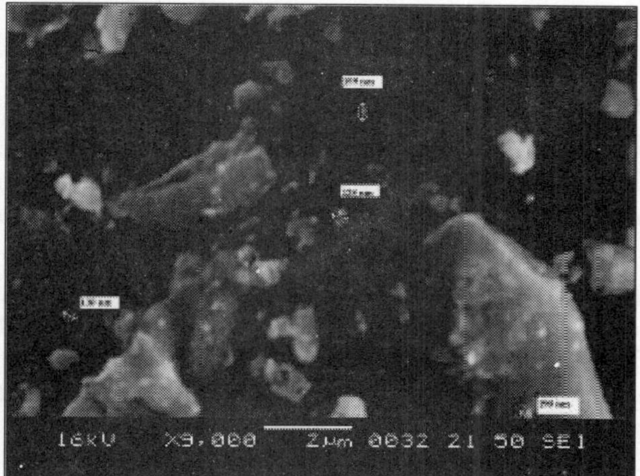

Figure 17.4: SEM Micrographs of Sr_2CeO_4 (200°C)

Figure 17.5: SEM Micrographs of Sr_2CeO_4 (400°C)

Figure 17.6: SEM Micrographs of Sr_2CeO_4 (600°C)

low magnification the particles appear agglomerated and at high enough magnification, the nature of the individual crystallites is clearly evident.

Photoluminescence Properties

This analysis gives information about an excitation and emission of electron of Sr_2CeO_4. The Photo fluoro spectra of Sr_2CeO_4 were taken on an Instrument RF-5301 in the scan range of 300.0nm to 700.0nm at high speed with high sensitivity with slit width 5nm, at Shri Shivaji Science College, Nagpur.

Excitation spectrum of Sr_2CeO_4 is shown in Figure 17.7. The spectrum displays a broad band with strong peaks, one at ~352 nm and the other at ~354 nm, 355nm the band shows the intensity of Sr_2CeO_4 phosphor material goes on increases with increasing temperature also the spectra could shift towards blue region of wavelength. The bands could be assigned to the different Ce^{4+}-O distances in the lattice [2]. The excitation band of the sample could be attributed to the transition $t_{1g} \rightarrow f$, where f is the lowest excited charge transfer state of the Ce^{4+} ion and t_{1g} is the molecular orbital of the surrounding ligands in the six-fold oxygen co-ordination [1].

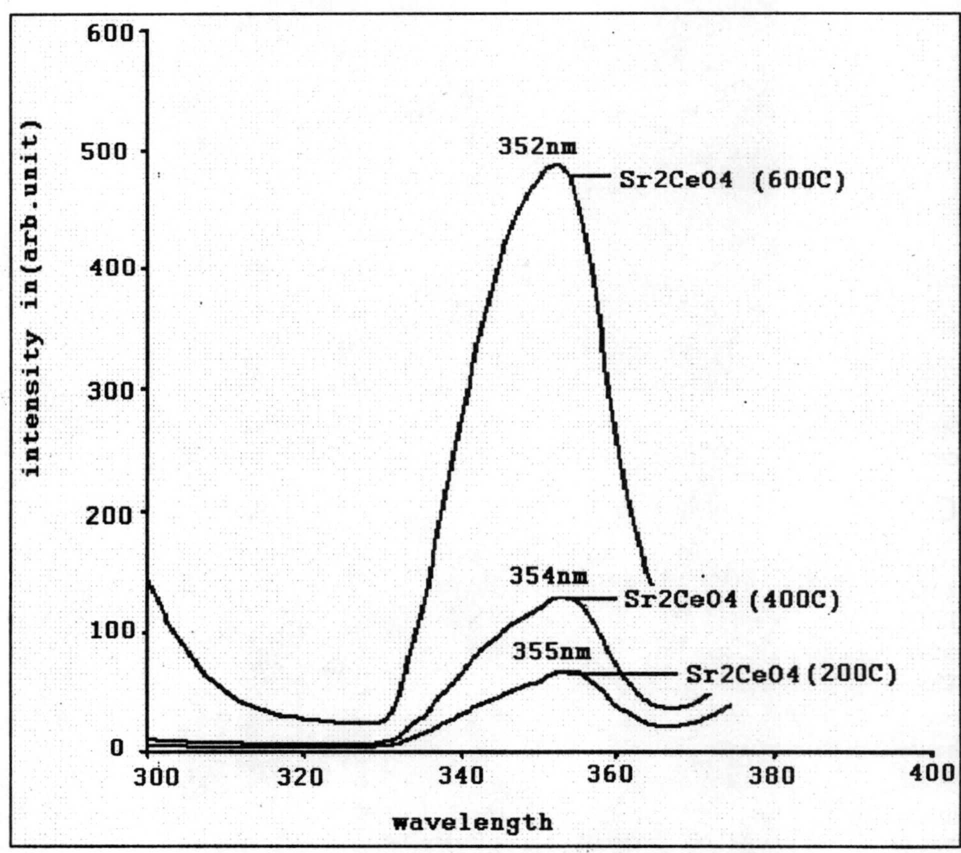

Figure 17.7: Excitation Spectrum of Sr_2CeO_4 Sample

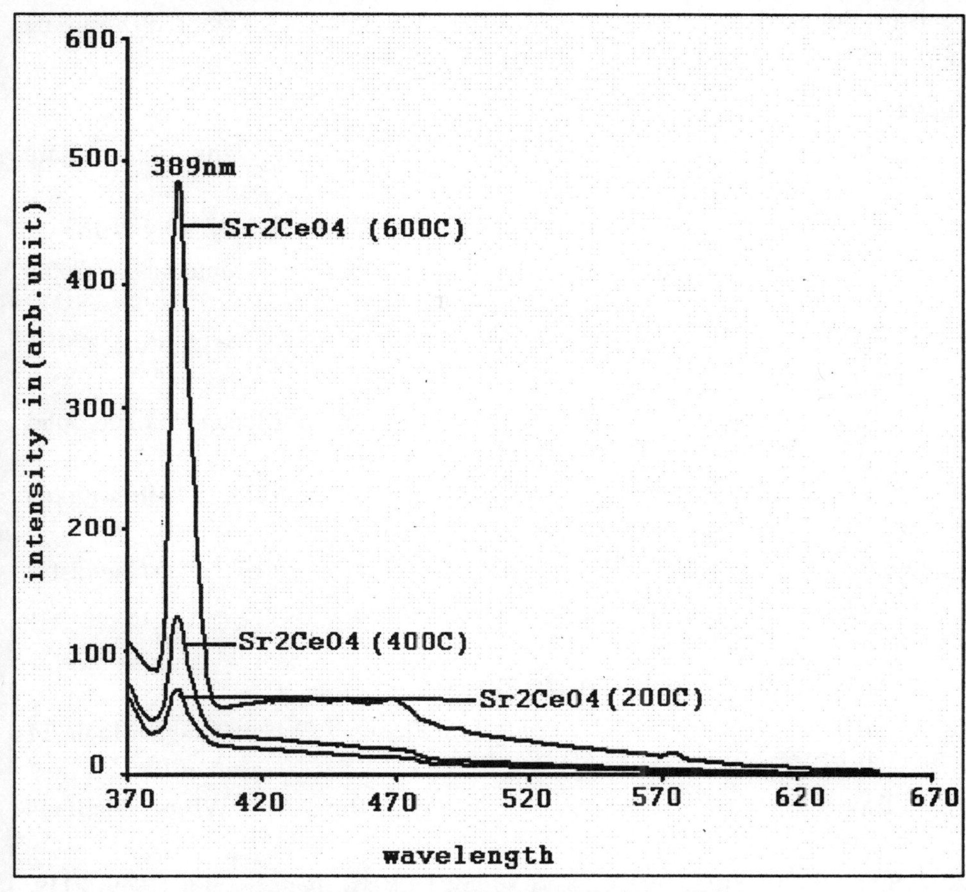

Figure 17.8: Emission Spectrum of Sr_2CeO_4

The emission spectrum of Sr_2CeO_4 excited at 352 nm is shown in Fig.8. The spectrum shows band in the region 370 – 420 nm with a peak around 389 nm. The emission band can be assigned to the $f \rightarrow t_{1g}$ transitions of Ce^{4+} ions.

Conclusion

The XRD pattern confirms the formation of Sr_2CeO_4 compound has been compared with JCPDS (file 00-0221422). The diameter of Sr_2CeO_4 was found to be 107.67 nm, 174.95nm and 111.45 nm with respective temperature 200°C, 400°C and 600°C. The scanning electronic microscope of Sr_2CeO_4 sample is shown in Figures 17.4–17.6. The sample exhibits grain like morphology with different sizes and shape. The excitation spectrum of Sr_2CeO_4 is shown in Figure 17.7. The spectrum displays a broad band with strong peaks, one at ~352 nm and the other at ~354 nm, 355 nm the band shows the intensity of Sr_2CeO_4 phosphor material goes on increases with increasing temperature also the spectra could shift towards blue region of wavelength. The emission spectrum of Sr_2CeO_4 excited at 352 nm is shown in Figure 17.8. The spectrum shows band in the region 370 - 420 nm with a peak around 389nm. The emission band can be assigned to the f→t_{1g} transitions of Ce^{4+} ions.

References

1. Zhang Chunxiang, Shi Jianshe, Yang Xujie, Lu Lude and Wang Xin, J. of Rare earths, 2010, 28, 513-518.
2. Chang-Hsin Lu, Change-Tao Chen, J. sol-gel sic. Technol. 2007, 43, 179-185.
3. N. Perea -Lopez, J. A. Gonzalez-Ortega, G. A. Hirata, Optical Materials, 29 (2006) 43.
4. A. Zukauskas, M.S. Shur, R. Gaska, Introduction to Solid-State Lighting, Wiley, New York, 2002.
5. J.S. Kim, Y.H. Park, J.C. Choi, H. L. Park, Gwang Chul Kim, Joong Hak Yoo, Solid State Communications 137, (2006), 187.
6. J. K. Park, C.H. Kim, S.H. Park, H. D. Park, S.Y. Choi, Appl. Phys. Lett. 84 (2004) 1647.
7. Jinyong Kuang, Yingliang Liu, Jianxian Zhang, Journal of Solid State Chemistry 179, (2006), 266.
8. J. Kuang, Y. Liu, Chem. Lett. 34,(2005),598.
9. D. Jia, W.M. Yen, J. Lumin. 101, (2003),115.
10. M. Yu, J. Lin, Z. Wang, J. Fu, S. Wang, H.J. Zhang, Y.C. Han, Chem. Mater. 14 (2002), 2224.
11. R. Ravichandran, S.T. Johnson, S. Erdei, R. Roy, W.B. White, Displays 19(1999)197.
12. B. Smets, J. Rutten, G.Hoeks, J. Verlijsdonk, J. Electrochem. Soc.136,(1989), 2119.
13. T. Katsumata, K. Sasajima, T. Nabae, S. Komuro, T. Morikawa, J. Am. Ceram. Soc. 81, (1998), 413.
14. J.L. Sommerdijk, A. Bril, J. Electrochem. Soc. 122,(1975),952.
15. K. Riwotzki, M. Haase, J. Phys. Chem. B 102,(1998),10129.
16. F. Gu, S.F. Shu, M.K. Lu, G.J. Zhou, S.W. Liu, D. Xu, D.R. Yuan, Chem. Phys. Lett. 380, (2003),185.
17. J. L. Scirrerdijk, A. Bril, F. M. J.H. Hoex-strik, Philips Res. Repts.32,(1977),149.
18. A. Zukauskas, M.S. Shur, R. Gaska, Introduction to Solid-State Lighting, Wiley, New York, 2002.
19. J.S. Kim, P.E. Jeon, J.C. Choi, H.L. Park, T.W. Kim, G.C. Kim, Appl. Phys. Lett. 85, (2004) 3696.

2013, Environmental Biotechnology *Pages 171–178*
Editors: **D.R. Khanna, A.K. Chopra, Gagan Matta, Vikas Singh & Rakesh Bhutiani**
Published by: **BIOTECH BOOKS, NEW DELHI**

Chapter 18

Use of Different Animal and Animal-Derived Products as Medicines by Tribal Peoples of Gadchiroli District, India

***A.A. Dhamani*[1]*, P.R. Chavhan*[2] *and S.D. Misar*[3]**

[1]*Department of Zoology, N.H.College, Bramhapuri – 441 206, Maharashtra*
[2]*Department of Zoology, Shri S.S. Science College, Ashti – 442 707, Maharashtra*
[3]*Department of Zoology, Janta Mahavidyalaya, Chandrapur – 442 401, Maharashtra*

The present ethnozoological study describes the traditional knowledge related to the use of different animals and animal-derived products as medicines by the peoples of villages Ashti, Dhanor, Lagam, and Bori in Gadchiroli district. Gadchiroli district is situated on the North-Eastern part of the Maharashtra, India and is well known for dense forest; having adjoin State borders of Andhra Pradesh and Chhattisgarh. Naxalites have taken shelter in the dense forest of this district. The field survey was conducted from October to December 2010 by performing interviews through structured questionnaires. The local and tribal peoples provided information regarding therapeutic uses of animals. A total of 15 animals and animal products were recorded which are used for different ethnomedical purposes, including jaundice, tuberculosis, paralysis, ear ache, constipation, asthma, weakness, arthritis, baldness, graying of hair, snake poisoning. The zootherapeutic knowledge was mostly based on domestic animals. We would suggest that this kind of neglected traditional knowledge should be included into the strategies of conservation and management of faunistic resources in the investigated area.

Keywords: *Tribal peoples, Medicine, Animals, Gadchiroli.*

Introduction

The use of complete range of natural resources including plant, animals is a common practice in traditional medicine. Animals and product derived from their organs have constituted part of inventory medicinal substances use in various cultures since ancient times (5, 12, 24, 28). The use of animals for medicinal purposes is a part of a body of traditional knowledge which is increasingly becoming more relevant to discussions on conservation biology public health policies, sustainable management of natural resources and biological prospection (3). Recent publication have shown the importance of zootherapy in various socio culture environments around the world, and examples of the use of animal derived remedies can currently be found in many urban, semi-urban and more remote localities in all parts of the world particularly in developing countries (1,16-18). In India the traditional knowledge system is fast eroding due to urbanization. So there is an urgent need to inventories and record all ethnobiological information among the different ethnic communities before the traditional culture are completely lost (33). A lot of work has been done in the Gadchiroli district on the medicinal plant and plant products and documented too, but there is definite scarcity of such knowledge when it comes to animal products. Thus there is an urgent need to make such study in the field of zootherapy. So keeping this aspect in view we have undertaken this study.

Study Area

The present work was carried out in the tribal community of Gond, Madiaya located in villages of Ashti, Dhannor, Lagam, and Bori in district Gadchiroli. This district is situated on the North-Eastern region of the Maharashtra State, India and is well known for dense forest; having State borders of Andhra Pradesh and Chhattisgarh. Naxalites have taken shelter in the dense forest of this district (Figure 18.1). The Gadchiroli district which covers the total area of about 14412 Sq.Kms. the geographical location of Gadchiroli is 18.43° to 21.50° N Latitude to 79.45° to 80.53°E Longitude. The District falls under assured and heavy rainfall zone. The rains are mainly received from South-West monsoon. The average rainfall is 1562mm. The climatic condition is extreme with temperature reaching 47.3°C in summer and 9.4°C in winter.

Forests are rich inTeak, Ain, Tendoo, Dhavada, Anjan, etc. Ain and Anjan are suitable for rearing Tussar silk worms. Similarly good quality bamboo is abundantly available. From the socio-culture point of view the Gadchiroli district exhibit great ethnic and cultural diversity.

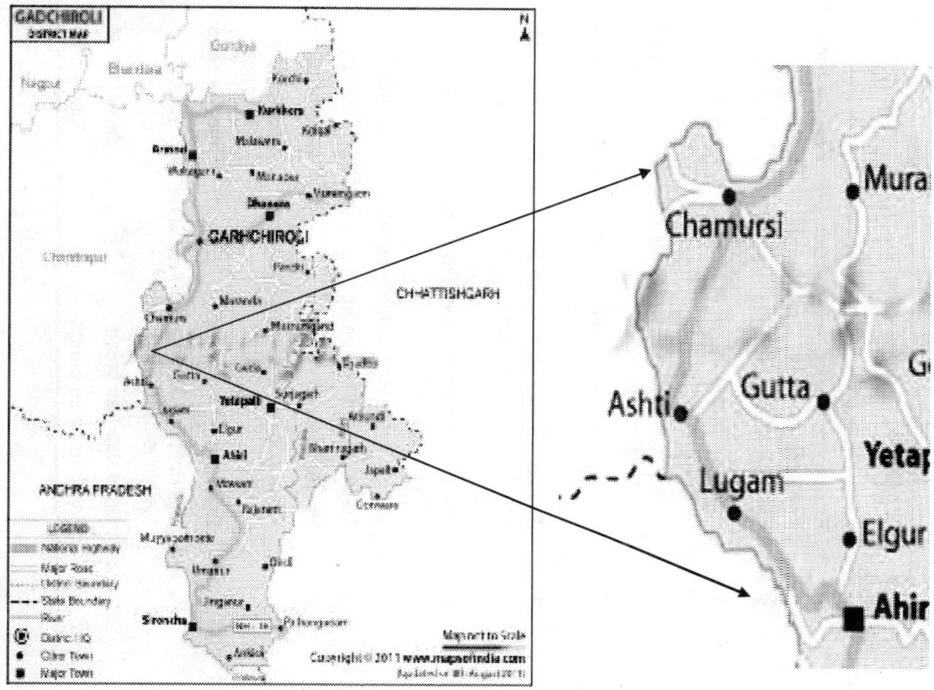

Figure 18.1: Map of Study Area

Methodology

Field survey was conducted from October to December 2010. During the first contact with the local population we have identified the peoples with specialized knowledge of medicinal use. (15). The information on the use of animals in traditional medicines was collected through interviews with tribal peoples mainly from elderly population who still retain the major portion of traditional knowledge in their communities. Survey data were gathered through interviews (35). They were asked about the ailments by animal based medicines and the manner in which the medicines were prepared and administrated. They were also asked regarding detailed information about mode of preparation and blending of animal products use as ingredients and were also asked about proper dose and length of medication. The name of animal and other information were documented. Since this kind of information indicates how a given medicine can be therapeutically efficient in terms of the right ingredients, the proper dose and right length of medication. According to them their knowledge of folk medicine was acquired mainly through parental heritage and experience about medicinal value of animal to heal themselves. The scientific name and species were identified using relevant and standard literature. (2, 20).

Result and Discussion

Animals have been used for medicinal purpose since colonial times in India [14] and they still play significant role in current folk healing practices. Various publications have shown the importance of zootherapy to traditional communities

Table 18.1

Sl.No.	English Name	Scintific Name	Common Name	Part Used	No of Mentioned	Method of preparation	Diseases
1.	Earthworm	*Pheretima postuma*	Gendur	Whole body	2	Earthworm are cooked in water for 10 min and eat	Snake bites
2.	Cockroach	*Blatta oriantalis*	Zurad	Whole body	2	Cockroach are eaten as raw	Asthma
3.	Crab	*Cancer pagurus*	Khekda	Whole body	3	Whole body is crushed in to and drink paste and boil in water	jaundice
4.	Honey bee	*Apis indica*	Madhmashi	Honey	4	Honey mixed turmeric powder and Tulsi leaves and eaten	Asthma, cold, mouth Ulcer
5.	Prawn	*Macrobrachium malcomsonii*	Zinga	Whole body dried powder	2	Dried powder eaten with water	Tuberculosis
6.	Scorpion	*Palamnaeus swammerdani*	Winchu	Whole body	3	The body is boiled in sarso oil and used for massaging	Rheumatic pain
7.	House sparrow	*Passer domesticus*	Chimni	Feaces	5	Dried fecal matter is used to cure asthma in children	Asthma
8.	Goat	*Capra indicus*	Bakari	Urine	3	Urine of goat administered orally	Tuberculosis
9.	Bat	*Megaderma lyra lyra*	Wat waghur	Flesh	3	Flesh cooked and eaten	Cough
10.	Human	*Homo sapiens*	Manus	Urine	8	Human urine consumed	Pile, internal injury
11.	Monitor	*Varanus monitor*	Ghorpad	flesh	5	Flesh cooked and make curry	Body pain and Joint pain
12.	Cow	*Bos indicus*	Gai	UrineMilk	10 15	Drink urine Drink milk	FeverWeakness
13.	Sheep	*Capra* sp	Mendhi	Milk	5	Milk+salt boil and drink and remain is used orally	Cough
14.	Pig	*Sus scrofa*	Dukar	Fat	10	Fat is used as cream	Muscle pain
15.	Mongoose	*Herpestes edwarssii*	Moongoose	Flesh	4	Fat is used as cream	Muscle pain

in various socio-culture environments in India [14]. The present study describe the traditional knowledge of treating various kinds of disease using different animal and their product by inhabitants (Gond, Madiaya) of villages in Gadchiroli district. The information of all local names of the animals, part or products used to cure and method of preparation were provided by the Mentions. In this study, we enlisted 15 animals' species, which are being used for 18 medicinal purposes. These animals are used as whole or body part or by product blood, milk, organ etc. for the treatment of various kinds of diseases such as asthma, paralysis, cough, weakness, snake poison etc. (Table 18.1). Since ancient times animals and their products have constituted part of inventory of medicinal substance used in various cultures. Different parts of single species provided the raw material to prepare different remedies, which are prescribed to treat various diseases. The possibility of using various remedies for the same disease is popularly valued, as it is rendered an adaptation to the availability for the animal possible [21]. On the other hand, different species were sometimes using to treat same illness. This strategy is important because many species have marked seasonality [26]. Zootherapeutical products are mainly used for the treatment of respiratory system diseases. Animals provide the raw materials for remedies used to treat physical and spiritual disease. The use of some Zootherapeutical resources is associated with popular briefs [21-25.27, 28]. In India nearly 15-20 percent of the ayurvedic medicine is based on animal derived substances. The Hindu religion has used 5 products (milk, urine, dung, curd and ghee) of the cow for purification since ancient times [31]. Different animals used by the Naga tribe of Nagaland [8], Ao tribe of Nagaland[11], Irular, Kurimba of Tamilnadu[32], Kachch (Gujarat)[6], Kanikar, Paliyar of Tamilnadu[29], Bhil, Gamit, Kokna etc of Maharashtra[19], Assam[4], Bhil of Rajasthan[30], Dibrugarh(Assam) [4] and Gond and Madiaya residing in this part of India. The use of urine of *Bos indicus* against the weakness due to fewer has reported amongst Gond and Madiaya tribes of Gadchiroli district and it has also been reported by Garasia tribe of Rajasthan [14] but the use of milk of *Bos indicus* has never been reported for weakness but it is reported for sexual health [14]. Rebari use honey for treatment of mouth ulcer, mouth infection and respiratory problem while tribal peoples of Tamilnadu use honey for cough and cold [29, 32]. Gond and madiaya tribes use the honey to improve digestive system of the young one but the saharia tribe of Rajasthan use honey to cure eye infection [13]. The use of honey is also reported for cough and cold, stomachache. The urine of Goat is reported to use to cure Tuberculosis by Gond and Madiaya people and it is also reported by Ao and Naga tribe of Nagaland [9] while the milk of this animal used to cure eye infection. Milk of sheep is used in cough by Gond and Madiaya community and the same is reported by Rebari community of Rajasthan. Gond and madiaya community are also found to use one animal product with plant derived material to cure particular diseases. They used honey, turmeric powder and Tulsi leaves (*Oscimum sanctum*) to cure cough. While the ash of peacock feather mixed with honey to cure asthma by Rebari community in Rajasthan. Further Gond and Madiaya community also use some non-domesticated animal species in traditional medicinal system. Fat is used in muscular pain [14].while the faeces of pig is used in neck tumor.

These examples confirm the knowledge and use of zootherapeutic in different parts of India and around the world. So the traditional knowledge should be included into the strategies of conservation and management of faunistic resources. Further studies are required to confirm the presence of bioactive compounds in these traditional remedies.

Acknowledgement

Authors are thankful to the people and the local medicine men from Gond and Madiaya tribes whose cooperation during the data collection could make the preparation of this manuscript possible.

References

1. A. A. C. Pinto and C. B. Maduro; 2003 "Produtos e subprodutosda medicina popular comercializados na cidade de Boa Vista,Roraima," *Acta Amaz^onica*, vol. 33, pp. 281–290.
2. Ali S; 1996 The book of Indian Birds. Bombay: Bombay. Natural History Society.
3. Alves RR, Rosa IL; 2005: Why study the use of animal products in traditional medicines? *J Ethnobiol Ethnomedicine* 2005, 1:5.
4. Dilip Kalita, Manashi Dutta, Nazim Islam Forid; 2005 Few plants and animal based folk medicines from Dibrugarh District, Assam.*Indian Journal of Traditional Knowledge* 2005, 4(1):81-85.
5. E. Lev; 2003 "Traditional healing with animals (zootherapy): medieval to present-day Levantine practice," *Journal of Ethnopharmacology*, vol. 85, no. 1, pp. 107–118. *Ethnobiology and Ethnomedicine*, vol. 5, article 12.
6. Gupta Leena, Silori CS, Mistry Nisha, Dixit AM; 2003: Use of Animals and Animal products in traditional health care systems in District Kachchh, Gujarat. *Indian Journal of Traditional Knowledge*. 2(1346356
7. H. P. Huntington; 2000 "Using traditional ecological knowledge in science: methods and applications," *Ecological Applications*, vol. 10, no. 5, pp. 1270–1274. *J Ethnobiol Ethnomedicine*, 1:5.
8. Jamir, N.S. and P. Lal; 2005. Ethnozoological practices among Naga tribes. *Indian Journal of Traditional Knowledge* 4(1): 100-104.
9. Kakati LN, Bendang Ao, Doulo V; 2006: Indigenous Knowledge of Zootherapeutic Use of Vertebrate Origin by the Ao Tribe of Nagaland. *J Hum Ecol*, 19(3163-167)
10. Kakati LN, Doulo V; 2002 Indigenous knowledge system of zootherapeutic use by Chakhesang tribe of Nagaland, India. *J HumEcol* 13(6): 419-423.
11. M. Mahawar and D. P. Jaroli; 2008 "Traditional zootherapeuticstudies in India: a review," *Journal of Ethnobiology and Ethnomedicine*, vol. 4, article 17.
12. M. O. Adeola; 1992 "Importance of wild animals and their parts in the culture, religious festivals, and traditional medicine, of Nigeria," *Environmental Conservation*, vol. 19, no. 2, pp. 125–134,
13. Mahawar, M.M. and D.P. Jaroli; 2007. Traditional knowledge on zootherapeutic uses by the Saharia tribe of Rajasthan, India. *Journal of Ethnobiology and Ethnomedicine* 3: 25

14. Mahawar, M.M.and D.P.Jaroli; 2006. Animals and their products utilized as medicines by the inhabitants surrounding the Ranthambhore National Park, India. *Journal of Ethnobiology and Ethnomedicine* 2: 46.
15. N. Alves, I. L. Rosa, and G. G. Santana; 2007. "The role of animal-derived remedies as complementary medicine in Brazil," *Bioscience*, vol. 57, no. 11, pp. 949–955.
16. N. Walston; 2005 "An overview of the use of Cambodia's wild plants and animals in traditional medicine systems," *TRAFFIC Southeast Asia, Indochina*.
17. O. A. Sodeinde and D. A. Soewu; 1999 "Pilot study of the traditional medicine trade in Nigeria," *Traffic Bulletin*, vol. 18, pp. 35–40.
18. P. E. V´azquez, R. M. M´endez, O. G. R. Guiasc ´ on, and E. J.N. Pi˜nera; 2006 "Uso medicinal de la fauna silvestre en los Altosde Chiapas, M´exico," *Interciencia*, vol. 31, no. 7, pp. 491–499.
19. Patil SH; 2003 Ethno-medico-zoological studies on Nandurbar districtof Maharashtra. *Indian Journal of Traditional Knowledge*.2(3297-299) pp. 82–103.
20. Prater SH; 1996. The Book of Indian Animals. Bombay. Bombay Natural History Society.
21. R. R. N. Alves and I. L. Rosa; 2006 "From cnidarians to mammals: the use of animals as remedies in fishing communities in NE Brazil," *Journal of Ethnopharmacology*, vol. 107, no. 2, pp. 259–276.
22. R. R. N. Alves and I. L. Rosa; 2007 "Zootherapeutic practices among fishing communities in North and Northeast Brazil: a comparison," *Journal of Ethnopharmacology*, vol. 111, no. 1,
23. R. R. N. Alves and I. L. Rosa; 2007 "Zootherapy goes to town: the use of animal-based remedies in urban areas of NE and N Brazil," *Journal of Ethnopharmacology*, vol. 113, no. 3, pp. 541–555.
24. R. R. N. Alves, I. L. Rosa, and G. G. Santana; 2007 "The role of animal-derived remedies as complementary medicine in Brazil," *BioScience*, vol. 57, no. 11, pp. 949–955.
25. R. R. N. Alves, H. N. Lima, M. C. Tavares, W. M. S. Souto, R.R. D. Barboza, and A.; 2008 Vasconcellos, "Animal-based remedies as complementary medicines in Santa Cruz do Capibaribe, Brazil," *BMC Complementary and Alternative Medicine*, vol. 8,article 44.
26. R. R. N. Alves, L. E. T. Mendonc¸a, M. V. A. Confessor, W.L. S. Vieira, and L. C. S. Lopez; 2009 "Hunting strategies used in the semi-arid region of northeastern Brazil," *Journal of Ethnobiology and Ethnomedicine*, vol. 5, article 12.
27. R. R. N. Alves, N. A. L´eo Neto, G. G. Santana, W. L. S. Vieira, and W. O. Almeida; 2009 "Reptiles used for medicinal and magic religious purposes in Brazil," *Applied Herpetology*, vol. 6, no.3, pp. 257–274.
28. R. R. N. Alves, W. L. Da Silva Vieira, and G. G. Santana; 2008" Reptiles used in traditional folk medicine: conservation implications," *Biodiversity and Conservation*, vol. 17, no. 8, pp.2037–2049,

29. Ranjith Sing, A.J.A. and C. Padmalatha; 2004. Ethnoentomological practices in Tirunelveli district, Tamil Nadu. *Indian Journal of Traditional Knowledge* 3(4): 442-446.

30. Sharma SK; 2002 A Study on Ethnozoology of Southern Rajasthanin Ethnobotany. In *Ethnobotany* Edited by: Trivedi PC. Jaipur: Aavishkar Publisher. 239-253.

31. Simoons FJ; 1974 The purification rule of the five products of the cow in Hinduism. *Ecology of Food and Nutrition*, 3:21-34.

32. Solavan A, Paulmurugan R, Wilsanand V, Ranjith Sing AJA; 2004 Traditional therapeutic uses of animals among tribal population of Tamil Nadu. Indian Journal of Traditional Knowledge. 3(2):206-207.

33. Trivedi PC: Ethnobotany; 2002 An overview. In *Ethnobotany* Edited by: Trivedi PC. Jaipur: Aavishkar publisher: 1.

2013, Environmental Biotechnology *Pages* ***179–185***
Editors: **D.R. Khanna, A.K. Chopra, Gagan Matta, Vikas Singh & Rakesh Bhutiani**
Published by: **BIOTECH BOOKS, NEW DELHI**

Chapter 19

Environmental Influences on Larval Duration in Silkworm Bivoltine Hybrid $NB_4D_2 \times SH_6$, *Bombyx mori* L.

Amardev Singh[1] *and Shamim Ahmed Bandey*[2]
[1]*Lecturer, Department of Sericulture,*
[2]*Assistant Professor, Department of Zoology,*
Govt. Degree College, Poonch, J&K

The rearing of silkworm is extremely vulnerable to change in environmental conditions especially in larval stages. Fluctuation in the larval stages concerning temperature and relative humidity prolonged the larval duration and also reduce yields of desirable crops while encouraging out breaks of diseases and many other problems. The results of the findings clearly depicted that the batches reared under optimum environmental conditions exerted better performance in the parameters studied such as single cocoon weight, single shell weight, shell per cent, larval durations *etc.*, over the batches which were reared under the experimental room without providing optimum environmental conditions.

Keywords: *Relative humidity, Cocoon production, Larval duration, Temperature.*

Introduction

The economical insect, silkworm (*Bombyx mori* L.) is a domesticated over 5000 years in different parts of the world (Nagaraju and Goldsmith, 2002). This insect has been extensively utilized as model organism in biological studies as well as for economic gains. Almost all-commercial silk is made from cocoons spun by silkworms of the genus *Bombyx* (Lee, 1999). Silkworm, in all stages of its growth, excluding egg stage, is characterized as a stenothermal and particularly a midthermal animal, according to its demands for environmental temperatures *i.e.*, 22-28pC Alexandros and Harizanis, (2004).The environmental conditions do influence the larval growth and also in attaining the cocoons productivity. Under the normal conditions of rearing the silkworms grow healthy and complete its larval duration within the specified period *i.e.*, 22-24 days, as a result the productivity of the cocoons enhanced significantly. Climatic fluctuations particularly during rearing of silkworm affect the silkworm crops and also prolonged the larval duration. Even though low temperatures have some merits they must be avoided because they decrease the rate of growth and rearing lasts longer, affecting the cost of rearing to a great extent (Ifantidis 1982; Patil and Gowda, 1986; Upadhyay and Misha, 1994). Since silkworms have been domesticated for many centuries, they are by nature delicate and are very sensitive and react very violently to the environmental conditions, particularly temperature and humidity. In silkworm rearing, it is paramount to maintain the optimum environmental conditions to ensure for maximum productivity of cocoons. Among the various environmental factors which influence the cocoons crops the most vital ones are the atmospheric temperature and humidity prevailing at the time of rearing silkworms. Silkworms are cold blooded insect; as such the temperature has direct influence on the various physiological activities of the system. Low and high temperature has adverse affect on the silkworm growth and development. On the other hand, humidity also plays a decisive role in silkworm rearing. Indirectly, humidity influences the rate of withering of leaf in the rearing beds and under too dry conditions, the leaves wither very fast and become unsuitable as silkworm feed, whereas too humid conditions especially in the case of late age worms build up bed humidity and create conditions that favour outbreak of diseases. Keeping in view, an experiment was carried out on the environmental influence on larval duration in silkworm bivoltine hybrid $NB_4D_2 \times SH_6$, *Bombyx mori* L.

Materials and Methods

The experiment was carried out at Department of Sericulture, Govt. Degree College, Poonch (J&K). Silkworm hybrids $NB_4D_2 \times SH_6$, was reared @200/tray with four treatments, and one control with four replications for each one, in two different rooms in order to compare the effect of temperature and humidity conditions for the silkworm rearing. The first room was considered as the control, where the silkworm rearing was carried out by maintaining the optimum conditions required to rear the silkworm (Table 19.1).The second room was considered as the experiment room, where no such optimum conditions was maintained in the entire rearing. The average temperature and humidity recorded in each instars in the experiment room given in (Table 19.2). A thermometer was installed to record the temperature and relative humidity in both the rooms. One week after mounting the cocoons were harvested and various cocoons parameters were recorded such as single cocoon weight, single shell weight, shell per cent, total larval and moulting durations. The observed data were subjected to one way ANOVA using SPSS package (7.5) for windows (Berkowitz and Allaway, 1998) and results are presented in the Figures 19.1–19.5.

Table 19.1: Normal Temperature and Relative Humidity Required during Rearing

Instars	*Temperature Required*	*Relative Humidity Required*	*Total Larval Period (Days)*	*Moulting Durations (h)*
1st	27-28°C	85-90 per cent	3½ days	20h
2nd	27-28°C	85-90 per cent	2½ days	24h
3rd	25-26°C	75-80 per cent	3½–4days	24h
4th	24-25°C	75-80 per cent	4½ days	30h
5th	24°C	65-70 per cent	6–7 days	Spinning (1-2 days)
	Total		**18 days**	**98h**

Table 19.2: Recorded Temperature and Relative Humidity in the Experimental Room during Rearing

Instars	*Average Temperature Recorded*	*Average Relative Humidity Recorded*	*Total Larval Period (Days)*	*Moulting Durations (Days)*
1st	24°C	75 per cent	6 days	1½ days
2nd	22°C	60 per cent	6½ days	2 days
3rd	21°C	62 per cent	5½ days	2 days
4th	19°C	55 per cent	6½ days	2½ days
5th	18°C	58 per cent	8½ days	Spinning (2½ days)
	Total		**34days**	**8days**

Results and Discussion

Several factors play an important role during the rearing of silkworm for the successful cocoon crops production (Benchamin and Jolly, 1986). Among them,

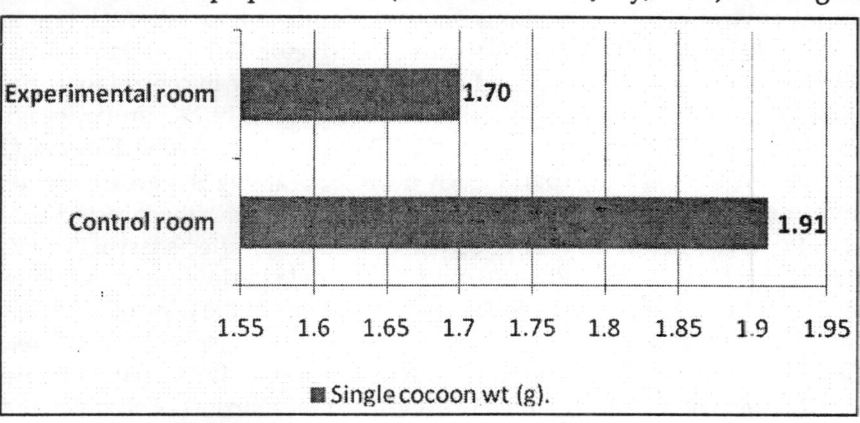

Figure 19.1: Environmental Effect on Single Cocoon Wt. (g)

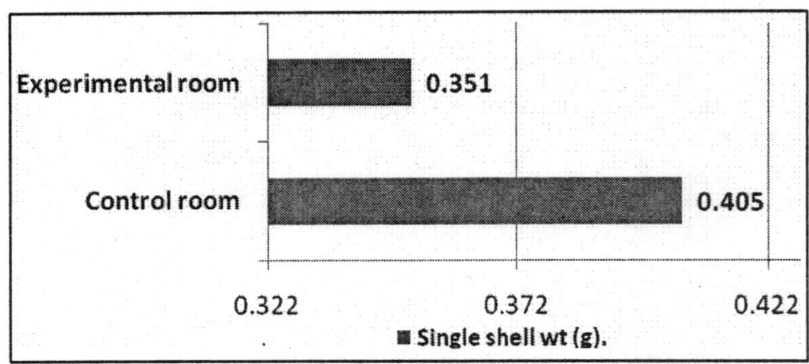

Figure 19.2: Environmental Effect on Single Shell Wt. (g)

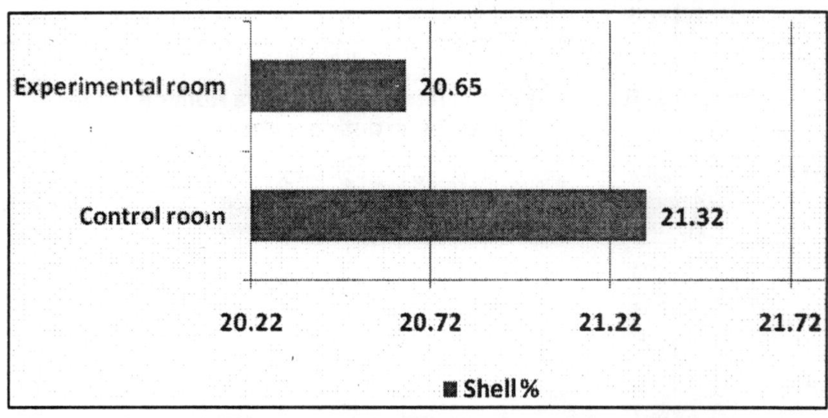

Figure 19.3: Environmental Effect on Single Shell per cent

environmental conditions especially temperature and relative humidity do influence the cocoon production (Mathur *et al.*, 1997). Some researchers (Mutswmura, 1975; Yokoyama, 1962) reported adverse effect of decline in relative humidity during larval rearing on the physiology of the silkworms which was resulted in poor cocoon production. The results of the present finding exhibited that the batches reared under optimum environmental conditions showed significant variation (1.91g) single cocoon weight over the experimental room (1.70g) (Figure 19.1). The data with regard to single cocoon weight also depicted significant difference in control room batches (0.405g) against the experimental room (0.351g) (Figure 19.2). Similar, trend was noticed in shell percentage also (Figure 19.3).As reported by Mubashar *et al.* (2011) that the life cycle of silkworm particularly during larval stages is greatly influenced by environmental stress, relative humidity affects all stages of the insects. Deviation in humidity levels below and above certain critical limits affects larval growth and development (Bursell, 1970; Rockstein, 1974).The data pertaining to moulting and larval durations also exhibited considerable variations in the control room batches the total moulting and larval duration was (8 and 24 days) respectively, whereas the results in experimental room was quite different (8 and 34 days) total moulting and larval duration respectively (Figures 19.4 and 19.5).Morohoshi (1979) stated that the silkworm gets affected extremely easily by the environmental conditions during

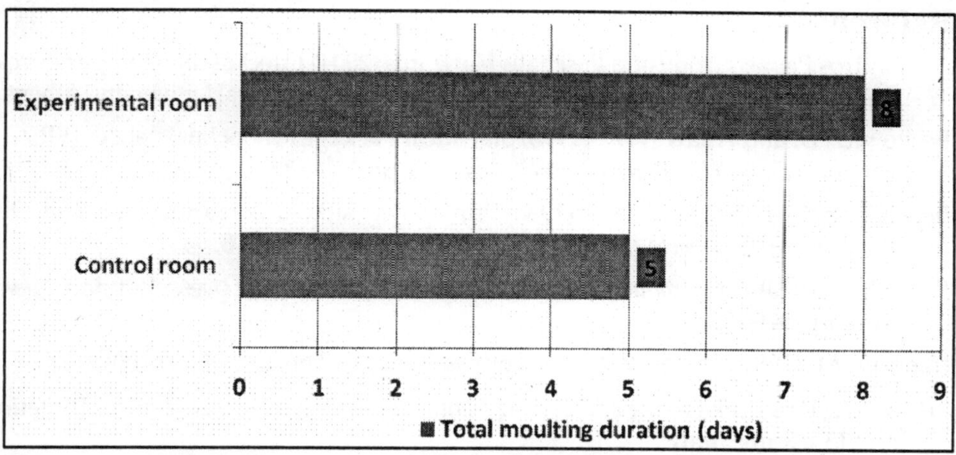

Figure 19.4: Environmental Effect on Total Moulting Duration (Days)

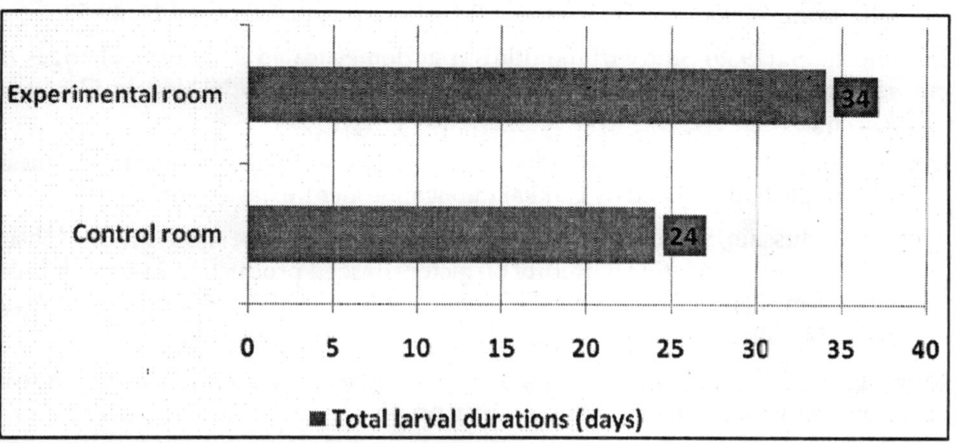

Figure 19.5: Environmental Effect on Total Larval Durations (Days)

different stages of development. Some workers (Sharma *et al.*, 1988; Sarker *et al.*, 1995; Zaman *et al.*, 1996) are opinion that seasonal variations in temperature and relative humidity and feeding of different mulberry leaves of different varieties influences the performance of rearing. Benchamin and Jolly (1986) showed that temperature plays a major role on growth and productivity of silkworm, as the silkworm is poikilothermic (cold blooded insect) any changes in the larval period has profound effect on the silkworm growth and development which directly influences the cocoon production to a great extent. According to Morohoshi (1939) findings that variation occurs in the grown silkworm larval stage due to the temperature and humidity impact (38° C, 90 per cent, 1 day-night).

Acknowledgment

Authors are grateful to the Deputy Director, State Sericulture Development Department, Poonch (J&K) for providing Dfls for this experimental work.

References

Alexandros Papasotitriou and Paschalis Harizanis (2004) Effect of high temperatures during silkworm rearing. B.Sc thesis submitted at Agricultural University, Athens.

Berkowitz, D. and Allaway, A. (1998) Statistical Package for Social Sciences (SPSS), version 7.5 for Windows NT/Windows 95, pp.130-132.

Benchamin, K. Y. and Jolly, M. S. (1986) Principles of silkworm rearing.*Proc. Sem. On Problems and Prospects of sericulture,* Vellore, India.pp.63-108.

Bursell, E. (1970) *An Introduction to Insect Physiology*. Academic press. London, New York.pp.241-250.

Ifantidis, M. (1982) *The silkworm (Biology and Rearing)*. Thessaloniki.p.100.

Lee, Y. W. (1999) *Silk Reeling and Testing Manual*.FAO Agricultural Services Bulletin NO.136. Rome, Italy.

Mathur, J. P. Prassad, A. and Tayagi, H. R. S. (1997) Performance of parent races of silkworm, *Bombyx mori* in southern Rajasthan. *Indian. J.Entomol.***11,** 49-52.

Morohoshi, S. (1939) Study of moultinism in domesticated silkworm, changes in moultinism due to humidity and temperature shock. Kyuudai Nogaku Geizatsu. 8, 276-281.

Morohoshi, S. (1979) *Development Physiology of Silkworms*. Translated from Japanese into English, published by Gakkai Publication Centre, Japan.p.11.

Mubashar Hussain, Shakil Ahmed Khan, Muhammed Naeem and Atul-ul-Mohsin (2011) Effect of relative humidity on factors of seed production in some inbred silkworm (*Bombyx mori*) Lines. *International Journal of Agriculture and Biology*. pp.1814-9596.

Mutswmura (1975) Silkworm Rearing. *Text book of Tropical Sericulture*. Japan Overseas Cooperative volunteers.Tokoyo, Japan, pp. 457.

Nagaraju, J. and Goldsmith, M. R. (2002) Silkworm genomics-progress and prospects. *Cur.Sci.* **83,** 415-425.

Patil, C. M. B. L. and Vishweshwara Gowda (1986) Environmental adjustment in sericulture. *Indian Silk,* **25(3),** 11-14.

Rockstein (1974) *The Physiology of Insect*. Vol.2, 2nd edition. Academic press. London, New York.pp.47-49.

Sarker, A. A. Hague, M. A. Rab and Absar, N. (1995) Effect of feeding mulberry (*Morus sp.*) leaf supplemented with different nutrients to the silkworm (*Bombyx mori* L.) *Curr.Sc.* **69,**185-188.

Sharma, D. K. Dutta, S. Khanikar, D. P. and Dutta, L. C. (1988) Effect of larval food plants on certain life parameters of Eri silkworm, *Philosamia ricini, Hult.J.Agri.Sci.Soc.* North East India.11:5-7.

Upadhyay, V. B. and Misha, A. B. (1994) Influence of temperature on the passage of food through the gut of multivoltine *Bombyx mori* L. *Indian Journal of Sericulture.* **33(2),** 183-185.

Yokoyama (1962) *Synthesized Science of Sericulture,* Central Silk Board, Bombay, India.pp.210-233.

Zaman, A.Qadar, M. A. Islam Baeman, A. C. Aslam, A. C. and Islam, M. (1996) Effect of feeding tukra affected mulberry leaves economic characters of silkworm, *Bombyx mori* L. *Pakistan. J. Zoology.* **28.**169-171.

2013, Environmental Biotechnology *Pages* ***187–192***
Editors: **D.R. Khanna, A.K. Chopra, Gagan Matta, Vikas Singh & Rakesh Bhutiani**
Published by: **BIOTECH BOOKS, NEW DELHI**

Chapter 20

Effect of Distillery Effluent on Growth Behaviour of *Phyllanthus niruri*

Chanchal Sharma and Sudanshudhar diwaide
Botany Department, Government Science and Commerce College, Benazir, Bhopal, Madhya Pradesh

A field work was undertaken to study the Physico-chemical and biological parameters of treated distillery effluent and the effect of various concentrations (0, 5, 10, 20, 40, 60, 80, 100 per cent) on plant growth *i.e.* seed germination, root length, shoot length, length of leaves, chlorophyll content in the *Phyllanthus niruri*. The effluent could not be directly applied to the field because of its excessive TDS, BOD and COD values and large quantities of soluble and suspended organic matter. Effluents was filtered through muslin cloth and diluted 2.5 times and there after various dilutions were prepared by diluting with tap water

Introduction

Phyllanthus niruri is belonging to the family Euphorbiaceae. The plant is distributed in the tropical and subtropical region of the country. It contained 300 genera and 7500 species. In India, family Euphorbiaceae is represented by 70 genera and 450 species. *Phyllanthus niruri* is one of the important species of this family due to its medicinal importance. The plant is annual herb up to 60 cm. in height with glabrous branched stem contains numerous leaves which are sub- sessile and disticous. Flower is yellow in colour. It has capsuler fruits, 2.5 mm in diameter. The whole plant is used in medicine for treatment of several diseases.

Water is most essential natural resource and is responsible for existence of life on this planet. It is worthwhile to mention here that the outcome of global industrialization has caused severe scarcity of water due to heavy consumption. It is also caused serious environmental hazards. The disposal of industrial effluents is one of the challenging problems being faced by the environmentalist. Consequently, utilization of industrial waste specifically of distillery industries can be used as soil amendment. This proposition has generated immense interest in recent times as is evident from myriad of studies carried out along these lines. As reported by Swaminatan and Vaidheeswaran (1991) effluent from dye industry increased the seed germination and chlorophyll content of groundnut seedling, at low concentration. Valdes *et al.* (1994, 1998), showed that treated wastewater contains various mineral nutrients, which are quite useful for the plant growth. Bishnoi and Gautam (1991) studied the effect of various concentrations of dairy effluents on seed germination and seedling growth of some kharif crops. It was noticed that with increasing concentration, the present seed germination decreased gradually. They concluded that the diluted effluent could be as liquid fertilizer as it promote seedling f=growth. The distillery effluent is mixture of organic and inorganic nutrients and has been reported to have a beneficial effect on seed germination Subramani *et al.* (1999). The effluent in the lower concentration enhanced the growth of corn rice crops, Kumar and Bhargava (1998) cautioned on the deleterious effects of higher concentration of effluents as it caused the decreasing effects on the growth of crops. This gave a new direction to the studies *i.e.* effluents should be used after dilutions.

In the present study an attempt has been made to assess the suitability of various dilutions of effluents released by distillery as a potential liquid fertilizer. The distillery industry is closely linked with the fermentation industry.

A field experiment was designed to know the effect of differentiation concentration (0-100 per cent) of distillery effluent on seed germination and various growth parameters of Phyllanthus niruri.

Material and Methods

The present study was conduct on effect of distillery effluent on growth behavior of Phyllanthus niruri. Effluent samples were collected from the outlet point of a tank of ETP of distillery in pre cleaned and dried ploy jars which were sealed immediately and labeled. Samples were analyzed for various Physico-chemical and biological parameter as per standard method of APHA (1998) for microbial studies, effluent samples were collected aseptically in pre sterilized glass bottles. They were immediately inoculated on suitable culture medium. The laboratory experiment were conducted to evaluate of different dilutions of effluent *viz.*, 0, 5, 10, 20, 40, 60, 80, and 100 per cent on growth of medicinal plant *Phyllanthus niruri.*

The soil was irrigated with different dilution of distillery effluent. Twenty healthy seeds of *Phyllanthus niruri* were sown in soil. Germination of seeds was observed and calculated after about 10 days. Soil was irrigated with equal volumes of different dilutions of effluents. Growth parameters such as seed germination, length of leaves, shoot length, branching number per plant height and chlorophyll content were recorded after a period of one month and two months of germination of the seeds.

Results and Discussion

Properties of the Effluent

The Physico-chemical and biological analysis of the distillery effluent is given in Table 20.1. Effluent of distillery was reddish brown in colour with unpleasant odour. Temperature of distillery effluent was 22.7pC. The average pH value of the distillery effluent was 7.70. The dissolved oxygen (D.O.) in the distillery effluent was 0.7mg/ml. The low value of D.O. is possibly due to high organic load. The value of total solids in distillery effluent is 3630 mg/l. The value of BOD in distillery was found to be 580.0. This indicates high organic load. The COD value of the distillery effluent was 36000mg/microbiological parameters were bacterial count >300 CFU/ml fungi one and Azotobacter nil CFU/ml

Table 20.1: Physico-chemical and Biological Characteristics of Distillery Effluent

Sl.No.	*Parameter*	*Units*	*Observation*
	Physical Parameters		
1.	Temperature	°C	22.70
2.	pH	–	7.70
3.	Conductivity	dS/m	10.11
4.	Salinity	ppt	5.70
5.	Total Solids	mg/L.	3630.00
6.	Total Dissolved Solids	mg/L.	667.00
7.	Total Suspended Solids	mg/L	2963.00
	Chemical Parameters		
8.	Chlorides	mg/L	1250.00
9.	Sulphates	mg/L	646.05
10.	Total Hardness	mg/L	2055.00
11.	Total Nitrogen	Percentage	0.05
12.	Total Phosphorous	ppm	3.90
13.	Dissolved Oxygen	mg/L	0.70
14.	B.O.D.	mg/L	580.00
15.	C.O.D.	mg/L	36000.00
16.	Total Alkalinity	mg/L	3700.00
17.	Sodium (Na)	ppm	65.40
18.	Potassium (K)	ppm	260.40
19.	Calcium (Ca)	ppm	546.60
	Biological Parameters		
20.	Standard Plat Count of–Bacteria	CFU/ml	>300
21.	Standard Plat Count of–Fungi	CFU/ml	1.00
22.	Azotobactor	CFU/ml	Nil

A perusal of the result given in the Table 20.1 reveals that conductivity, salinity, TDS total solids, total suspended solids, D.O., B.O.D, C.O.D were above the permissible limits. Effluent is rich in potassium that is a major plant nutrient besides containing minor amounts of two other major nutrients, nitrogen and phosphorus. Smaller amounts of other minor plant nutrients such as calcium, magnesium, sulphate and chloride are present. However pH and nutrients were with in permissible limits suggesting thereby that this effluent can be a good liquid fertilizer.

Effect of Different Dilutions of Effluent on Growth of *Phyllanthus niruri*

The results regarding the impact of effluent on growth of Phyllanthus niruri plant are presented in Table 20.2. Various parameters such as seed germination, length of leaves, shoot length, root length, branching no. per plant height and chlorophyll content per plant were observed at a period of one month and two month after seed germination. The seed germination varied from 70 per cent to 90 per cent with control as well as with all dilution used of the effluents. Length of the leaves was maximum *i.e.* 7.1 cm.recorded with 20 per cent effluent dilution, while 3.4 cm in the control. Shoot length varied from 11.5 cm. to 31 cm. treated with different dilution and 26.2 cm in the control. However maximum shoot length *i.e.* 31 cm. was observed with 60 per cent effluent dilution.

After a gap of two months of seed germination the length of leaves of varied from 4.1 cm. to 7.1 c. in the dilution of 80 per cent and 10 per cent respectively while it was 4. 0 cm. in the control. However maximum length *i.e.*7.1 m. was noticed with 10 per cent effluent dilution.

The shoot length varied from 20.5 cm. to 38.4 cm. in all treatments. However minimum 20.5 cm. was noticed with 100 per cent dilution and appreciable shoo length 38.4 c m. was observed with 80 per cent effluent dilution while it was 25.1 in the control. The branching per plant varied 2 to 4 which was insignificant including treated and control experiments. The root length was maximum 13.0 cm. with 60 per cent effluent dilution and minimum 7.2 cm. with 100 per cent dilution. The plant height was 38.5 cm. with the treatment of 60 per cent and minimum 26. 6 cm. with 20 per cent treatment. However it was 28.3 cm. in case of control experiment and it increased considerably in all effluent dilutions.

A perusal of the result given in the Table 20.1 reveals that conductivity, salinity, TDS total soli ds, total suspended solids, D.O., B.O.D, and C.O.D were above the permissible limits. Effluent I s rich in potassium that is a major plant nutrient besides containing minor amounts of two other major nutrients, nitrogen and phosphorus. Smaller amounts of other minor plant nutr- ients such as calcium, magnesium, sulphate and chloride are present. However pH and nutrients were with in permissible limits suggesting thereby that this effluent can be a good liquid fertilize

Conclusion

A careful analysis of the parameters evaluated show that in general there was a rising trend in various parameters such as chlorophyll content, Plant height in the

Table 20.2: Effects of Various Soil-Effluent Mixtures on Seed Germination and Growth of *Phyllanthus niruri* on Laboratory Scale from Distillery

Sl.No.	Parameter		Units	Dilutions							
				5%	10%	20%	4 0%	60%	80%	100%	Control Soil
1.	Percent germinational (per cent)		per cent	80	80	80	80	90	80	70	80
2.	After one month of	Length of leaves	Cm	4.1	5.7	7.1	4.4	4.9	4.4	3.8	3.4
3.	germination	Shoot length	Cm	16	23.3	17.1	1 1.5	31	28.6	25.5	26.2
4.	After two months of	Length of leaves	Cm	4.7	7.1	6.8	4.6	6	4.1	6	4.0
5.	sprouting	Shoot length	Cm	31.2	30.6	26.4	32.6	35.3	38.4	20.5	25.1
6.		Branching	Number	4	3	3	3	4	4	3	3
7.	Root length		Cm	10.2	11.4	10.0	10. 5	13.0	11.1	7.2	10.0
8.	Total growth of plants		Cm	31.5	32.2	26.6	33. 3	38.5	29.6	27.9	28.3
9.	Chlorophyll contents (Absorbance value)		620 mm	0.484	0.49	0.535	0.58 6	0.737	0.563	0.57	0.42

plant grown in soil treated with 5, 10, 20, 40, 60 per cent effluents maximum value being in the last case after which the re was downward trend in the plants grown with 80 and 100 per cent effluents. Length of leaves, shoot length, root length and total growth of plants showed fluctuations, though overall results we re better with the treatment of 60 per cent dilution. Thus it may be concluded that the effluents with 60 per cent dilution have the maximum potential for healthy growth of *Phyllanthus niruri* and is recommended to be used as a potential liquid fertilizer.

References

APHA, (1998) standard method for examination of water and wastewater, (20th ed.) American Public Health Association, Washington, D.C.

Bishonai, S. and Gautam, D.D. (1991) Effect of diary effluents on seed germination and seedling growth of some crop plants. Inst. J. Eco. Environ. Sci 17(1):61-7

Chopora and Kanwar (1986) Analytical Agriculture Chemistry, Kalayani Publisher.

D.W. Westcot, and R.S. Ayers. 1984. Irrigation Water Quality Criteria.

Joshi, H.C., *et al.*, 1994. Environmental issues related with distillery effluent utilization in agriculture in India. Asia Pac. J.Env. Develop. 1:92-103.

Joshi, H.C., *et al.*, 1998. Agro cycling- A pollution control strategy for distilleries in India.

Kumar, R.K.T., *et al.*, 1990. Effects of chemical factory effluent on germination and growth of guar: *Cyamopsis tetragonoloba* L. Taubvar. PNB. Adv. Plant Sci., 3(1):34-42.

Kumar, R., Bhargava, A.K. (1998). Effect of sugar mill effluent on the vegetative growth and yield of *Tritium aestivums* CV.UP 2003. Ad. Plant Sci.11 (2):221-227.

Olsen, S.R., Cole, C.V., Watanabe, F.S., and Dean, L.A.(1954). Estimation of available phosphorus in soils by extraction with sodium bicarbonate. U.S. Department of Agriculture Circular.

Singh and Mishra (2004) Effects of fertilizer factory effluent on soil and crop productivity, Springer Link Journal, DOI 10.1007/BF00294199 pages 309-320.

Subramani A., p. Sundermoorti, S. Saravanan, M. Silvarju and A.S. Lakshmanchary. 1999. Impact of biologically treated distillery effluent on growth behavior of Green gram (*Viniga radiata*). In: Jr. of Industrial Pollution Control. 15(2):281-286.

Swaminathan K, Vaidheeswaran P (1991). Effect of dyeing factory effluent on seed germination and seeding development of groundnut (*Arachis hypogea*) 1. Environ. Biol. 12(3):253-258.

2013, Environmental Biotechnology *Pages* ***193–203***
Editors: **D.R. Khanna, A.K. Chopra, Gagan Matta, Vikas Singh & Rakesh Bhutiani**
Published by: **BIOTECH BOOKS, NEW DELHI**

Chapter 21

Protective Effect of *Emblica officinalis* on Nicotine Toxicity to Rat Liver

J. Vadivelu[1] *and S. Dawood Sharief*[2]
[1]*Ph.D Research Scholar, School of Environmental Science,*
[2]*Associate Professor,*
Department of Zoology, The New College, Chennai – 14

Emblica officinalis (Amla) is widely used in the Indian system of medicine and is believed to increase defense mechanism against diseases. It is one of oriental traditional medicine used for hepatic disorders from time immemorial. Nicotine is the most abundant component in cigarette smoke, and nicotine is first metabolized in the liver. Present study was carried out to investigate the protective effect of *Emblica officinalis* on nicotine induced toxicity in liver of rats.

Male wistar rats (Group-II, Group-III and Group-IV) were treated with oral nicotine diluted with drinking water for 32 days, while control (Group-I) were treated with drinking water simultaneously. After 32 days Group-III, Group-IV was administered with two different concentrations of *Emblica officinalis* (250, 500 mg/body weight, diluted with drinking water) for 7 days. Group-II served as a toxicity group (only with nicotine). Rats were sacrificed 24 hours after last day of administration (40th day), the blood was analyzed for liver marker enzymes, and the liver was used for histopathological studies.

A significance increase in the liver marker enzymes, such as ALT, AST, and ALP (Alanine Aminotransferase, Aspartate Aminotransferase, Alkaline Phosphataseand) and decreased the activity of GGT (Gamma Glutamyl transferase) was recorded, Total Protein and elevated levels of cholesterol, triglycerides in nicotine control group was observed. In the two treated group (Group-III and Group-IV), the effect of *Emblica officinalis* was more significant in animals treated with 500 mg/kg dose. The results suggest that *Emblica officinalis* exerts the protective effect of nicotine toxicity in liver. The results are supported by histopathological observation of liver, and are discussed in the light of recent literature.

Keywords: *Emblica officinalis, Nicotine, Alanine Aminotransferase, Aspartate Aminotransferase, Glutamyl transferase.*

Introduction

Nicotine

Cigarette smoke has enormous negative health consequences worldwide, and the use of tobacco is still rising globally (CDC, 2000). Although approximately 4000 components occur in the cigarette, nicotine is the alkaloid most active in the tobacco. Nicotine is an amine composed of pyridine and pyrrolidine rings (Trushin and Hecht, 1999). It has been shown that nicotine can cross the biological membranes including the blood brain barrier (Balfour *et al.,* 2000).The actions of nicotine have been extensively investigated in human, in animal, and in a variety of cell systems (Gryglewski, 1968; Ashakumary and Vijayammal, 1997; Benowitz and Gourlay, 1997; Benowitz *et al.*, 2002; Pausova *et al.*, 2003; Cooke and Bitterman, 2004; Valenca *et al.*, 2004). The predominant effects of nicotine in the whole intact animal or human consist of an increase in pulse rate, blood pressure, increase in plasma free fatty acids and lung injury (Kavitharaj and Vijayammal, 1999; Liu *et al.*, 2001; Benowitz *et al.*, 2002). In addition, nicotine has also been found to disturb the antioxidant defense mechanisms in rats fed a high fat diet (Senthilkumar *et al.*, 2004; Kalpana *et al.*, 2005; Perlemuter *et al.*, 2005). Nicotine has also been studied as an experimental therapy for Parkinson's disease, Alzheimer's disease, and ulcerative colitis (Westman *et al.*, 1995; Baron, 1996; Birtwistle and Hall, 1996).

Nicotine is metabolized by various pathways, of which cotinine is the primary product of the C-oxidation pathway of nicotine biotransformation (Nakajima *et al.*, 1996; Wang *et al.*, 2005). While the liver is considered to be the major site of nicotine biotransformation, metabolism also occurs in the lung and kidney (Nakajima *et al.*, 1996; Trushin and Hecht, 1999). Nicotine is metabolized by CYP forms to a nicotine-10(50)-minimum ion which is converted to cotinine by a cytosolic aldehyde oxidase enzyme (Price *et al.*, 2004; Ochiai *et al.*, 2006). In rat liver nicotine is metabolized by CYP1A2, CYP2B1, CYP2C11 and other CYP forms; CYP2B1 also being constitutively expressed in rat lung (Ochiai *et al.*, 2006).Fatty liver or steatosis refers to a histopathological condition characterized by an excessive accumulation of lipids, primarily triglycerides, within hepatocytes (Wang *et al.*, 2006).

Emblica officinalis

Emblica officinalis (*Phyllanthus emblica* L.) as an euphorbiaceous plant is widely distributed in subtropical and tropical areas of India, China, Indonesia, and Malaysia. It has abundant amounts of vitamin C and superoxide dismutase (Verma and Gupta,

2004) and is used in many traditional medicinal systems. Emblica fruit is reported to have hypolipidemic (Anila and Vijayalakshmi, 2000; Jacob *et al.*, 1988; Mathur *et al.*, 1996; Thakur *et al.*, 1988) and hypoglycemic activities (Abesundara *et al.*, 2004; Anila and Vijayalakshmi, 2000), and acts as a important constituent of many hepatoprotective formulations available (Antarkar *et al.*, 1980; De *et al.*, 1993; Panda and Kar, 2003). It is also used as antimicrobial agent (Dutta *et al.*, 1998; Godbole and Pendse, 1960; Rani and Khullar, 2004), anticancer (Jeena *et al.*, 2001; Zhang *et al.*, 2004), and anti-inflammatory agent (Asmawi *et al.*, 1993; Lampronti *et al.*, 2004; Perianayagam *et al.*, 2004).

It was reported that emblica has a strong antioxidant activity (Bafna and Balaraman, 2004; Anila and Vijayalakshmi, 2003; Jose and Kuttan, 1995), which may be partially due to the existence of flavonoids and several gallic acid derivatives including epigallocatachin gallate (Anila and Vijayalakshmi, 2002; Sabu and Kuttan, 2002). The aim of the present study was to investigate the protective effect *Emblica officinalis* on nicotine induced toxicity in rat liver.

Materials and Methods

Animals

Male albino rats (*rattus norvegicus*) ranging in body weight from 180-200 gms were obtained from the King Institute, Guindy, Chennai and maintained according to the guidelines of CPCSEA (No: 324), under the supervision of Animal Ethical Committee were used for the experiment. They were acclimatized to laboratory conditions prior to use and fed with pelletted chow (supplied by Poultry Research Station, Chennai) and water provided *ad libitum*.

Chemicals

Nicotine ((-) - nicotine ([-]-1methyl-2-[3-pyridyl]-pyrrolidine, Sigma, Chennai, India)) Nicotine solution was prepared daily. Special drinking bottles were used to avoid nicotine solution exposition to light.

Plant Material

Emblica officinalis was procured from local market and fruit of *Emblica officinalis* was separated, shade dried, grounded with mortar and pestle and sieved to get fine powder.

Experimental Design

The rats were randomly distributed into four different groups of six animals each under identical conditions and were grouped as follows:

Group I Served as control animals and was given drinking water.

Group II Animals received nicotine (5 mg/kg bwt) in drinking water for 32 days. (Samuel Santos Valenca *et al.*, 2008.)

Group III Animals received *Emblica officinalis* (250mg/bwt) in drinking water for 7 days (after 32 days of nicotine administration)

Group IV Animals received *Emblica officinalis* (500mg/b.wt) in drinking water for 7 days (after 32 days of nicotine administration)

At the end of the experimental period (39 days) all the animals were anaesthetized and sacrificed by cervical dislocation after an overnight fast. Blood was collected in heparinised tubes and centrifuged at 5000 rpm for 10 min. Plasma was separated by aspiration, transferred into eppendorfs tubes and stored at –20°C for analysis.

Analysis of Liver Marker Enzymes and Biochemical Parameters

The activities of Alanine Aminotransferase (ALT) (Bergmeyer *et al.*, 1980 and Tietz, 1987), Aspartate Aminotransferase (AST) (Bergmeyer *et al.*, 1978 and Tietz, 1987), Alkaline Phosphatase (ALP) (Bessey *et al.*, 1946; Bowers and McComb, 1975), Gamma Glutamyl transferase (GGT) (Rosalki and Tarlow, 1975) were assayed and the levels of Total Protein (Lowry *et al.*, 1951), cholesterol (Allain *et al.*, 1974) triglycerides (Denmark, 1960 and Tietz, 1987) were estimated in plasma.

Results

The activities of biomarker enzymes in the plasma are shown in Table 21.1. In nicotine treated rats, the activities of ALT, AST, and ALP were significantly increased and GGT was decreased when compared to the control. Administration of *Emblica officinalis* to nicotine treated rats at two different doses significantly decreased the activities of these enzymes (ALT, AST, and ALP) and increased the activity of GGT when compared to the nicotine treated animals. The dose of 500 mg/kg body weight of *Emblica officinalis* was found to be most effective.

Table 21.1: Changes in the activities of marker enzymes in the plasma (Values are mean ± SD from 6 rats in each group)

Groups	*ALT*	*AST*	*ALP*	*GGT*
Control	42.13 ± 2.46	76.06 ± 4.32	81.34 ± 3.26	8.3 ± 0.38
Nicotine	79.04 ± 4.62	138.12 ± 12.61	167.36 ± 8.42	3.4 ± 0.32
N + EO (250mg/b.wt)	47.32 ± 3.12	82.42 ± 4.26	94.64 ± 4.61	6.4 ± 1.27
N + EO (500mg/b.wt)	44.23 ± 3.26	78.26 ± 3.68	83.72 ± 6.48	7.6 ± 0.23

N: Nicotine; EO: *Emblica officinalis*; ALT: Alanine Aminotransferase; AST: Aspartate Aminotransferase; ALP: Alkaline Phosphatase; GGT: Gamma Glutamyl Transferase.

The levels of cholesterol, triglycerides and total protein in plasma and liver tissue of experimental animals is shown in Table 21.2. The levels of cholesterol and triglycerides in plasma of experimental animals increased and total protein was decreased significantly in the nicotine treated group. Significance protection was seen in the *Emblica officinalis* supplemented nicotine treated animals, but the 500 mg/kg body weight dose was more effective than the 250 mg/kg body weight dose tested.

There was significance reduction in the body weight in the nicotine treated animals when compared to the control rats shows (Table 21.3). Treatment with *Emblica officinalis* at a dose of 500 mg/kg reversed this decrease.

Table 21.2: Changes in the Level of Cholesterol, Triglycerides and Total Protein (Values are mean ± SD from 6 rats in each group)

Groups	*Cholesterol*	*Triglycerides*	*Total Protein*
Control	81.32 ± 7.68	72.36 ± 5.46	6.9 ± 0.20
Nicotine	154.32 ± 9.78	98.42 ± 7.26	6.1 ± 0.27
N + EO (250mg/b.wt)	87.64 ± 6.72	77.62 ± 6.48	6.3 ± 0.33
N + EO (500mg/b.wt)	82.62 ± 8.64	73.56 ± 8.23	6.7 ± 0.20

N: Nicotine; EO: *Emblica officinalis.*

Table 21.3: Changes in the Body Weight of Control and Experimental Animals (in grams) (Values are mean ± SD from 6 rats in each group)

Days	*Control*	*Nicotine*	*N+EO (250mg/b.wt)*	*N+EO (500mg/b.wt)*
1st day	165.33 ± 1.28	159.00 ± 1.78	154.23 ± 1.56	150.23 ± 1.81
32nd day	186.45 ± 1.03	98.00 ± 3.06	98.34 ± 4.23	96.42 ± 1.28
39th day	201.23 ± 1.08	–	132.42 ± 3.64	144.46 ± 2.1

N: Nicotine; EO: *Emblica officinalis.*

Discussion

Nicotine, the major component of cigarette smoke causes oxidative damage to liver, kidney, lungs, brain and heart; it is a potential oxidant, which is capable of producing free radicals and reactive oxygen species (Yildiz, *et al.*, 1998). The nicotine induced free radicals react with biomembranes causing oxidative destruction of polyunsaturated fatty acids and forming cytotoxic aldehydes by lipid peroxidation (Yildiz, *et al.*, 1999). Lipid peroxidation has been implicated in pathogenesis of a number of diseases (Morel, *et al.*, 1970). An increase in the activity of marker enzymes like ALT, AST, ALP in nicotine treated rats indicates tissue damage. This increase in the ALP activities is the indicative of cellular damage due to loss of functional integrity of cell membranes (Wetscher, *et al.*, 1995). Oxidative damage to the cellular membranes produce marked changes in molecular organization of lipids resulting in increased membrane permeability. Further GGT levels decreased in the serum of nicotine induced rats.

Lipid peroxidation can be used as an index for measuring the extent of damage that occurs in membranes of tissues as a result of free radical generation (Jason, *et al.*, 1998). Lipid peroxidation was enhanced in liver of the nicotine treated rats. Nicotine is oxidized primarily in to its metabolite cotine in liver, generates free radicals in tissues and induces oxidative tissue injury (Wang, *et al.*, 2000). Thus, the damage to the tissues in the nicotine treated animals may be due to the excessive generation of free radicals.

In the present study the cholesterol level was elevated in the nicotine treated animals. The prevalence of hypercholesterolemia and triglyceridemia has been reported in heavy smokers (Masora, 1977). The increased level of cholesterol is attributed to the increased activity of 3-hydroxy-3-glutaryl coA reductase (HMG-coA reductase) and increased incorporation of labeled acetate in to cholesterol (Brunzell, *et al.*, 1983). Nicotine decreased the activity of lipoprotein lipase resulting in elevated levels of triglycerides. This enzyme involved in the uptake of circulating triglycerides rich lipoprotein, chylomicrons or VLDL by the extra hepatic tissues (Huttunen, *et al.*, 1976). Chromaffin cells of adrenal medulla synthesize catecholamines by the stimulation of nicotine and adipose tissue lipolysis is carried out by catacholamines, which in turn elevates the levels of cholesterols, triglycerides and also increases fatty acids (Cryer, 1981).

Total protein content is slight decreased in nicotine treated rats that were found to recover on treatment with *Emblica officinalis*. This might be due to phytochemical compound present in the natural products in reducing or detoxifying the effect of nicotine. According to Bandyopadhyay *et al.* (1999), total protein serve as a source of nutrition for the tissue and the liver function. The site specific oxidative damage of some of the susceptible amino acids is now regarded as the major cause of metabolic dysfunction during pathogenesis. Extensive liver injury may lead to decreased blood levels of total proteins synthesized exclusively by hepatocytes.

Emblica officinalis is rich in Vitamin C compound and antioxidant activity of the fruit (Scartezzini, *et al.*, 2005). The antioxidant property of *Emblica officinalis* extract are many times more than that of water soluble vitamin E (Maulik *et al.*, 1996, 1997) *Emblica officinalis* is a constituent of various liver tonics used against acute viral hepatitis and other liver disorders (Antarkar *et al.*, 1980; Handa *et al.*, 1986). Antioxidants such as vitamin E ellagic acid (Thresiamma and Kuttan, 1996) and curcumin (Nishigaki *et al.*, 1992) have been reported to protect liver injury and fibrosis induced by hepatotoxins. The hepatoprotective effect of *Emblica officinalis* extracts were related mostly to their reported antioxidant properties (Jeena and Kuttan, 1995). *Emblica officinalis* fruit extract neutralizes the oxidizing potentials of reactive oxygen species generated thereby maintaining cell membrane integrity and viability.

The phytochemical analysis of the *Emblica officinalis* fruit revealed the presence of saponins, tannins, anthraquinones, coumarins, sterols and/or triterpenes. Additionally, saponins, and tannins are known to affect the integrity of mucus membranes (Oliver, 1960). Tannins also being stringent may have precipitated micro proteins on the site of ulcer thereby forming an impervious protective pellicle over the lining to prevent absorption of toxic substances and resist the attack of proteolytic enzymes (John and Onabanjo, 1990; Nwafor *et al.*, 1996; Nwafor *et al.*, 2000). Thus, the present study shows that *Emblica officinalis is* an effective scavenger of free radicals. This property helps to protect the liver from nicotine induced toxicity.

Nicotine, an active ingredient in tobacco is known to influence body weight (Frankhan and cabanac, 2003). In the present study the initial body weight of the animals was 160-180 g. treatment with nicotine decreased the body weight. The weight loss during nicotine treatment may be due to less food intake, stimulation of the metabolic rate (Frankish, *et al.*, 1995), activation of lipoprotein lipase (Winders and Grunberg, 1990) and suppression of glycolysis (Maritz, 2003).

Supplementations of *Emblica officinalis* at a dose of 500 mg/kg body weight showed a significant improvement in body weight. This may be due to antioxidant property of *Emblica officinalis.* Thus, *Emblica officinalis* showed its protective nature against nicotine induced toxicity on rat liver.

References

Abesundara, K.J.M., Matsui, T., Matsumoto, K., 2004. a-glucosidaseinhibitory activity of some Sri Lanka plant extracts, one of which, *Cassia auriculata*, exerts a strong anti hyperglycemic effect in rats comparable to the therapeutic drug acarbose. Journal of Agricultural and Food Chemistry 52, 2541–2545.

Anila, L., Vijayalakshmi, N.R., 2000. Beneficial effects of flavonoids from Sesamum indicum, Emblica officinalis and Momordica charantia. Phytotherapy Research 14, 1–4.

Anila, L., Vijayalakshmi, N.R., 2002. Flavonoids from Emblica officinalis and Mangifera indica– effectiveness for dyslipidemia. Journal of Ethnopharmacology 79, 81–87.

Anila, L., Vijayalakshmi, N.R., 2003. Antioxidant action of flavonoids from Mangifera indica and Emblica officinalis in hypercholesterolemic rats. Food Chemistry 83, 569–574.

Antarkar, D.S., Ashok, B.V., Doshi, J.C., Athavale, A.V., Vinchoo, K.S., Natekar, M.R., Thathed, P.S., Ramesh, V., Kale, N., 1980. Doublebind clinical trial of Arogyawardhani–an ayurvedic drug in acute viral hepatitis. Indian Journal of Medical Research 72, 588–593.

Ashakumary, L. and Vijayammal, P. L. Effect of nicotine on lipoprotein metabolism in rats. *Lipids, 32*:311-5, 1997.

Asmawi, M.Z., Kankaanranta, H., Moilanen, E., Vapaatalo, H., 1993. Anti-inflammatory activities of Emblica officinalis Gaertn. Leaf extracts. Journal of Pharmacy and Pharmacology 45, 581–584.

Bafna, P.A., Balaraman, R., 2004. Anti-ulcer and antioxidant activity of DHC-1, a herbal formulation. Journal of Ethnopharmacology 90, 123–127.

Balfour, D.; Benowitz, N.; Fagerstrom, K.; Kunze, M. and Keil, U. Diagnosis and treatment of nicotine dependence with emphasis on nicotine replacement therapy. A status report. *Eur. Heart J., 21*:438-45, 2000.

Baron, J. A. Beneficial effects of nicotine and cigarette smoking: the real, the possible and the spurious. *Br. Med. Bull., 52*:58-73, 1996.

Benowitz, N. L. and Gourlay, S. G. Cardiovascular toxicity of nicotine: implications for nicotine replacement therapy. *J. Am. Coll. Cardiol., 29*:1422-31, 1997.

Benowitz, N. L.; Hansson, A. and Jacob, P., 3rd. Cardiovascular effects of nasal and transdermal nicotine and cigarette smoking. *Hypertension, 39*:1107-12, 2002.

Benowitz, N. L.; Hansson, A. and Jacob, P., 3rd. Cardiovascular effects of nasal and transdermal nicotine and cigarette smoking. *Hypertension, 39*:1107-12, 2002.

Bergmeyar,H.U., Bowers,G.N.Jr., 1980. IFCC 9International Federation of Clinical Chemistry) methods for the measurement of catalytic concentration of enzymes. Part 3. IFCC method for alanine aminotransferase. Journal of Clinical Chemistry 24, 720-721.

Bessey, O.A., Lowery, O.H., Brock, M.J., 1946. A method for the rapid determination of alkalin phosphatase with five cubic millimeters of serum.Journal of Biological Chemistry 164, 321-329.

Birtwistle, J. and Hall, K. Does nicotine have beneficial effects in the treatment of certain diseases? *Br. J. Nurs., 5*: 195- 202, 1996.

Biswas, S., Talukder, G., Sharma, A., 1999. Protection against cytotoxic effects of arsenic by dietary supplementation with crude extract of Emblica officinalis fruit. Phytotherapy Research 13, 513–516.

Brunzell J D, Miller N E, Alaupovic P, St. Hilaire R J and Wang C S, Familial Chylomicronemia due to circulation inhibitor of lipoprotein lipase activity, J Lipid Res,24 (19830 12.

CDC (Centers for Disease Control and Prevention). State specific prevalence of current cigarette smoking among adults and the proportion of adults who work in a smoke free environment–United States, 1999. *MMWR. 2000:49*:978-82, 2000.

Cooke, J. P. and Bitterman, H. Nicotine and angiogenesis: a new paradigm for tobacco-related diseases. *Ann. Med., 36*:33-40, 2004.

Cryer P E, Disease of the adrenal medulla and sympathetic nervous system, Clin Sci Mol Med, 50 (1976) 249.

De, S., Ravishankar, B., Bhavsar, G.C., 1993. Plants with hepatoprotective activity–a review. Indian Drugs 30, 355–363.

Dhir, H., Roy, R.K., Sharma, A., Talukdar, G., 1990. Modification of clastogenicity of lead and aluminium in mouse bone marrow cells by dietary ingestion of Phyllanthus emblica fruit extract. Mutation Research 241, 305–312.

Dutta, B.K., Rahman, I., Das, T.K., 1998. Antifungal activity of Indian plant extracts. Mycoses 41, 535–536.

Ghosal S, Tripatty VK, Chauhan S (1996) Active constituents of *Emblica officinalis:* I: the chemistry and antioxidative effect of two new hydrolysable tannins, Emblicannin A and B. Indian Journal of Chemistry 35B: 941-948

Godbole, S.H., Pendse, G.S., 1960. Antibacterial property of some plants. Indian Journal of Pharmacy 22, 39–42.

Gryglewski, R. Participation of nicotine in damage caused by addiction to tobacco smoking. *Przegl. Lek.*, 24:514-6, 1968.

Jacob, A., Pandey, M., Kapoor, S., Saroja, R., 1988. Effect of the Indian gooseberry (amla) on serum cholesterol levels in men aged 35–55 years. European Journal of Clinical Nutrition 42, 939–944.

Jason D M, Balz F, Atkinson W L, Michael J G, Sean M L, Yu S, William E S, John A O and Jackson R L, Increase in circulating products of lipid peroxidation F2-Isoprostanes in smokers – Smoking as a cause of oxidative damage, The New England J Med, 18 (1998) 332.

Jeena KJ, Kuttan R (1995) Antioxidant activity of *Emblica officinalis.* Journal of Clinical Biochemistry and Nutrition 19: 63-70

John TA, Onabanjo AO (1990) Gastroprotective effects of an aqueous extract of *Entandrophragma utile* bark in experimental ethanol-induced peptic ulceration in mice and rats. Journal of Ethnopharmacology 29: 87-93

Jose, J.K., Kuttan, R., 1995. Antioxidant activity of *Emblica officinalis.* Journal of Clinical Biochemistry Nutrition 19, 63–70.

Kalpana, C.; Rajasekharan, K. N. and Menon, V. P. Modulatory effects of curcumin and curcumin analog on circulatory lipid profiles during nicotine-induced toxicity in Wistar rats. *J. Med. Food, 8*:246-50, 2005.

Kavitharaj, N. K. and Vijayammal, P. L. Nicotine administration induced changes in the gonadal functions in male rats. *Pharmacology, 58*:2-7, 1999.

Lampronti, I., Khan, M.T.H., Bianchi, N., Borgatti, M., Gambar, R., 2004. Inhibitory effects of medicinal plant extracts on interactions between DNA and transcription factors involved in inflammation. Minerva Biotecnologica 16, 93–99.

Liu, R. H.; Kurose, T. and Matsukura, S. Oral nicotine administration decreases tumor necrosis factor-alpha expression in fat tissues in obese rats. *Metabolism, 50*:79-85, 2001.

Masora E J, Lipids and lipid metabolism, Annu Rev Physiol,39 (1977) 301.

Mathur, R., Sharma, A., Dixit, V.P., Varma, M., 1996. Hypolipidaemic effect of fruit juice of Emblica officinalis in cholesterol fed rabbits. Journal of Ethnopharmacology 50, 61–68.

Moral d W, Hasler J and Clistom G M, Low density lipoprotein cytotoxicity indused by free radicals peroxidation of lipids, Lipids, 24 (1963) 1970.

Nakajima, M.; Yamamoto, T.; Nunoya, K.; Yokoi, T.; Nagashima, K.; Inoue, K.; Funae, Y.; Shimada, N.; Kamataki, T. and Kuroiwa, Y. Role of human cytochrome P4502A6 in C-oxidation of nicotine. *Drug Metab. Dispos., 24*:1212-7, 1996.

Nakajima, M.; Yamamoto, T.; Nunoya, K.; Yokoi, T.; Nagashima, K.; Inoue, K.; Funae, Y.; Shimada, N.; Kamataki, T. and Kuroiwa, Y. Role of human cytochrome P4502A6 in C-oxidation of nicotine. *Drug Metab. Dispos., 24*:1212-7, 1996.

Nwafor PA, Effraim KD, Jacks TW (1996) Gastroprotective effects of aqueous extract of *Khaya senegalensis* bark on Indomethacin-induced ulceration in rats. West African Journal of Pharmacology and Drug Research 12: 46-50

Nwafor PA, Okwuasaba FK, Binda LG (2000) Antidiarrhoeal and antiulcerogenic effects of methanolic extract of *Asparagus pubescens* root in rats. Journal of Ethnopharmacology 72: 4 2 1 ^ 27

Ochiai, Y.; Sakurai, E.; Nomura, A.; Itoh, K. and Tanaka, Y. Metabolism of nicotine in rat lung microvascular endothelial cells. *J. Pharm. Pharmacol.,. 58*:403-7, 2006.

Oliver B (1960) Medicinal Plants in Nigeria. Nigeria College of Arts and Science and Technology. Ibadan. 358

Panda, S., Kar, A., 2003. Fruit extract of *Emblica officinalis* ameliorates hyperthyroidism and hepatic lipid peroxidation in mice. Pharmazie 58, 753–761.

Pausova, Z.; Paus, T.; Sedova, L. and Berube, J. Prenatal exposure to nicotine modifies kidney weight and blood pressure in genetically susceptible rats: a case of geneenvironment interaction. *Kidney Int., 64*:829-35, 2003.

Perianayagam, J.B., Sharma, S.K., Joseph, A., Christina, A.J.M., 2004. Evaluation of anti-pyretic and analgesic activity of *Emblica officinalis* Gaertn. Journal of Ethnopharmacology 95, 83–85

Perlemuter, G.; Davit-Spraul, A.; Cosson, C.; Conti, M.; Bigorgne, A.; Paradis, V.; Corre, M. P.; Prat, L.; Kuoch, V.; Basdevant, A.; Pelletier, G.; Oppert, J. M. and Buffet, C. Increase in liver antioxidant enzyme activities in non-alcoholic fatty liver disease. *Liver Int., 25*:946-53, 2005.

Price, R. J.; Renwick, A. B.; Walters, D. G.; Young, P. J. and Lake, B. G. Metabolism of nicotine and induction of CYP1A forms in precision-cut rat liver and lung slices. Toxicol. *In Vitro, 18*:179-85, 2004.

Rani, P., Khullar, N., 2004. Antimicrobial evaluation of some medicinal plants for their anti-enteric potential against multi-drug resistant *Salmonella typhi*. Phytotherapy Research 18, 670–673.

Sabu, M.C., Kuttan, R., 2002. Anti-diabetic activity of medicinal plants and its relationship with their antioxidant property. Journal of Ethnopharmacology 81, 155–160.

Samuel Santos Valenca, Lucas Gouveia, Wagner Alves Pimenta and Luís Cristovao Porto2008. Effects of Oral Nicotine on Rat Liver Stereology *Int. J. Morphol., 26(3)*:1013-1022.

Senthilkumar, R.; Viswanathan, P. and Nalini, N. Effect of glycine on oxidative stress in rats with alcohol induced liver injury. *Pharmazie, 59*:55-60, 2004.

Thakur, C.P., Thakur, B., Singh, B., Singh, S., Sinha, P.K., Sinha, S.K., 1988. The Ayurvedic medicines, Haritaki, Amla and Bahira reduce cholesterol induced atherosclerosis in rabbits. Indian Journal of Cardiology 21, 167–175.

Torel, J., Cillard, J., Cillard, P., 1986. Antioxidant activity of avonoids and reactivity with peroxy radical. Phytochemistry 25, 383–385.

Trushin, N. and Hecht, S. S. Stereo selective metabolism of nicotine and tobacco-specific N-nitrosamines to 4- hydroxy-4-(3-pyridyl) butanoic acid in rats. *Chem. Res. Toxicol., 12*:164-71, 1999.

Trushin, N. and Hecht, S. S. Stereoselective metabolism of nicotine and tobacco-specific N-nitrosamines to 4- hydroxy-4-(3-pyridyl)butanoic acid in rats. *Chem. Res. Toxicol., 12*:164-71, 1999.

Valenca, S. S.; de Souza da Fonseca, A.; da Hora, K.; Santos, R. and Porto, L. C. Lung morphometry and MMP-12 expression in rats treated with intraperitoneal nicotine. *Exp. Toxicol. Pathol., 55*:393-400, 2004.

Wang H, Ma L, Li Y and Cho C H, Exposure to cigarette smoke increase apoptosis in the gastric mucosa through a reactive active species mediated and p-53 independent pathway, Free Radic Biol Med, 28 (2000) 1295.

Wang, D.; Wei, Y. and Pagliassotti, M. J. Saturated fatty acids promote endoplasmic reticulum stress and liver injury in rats with hepatic steatosis. *Endocrinology, 147*:943- 51, 2006.

Wang, S. L.; He, X. Y. and Hong, J. Y. Human cytochrome p450 2s1: lack of activity in the metabolic activation of several cigarette smoke carcinogens and in the metabolism of nicotine. *Drug Metab. Dispos., 33*:336- 40, 2005.

Westman, E. C.; Levin, E. D. and Rose, J. E. Nicotine as a therapeutic drug. *N. C. Med. J., 56*:48-51, 1995.

Wetscher G J, Bagchi D, Perdikis G, Redmond E J, Hinder P R, Glaser L and Hinder R A, in vitro free radicals production in rat oesophageal mucosa by nicotine, Digest Dist Sci, 40 (1995) 853.

Yildiz D, Ercal N and Armstrong D W, Nicotine enantiomers and oxidative stress, Toxicology, 130 (1988) 155.

Yildiz D, Liu Y S, Ercal N and Armstrong D W,Comparison of pure nicotine and smokeless tobacco extract indused oxidative stress. Arch Environ Toxicol, 37 (1999)434.

2013, Environmental Biotechnology *Pages* **205–209**
Editors: **D.R. Khanna, A.K. Chopra, Gagan Matta, Vikas Singh & Rakesh Bhutiani**
Published by: **BIOTECH BOOKS, NEW DELHI**

Chapter 22

Antihyperlipidemic Activity of Celery Leaves of Rats

Renugopal Perumalraja and S. Dawood Sharief
School of Environmental Science, PG and Research Department of Zoology, The New College, Chennai – 14

India has about 45000 plant species; while medicinal properties have been assigned to several thousand. *Apium graveolens* (Wild Celery) belonging to family Apiaceae is an alimentary herb, used as a decorative herb in India in general and more particular in Punjab and Uttar Pradesh. Celery roots, leaves and seeds are used for the therapeutic purpose in treating and preventing many diseases.

In the present study the effect of celery on cholesterol induced rats was studied. Celery extract was orally administered by post treatment at the dose level of 100mg and 200mg/kg of body weight for 7 days. Administration of cholesterol induced rats resulted in increased total cholesterol, triglycerides, VLDL and decreased the HDL cholesterol. Administration of *celery* extract resulted in HDL cholesterol increase and decrease in the total cholesterol, triglycerides and VLDL in dose dependent manner. The above findings proved that 200mg/kg of body weight of celery was found to have more protective effect on lipid profiles in cholesterol induced rats. The results of the above studies are discussed in the light of recent literature.

Keywords: *Apium graveolens, Cholesterol, Triglyceride, LDL, HDL.*

Introduction

In recent times, focus on plant research has increased all over the world and large body evidence has corrected to immense potential of medicinal plants used in various traditional system. India has about 45,000 plant species; while medicinal properties have been assigned to several thousand. *Celery* (*Apium Graveolens*) belonging to family Apiaceae. Review of the literature indicates that *celery* has been cultivated for the last 3,000 years (Momin and Nair 2001). It is rich source of Vitamin C and dietry fibre, potassium, folate, manganese, vitamin B6, calcium, vitaminB1, vitamin B2, magnesium, vitamin A, phosphorus and iron were also present (Mitra *et al.*, 2001).

According to Zahra *et al.* (2011), celery consists of twenty two volatile compounds. It has been extensively studied for its biological activities. Aqueous extract of *celery* stem caused significant reduction in serum total cholesterol level in hypercholesterolemic rats (Tsi and Tan, 2000). Nitrogenous compounds from essential oil of *celery* seed have been reported to have effect on the central nervous system (Al Hindawi *et al.*, 1989). *Celery* posses anti-inflammatory effect (Momin and Nair 2002), and celery juice showed protective effect when applied with doxorubicin (Kolarovic *et al.*, 2009). It is an aromatic biennial herb, almost the whole plant is used, including the roots, seeds, leaves, and oil. It is a bitter herb with a pleasant smell that relieves indigestion, reduces inflammation, and acts as a mild diuretic (Newall *et al.*, 1996). In Germany, *celery* preparations are used to treat loss of appetite, general and nervous exhaustion (Wren, 1988). Some pharmacological effects of *celery* have been reported, such as vasodilatory action in rat thoracic aorta (Ko *et al.*, 1991) and mosquito repellent (Tuetun *et al.*, 2005). Sultana *et al.* (2005) reported that celery is a potent plant against experimentally induced hepatocarcinogenesis in Wistar rats. In the present study the effect of *celery* leaves extract on cholesterol induced toxicity in rats was studied.

Materials and Methods

Plant Processing

Celery was purchased from the local market (Chennai). The celery leaves were shade dried, finely powdered and subjected to soxhlet extraction for 30 hours using methanol as solvent. The extract was filtered and concentrated to dryness under low temperature (40°C) and reduce pressure. The extract was used for the study.

Cholesterol

Cholesterol was purchased from Sisco research laboratories Pvt. Ltd. Bombay, India.

Animal Model

Animals were housed, fed and treated in accordance with the in house guidelines for animal protection to minimize pain and discomfort. Adult male wistar rats weighing about 180 - 200g each were used throughout the study. The animals were left for seven days to adapt to the room conditions (temperature, humidity, light and dark period, aeration, and caging).

Treatment

Male wistar rats were divided into four groups (n = 6). Group I served as control, administered with vehicle only (2ml of hydrogenated groundnut oil). Group II cholesterol control where in 400mg/kg body wt of cholesterol in 2ml of hydrogenated groundnut oil was administered for fourteen days, The treatment of Group III and IV rats were fed with 1ml of the *celery* extract for after seven consecutive days at doses of 100 and 200 mg/kg body wt respectively.

Collection of Blood

At the end of the experiment after 12-14 h of fasting blood samples were drawn from rats in plain tube, allowed to clot and were centrifuged to obtain serum. The serum was stored -20°C for biochemical analysis.

Determination of Serum Lipid Profile

Serum lipid profile, including: total cholesterol (TC) and triglycerides (TG) were calorimetrically determined (Allain *et al.*, 1974 and Wahlefeld 1974); high-density lipoprotein cholesterol (HDL-c) was calorimetrically determined, low-density lipoprotein cholesterol (LDL-c), and very low-density lipoprotein cholesterol (VLDL-c) were mathematically calculated (Richmond 1973).

Results and Discussion

Herbal preparation has been used in many parts of the world since ancient times. In recent years, their popular alternative to modern medicine has increased considerably even in developing countries (Maurya *et al.*, 2004).

On feeding normal rat with cholesterol significantly increases the level of Total cholesterol, TGL-c, LDL-c, VLDL-c and decrease the HDL-c. On feeding celery it increased the HDL-c and at the same time it remarkably decreases the Total cholesterol, TGL-c, LDL-c and VLDL-c when compared to the control experimental animal. Better results were obtained when the animals were fed with 200 mg/kg celery than the 100 mg/kg celery treated animal (Table 22.1 and Figure 22.1). The results suggest that the lipid lowering action of this natural product maybe reduction of lipid absorption in the intestine.

Table 22.1: Effect of Treatment with Methonolic Extract of Celery on Serum Lipid Profile Concentration in Rats

Parameters	*Control*	*Cholesterol Control*	*Celery (100 mg/kg)*	*Celery (200 mg/kg)*
Total cholesterol mg	71.0±1.41	86.25±1.50	77.75±4.42	61.5±2.88
Triglycerides mg	73.5±0.70	106.75±3.30	72.25±5.05	60.75±5.25
HDL cholesterol mg	22±1.43	21.25±2.5	32.5±2.38	37.0±1.82
LDL cholesterol mg	34±2.82	43.5±2.30	30.5±5.06	11.75±2.98
VLDL cholesterol mg	14±0.00	21±0.81	14±1.15	11.75±0.95

The exact mechanism by which the plant extract induce weight loss is not well known. However, several studies have shown that agents could cause body weight reduction through several proposed mechanisms. These include: stimulation of the mobilization, inhibition of lipoproteinslipase activity, increasing energy expenditure,

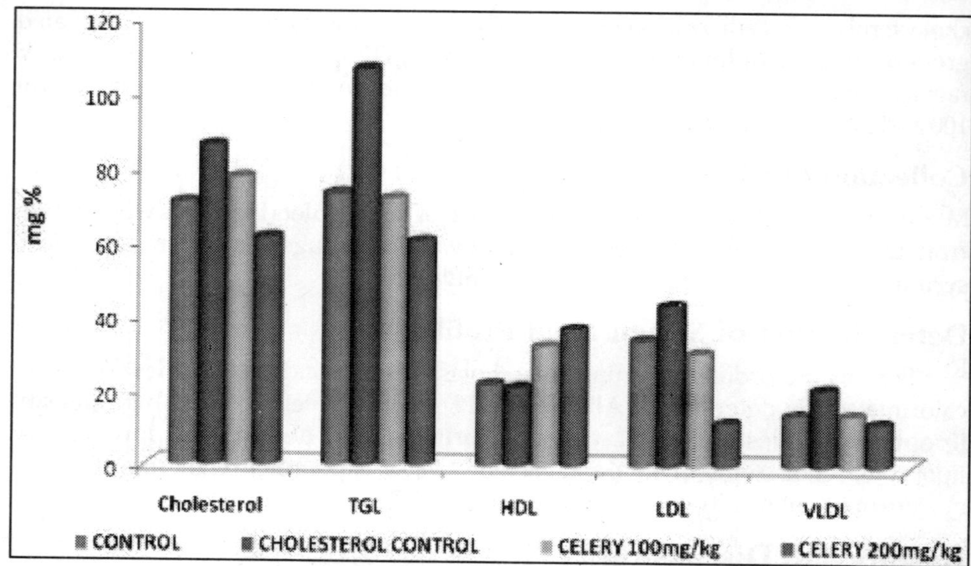

Figure 22.1: Effect of Treatment with Methonolic Extract of Celery on Serum Lipid Profile Concentration in Rats

inhibition of absorption of nutrients from the gastrointestinal tract, suppression of the appetite, and reduction of food intake (Dyer, 1994: Angelica, 1998).

The preliminary study shows that *celery* has antihyperlipidemic effect and could be of value in reducing serum Total cholesterol, Triglyceride, LDL-c, VLDL-c and increasing HDL-c.

References

Al Hindawi M, Al Deen I, Nabi M, Ismail M. Antiinflamatory activity of some Iraqi plants using intact rats. J Ethnopharmacol. 1989; 26:163-168.

Allain CC, Poon LS, Chan CS. Enzymatic determination of serum total cholesterol. Clin Chem. 1974; 20: 470–5.

Angelica L. Hormonal regulation of appetite and food intake. *Annals of Medicine.* 1998; **30:** 7 - 20.

Dyer R. Traditional treatment of obesity. *Baillieres Clinical Endocrinology and Metabolism.* 1994; 8: 661 – 688.

Ko FN, Huang, TH, Teng, CM. Vascodilatory action mechanisms of Apigenin isolated from *A. graveolens* in rat thoracic aorta. *Biochemical and Biophysical Acta.* 1991; **1115:** 69 - 74.

Kolarovic J, Popovic M, Mikov M, Mitic R, Gvozdenovic Lj. Protective effects of celery juice in treatments with doxorubicin. Molecules. 2009; 14:1627-1638.

Maurya R, Srivastavas S, Kalshreshtha DK, Gupta CM. Traditional remedies for fertility regulation. *Curr. Med. Chem.* 2004; **11:** 1350 - 1431.

Mitra SK, Venkataranganna MV, Gopumadhavn S, Anturlikar SD, Seshadri, Udupa UV. The protective effect of HD-03 in CCL4- induced hepatic encephalopathy in rats. Phytother Res. 2001; 15:493–6.

Momin R, Nair G. Mosquitocidal, Nematicidal and Antifungal Compounds from Apium graveolens L. Seeds J Agric Food Chem. 2001; 49:142–145.

Momin RA, Nair MG. Antioxidant, cyclooxygenase and topoisomerase inhibitory compounds form *A. graveolens* seeds, *Phytochemistry*. 2002; **9:** 312 - 318.

Newall CA, Anderson LA, Phillipson JD. *Herbal Medicines: A Guide for Health-Care Professionals*. London, Pharmaceutical Press. 1996; pp. 65–66.

Richmond N. Colorimetric determination of total cholesterol and high density lipoprotein cholesterol (HDL-c) Clin Chem. 1973; 19: 1350–6.

Sultana S, Ahmad S, Jahangir T, Sharma S. Inhibitory effect of Celery seeds extract on chemically induced hepatocarcinogenes modulation of all proliferation, metabolism and altered hepatic foci development. *Cancer Letters*. 2005; 221:11 - 20.

Tsi D, Tan B. The mechanism underlying the hypocholesterolaemic activity of aqueous celery extract, its butanol and aqueous fractions in genetically hipercholesterolaemic rico rats. Life Sci. 2000; 66:755-767.

Tuetun B, Choochote W, Rattanachan P, Chain E, Chaithong U, Jitpakdi A, Tippawangkosol P, Riyong D, Pitasawat B. Mosquito repellency of the seeds of Celery (*A. graveolens* L.). *Ann Trop Med Parasitol*. 2004; **98:** 217 - 407.

Wahlefeld AW. "Methods of Enzymatic Analysis". Academic Press. 1974. pp. 1831–5. Chapter 5.

Wren RC Walden, C.W. Daniel. *Potter's New Cyclopaedia of Botanical Drugs and Preparations*. Saffron Pharmaceutical Biology Downloaded from informahealthcare.com by University of Col. 1988; pp. 1–384.

Zahra Asadiyeh Shojaei, Ataaollah Ebrahimi and Mandana Salimi. Chemical composition of three ecotypes of wild celery. J. H. Sp. Medi. Plt. 2011; 17:62-68.

2013, Environmental Biotechnology *Pages* ***211–216***
Editors: **D.R. Khanna, A.K. Chopra, Gagan Matta, Vikas Singh & Rakesh Bhutiani**
Published by: **BIOTECH BOOKS, NEW DELHI**

Chapter 23

Pretreatment of *Moringa oleifera* Against Stannous Chloride Toxicity on Male Rats

K. Kothandaraman[1] *and S. Dawood Sharief*[2]
[1]*Ph.D Scholar, School of Environmental Sciences,*
[2]*Associate Professor,*
Department of Zoology, The New College, Chennai – 14

Moringa oleifera Lam (Moringaceae) is a highly valued medicinal plant, distributed in many tropical and subtropical countries. It is generally known in the developing world as a vegetable medicinal plant and a source of vegetable oil. Besides the plant is reported to have various biological activities, including regulation of thyroid hormone, rich antioxidant, antidiabetic, antigastric ulcers, anti-tumor and hypotensive agent. It is also used for treating various diseases, such as inflammation, cardiovascular and liver disease.Taking a clue from the above literature reports, the effect of stannous chloride on mammals is studies because stannous chloride ($SnCl_2$) is widely used in human life in soft drinks, in food manufacturing and biocides preparations. Therefore, the present experiment was carried out to determine the effectiveness of Moringa oleifera in alleviating the toxicity of stannous chloride ($SnCl_2$) on biochemical parameters in male white rats, (Rattus norvegicus) which is an animal model.

Animals were assigned to treat one of four treatment groups. Group-I served as control. Group-II was administrated with stannous chloride (200mg/kg b.wt.for 28 days orally). Group-III and IV were administrated with Moringa oleifera 250mg/kg, 500mg/kg b.wt for 7 days before the administration of stannous chloride. It was observed that stannous chloride significantly decreased, Aspartate transaminase (AST), Alanine transaminase (ALT), Alkaline Phosphatase (ALP), and Acid Phosphatase (ACP) in blood serum, when compared to control groups.

On the other hand Total Cholesterol (TC), Triglycerides (TG), Low-density lipoprotein (LDL), and Very low-density lipoprotein (VLDL) were significantly decreased in blood serum. Animals treated with 250mg/kg b.wt of Moringa oleifera showed slight increase in above components, while animal treated with 500mg/kg the enzyme parameters increased as significantly nearing the control values. Therefore the present results revealed the treatment with Moringa oleifera could minimize the toxic effects of stannous chloride. The results are discussed in the lights of recent literature.

Keywords: *Moringa oleifera, Stannous Chloride, Biochemical parameters, Lipid profile.*

Introduction

Stannous Chloride

Tin compounds have been widely used in various activities. Tin is extensively used in tin plated containers and alloys. Inorganic tin compounds are used in a variety of industrial processes for the strengthening of glass, soft drinks, as a base for colors, as catalysts, as stabilizers in perfumes and soap and a dental cryogenic agent. The stannous ion mainly the stannous chloride (Sncl2) salt form is widely used as a reducing agent to label radiotracers with technetium-99m [(99m) Tc]. These radiotracers can be employed as radiopharmaceuticals in nuclear medicines procedures. In this case there is no doubt about absorption of this complex, because it is intravenously administrated in humans, (Assis *et al.*, 2002). The Sncl2 is capable to promote the generation of reactive oxygen species (ROS) that are responsible for the oxidative stress which can damage DNA (de Mattos *et al.*, 2000: Silva *et al.*, 2002). The stannous chloride was carried out to investigate the alterations in the levels of free radicals and some enzyme activates and in different tissues of male rabbits. (-Demerdash, and Yousef *et al.*, 2005).

Moringa oleifera

Moringa oleifera (Lam) is a rapidly growing tree belonging to the family Moringaceae and popularly known as "arbol de rabano", horseradish tree and drumstick tree (Ramachandran *et al.*, 1980: Kwaambawa and Maikokera, 2007). It is an indigenous tree to northern India and Pakistan. The leaves and fruits of Moringa tree are rich in protein content, carotenoids, minerals and ascorbic acid (Vitamin-C) and also are used as food supplements for human and animals (Makkar and Becker, 1997: Afuang *et al.*, 2003: Dongmeza *et al.*, 2006). In addition to its use as a food supplements. Most of its parts are widely used for many medical applications. Seeds extracts of Moringa have established anti- inflammatory, antispasmodic, and diuretic antioxidant activities (Rao *et al.*, 1999). In many countries root, bark, gum, leaf, fruit (pods), flowers, seed and seed oil have been used for ailments in the indigenous medicine of south Asia, including the treatment of inflammation and infectious diseases along with cardiovascular, gastrointestinal, hematological and hepato renal disorders (The Wealth of India., 1962., Singh and Kumar, 1999: Mormitsn *et al.*, 2000:

Siddhuraju and Beeker, 2003). Moringa oleifera substantially minimizes the toxic effects of some environmental pollutants (Flodin 1990: Salem *et al.*, 2001). Therefore the present study was carried out to know the capability of Moringa oleifera to minimize the harmful effects of stannous chloride on lipid profile and biochemical parameters of rats.

Materials and Methods

Chemicals

Stannous chloride: AR, 97.0 per cent [Tin (11) chloride] ($SnCl_2.2H_2O$) was purchased from S.D fine chem. Ltd. Mumbai – 400 025.

Plant Material

Moringa oleifera(MO) was procured from local market and seeds of Moringa oleifera was separated, shade dried, grounded with mortar and pestle and sieved to get fine powder.

Experimental Animals and Treatment

Adult male albino rats of Wister strain weighing 180-190gms were obtained from the Animal house Tamil Nadu University of Veterinary and Animal Sciences, Chennai. Animals were maintained according to the principle and guidelines of the CPCSEA (No: 324) under the supervision of Animal Ethical Committee in accordance with the Indian National Law on animal care and use. They were fed with commercial pelleted chow obtained from Poultry Research Station Chennai. Soaked Bengal gram and water was provided ad libitum.Animals were assigned to treat one of four treatment groups. Six animals were housed in each group. Group-I served as control fed with animal feed and water was provided ad libitum. Group-II was administrated with Stannous Chloride (200mg/kg b.wt. For 28 days orally). Group-III and IV were administrated with Moringa oleifera 250 mg/kg, 500mg/kg b.wt for 7 days before the administration of Stannous Chloride. The animals were starved prior to the day of the experiment; they were sacrificed by cervical dislocation. Blood samples were collected separately from individuals and stored for further analysis.

Serum Biochemical Parameters

The activities of.Aspartate transaminotransferase.(AST) and alanine transaminotransferase (ALT) activities were determined (Bergmeyar *et al.*, 1980). alkaline phosphatase (ALP) and acid phosphatase (ACP) activity (Bessey *et al.*, 1985) cholesterol and triglycerides (Allain *et al.*, 1993). Low density lipoprotein (LDL) and very low density lipoprotein (VLDL) were determined following the procedure of Warnick *et al.* (1983) and Bergmeyer (1985)

Results

The Treatment with $SnCl_2$ resulted in a significant decrease in serum aspartate amino transferase (AST), alanine trans aminase (ALT), acic and alkaline phosphatase compared to control animals (Table 23.1). The present study indicated that treatment with Moringa oleifera alone had no significant effect on the activities of these enzymes.

while the prescence of Moringa oleifera with $SnCl_2$ countracted its toxic effect.on the other hand serum total cholesterol (TC), triglycerides (TG), low-density lipoprotein (LDL) and Very low density lipoproteins (VLDL) concentrations were significantly increased by $SnCl_2$ treatment as compared with control (Table 23.2). Treatment with Moringa oleifera alone decreased TC,TG, LDL and VLDL. As compared with control group. While Moringa oleifera did not affect the levels of Cholesterol, TG, LDL and VLDL. In addition, Moringa oleifera minimized the hazardous effect of $SnCl_2$ on serum lipids.

Table 23.1: Changes in the Activities of Serum of Male Rats Treated with *Moringa oleifera* (Mo), Stannous Chloride ($SnCl_2$) and MO + $SnCl_2$)

Enzymes (U/l)	*Control*	*$SnCl_2$*	*$SnCl_2$+MO (250mg/kg)*	*$SnCl_2$+MO (5000mg/kg)*	*MO (500mg/kg)*
AST	36.6±6.05	19.6±3.2	27.5±5.2	36.6±1.6	37.8±4.5
ALT	50.5±7.17	39.1±5.8	43.1±2.9	49.8±3.7	54.0±4.8
ALP	108.3±14.7	34.1±5.8	74.1±10.2	88.8±9.2	106.5±10.2
ACP	53.5±1.2	0.5±1.1	2.3±0.4	3.6±0.6	4.7±0.6

Table 23.2: Serum Lipid and Lipoprotein Profile of Male Rats Treated with *Moringa Oleifera* (Mo), Stannous Chloride ($SnCl_2$) and (Mo+$SnCl_2$)

Lipids (mg/dl)	*Control*	*$SnCl_2$*	*$SnCl_2$+MO (250mg/kg)*	*$SnCl_2$+MO (5000mg/kg)*	*MO (500mg/kg)*
Cholesterol	195±10.4	37.0±6.3	121.6±23.1	158.3±14.7	198.3±14.7
Triglycerides	195±18.7	20.8±10.6	118.3±23.1	102.5±10.8	166±16.3
LDL	150±16.7	15.8±26.3	73.3±12.1	111.6±21.3	120.0±14.1
VLDL	11.1±4.5	1.5±1.2	4.1±0.1	5.9±1.1	13.1±4.6

Discusssion

Several soluble enzymes of blood serum have been considered as indicators of the hepatic disfunction and damage. Transaminases (AST and ALT) are important and critical enzymes in the biological processes. The increase in the activities of AST and ALT in plasma of rabbits treated with $SnCl_2$ (Yousuf *et al.*, 2003). It is mainly due to the leakage of these enzymes from the liver cytosol into the blood stream (Navarro *et al.*, 1993). The activity of AST is significantly increases in such cases and escapes to the plasma from injured hepatic cells. In addition, ALT level is of value also indicating the existence of liver diseases, as this enzyme is present in large quantities in the liver. It increases in serum when cellular degeneration or destruction occurs in this organ (Hassoun and Stohs, 1995).

Phosphatases are important and critical enzymes in biological processes, they are responsible for detoxification, metabolisam and biosynthesis of energetic macromolecules for different essential functions. Any interference in these enzymes leads to biochemical impairment and lesions of the tissue and cellular function (Khan *et al.*, 2001). The increases in alkaline and acid phosphatases (ACP and ALP) in plasma caused by $SnCl_2$ are in accordance with the finding of (Yamaguchi *et al.*, 1981) in rats. They reported that the changes in the activities of these enzymes in $SnCl_2$-treated rats were regarded as the biochemical manifestation of the toxic action of inorganic tin. Also,(Ochmanski and Barabasz *et al.*, 2000) reported that stannous chloride may bind to DNA,RNA and inhibit the activities of acid and alkaline phosphates. In addition (Rahman *et al.*, 2000) suggested that increase in the activities of ALP and ACP in plasma might be due to increased permeability of plasma membrane or cellular necrosis and this showed the stress condition of the treated animals (Wang and Zhai *et al.*, 1988).

According to (Yousuf *et al.*, 2004) Ascorbic Acid against $SnCl_2$ decreased the induction of AST,ALT, ACP and ALP and maintained the levels of these enzymes to the normal values. Also, our previous studies showed the protective effects of AA against the harmful effects of aflotoxin B1 and aluminium on the activities of these enzymes. Also Sies and Stahl (1995) reported that AA can protect biomembranes enzyme against per oxidative damage of xenobiotics.

According to Simon and Hudes (2000), vitamin C involved is in the metabolisam of cholesterol to bile acids which may be implications for blood cholesterol levels and the incidence of gallstones.The previous study showed that Ascorbic acid caused in the lipid profile Also, Sahin *et al.* (2002) reported that greater dietary vitamin-C (100 and 200mg/kg of diet) resulted in a normaley in serum triglycerides and cholesterol concentrations. (Kronhausen *et al.*, 1989) monitored the cholesterol levels of people who took 1500mg of vitamin C a day. This was found encouraging in the conversion of cholesterol into bile acids which are then eliminated from the body in the feces. Similarly several animal studies indicated that vitamin-C contributes to this conversion by stimulating an enzyme that regulates to this conversion.The presence of *Moringa oleifera* with $SnCl_2$ induction of Cholesterol, trglycerides, LDL, and VLDL and maintaines the normal values of lipid profile.

Conclusions

From the present study, it can be concluded that use of Moringa oleifera and their combination have the capability to alleviate the harmfull effects of Stannous Chloride.

References

Assis, M.L., De Mattos, J.C., Caceres. M.R., Dantas, F.J., Asad, L.M., Asad, N.R., Bezerra, R.J., Caldeira-de-Araujo, A., Bernardo-Filho, M., 2002. Adaptive response to H2O2 protects against SnCl2 damage; the OxyR system involvement. Biochimie 84(4), 291-294.

Bergmeyar,H.U., Bowers,G.N.Jr., 1980. IFCC 9International Federation of Clinical Chemistry) methods for the measurement of catalytic concentration of enzymes. Part 3. IFCC method for alanine aminotransferase. Journal of Clinical Chemistry 24, 720-721.

Bessey, O.A., Lowery, O.H., Brock, M.J., 1946. A method for the rapid determination of alkalin phosphatase with five cubic millimeters of serum.Journal of Biological Chemistry 164, 321-329.

De Mattos, J.C., Dantas, F.J., Bezerra, R.J., Bernardo-Filho, M., Cabral-Neto, J.B., Lage, C., Leitaqo, A.C., Caldeira-de-Araujo, A., 2000. Damage induced by stannous chloride in plasmid DNA. Toxicology Letters 116, 159-163.

Hassoun, E.A., Stohs, S.J., 1995. Comparative studies on oxidative stress as a metchanism for the fetotoxic of TCDD, endrin and lindane in C57BL/6J and DBA/2J mice. Teratology 51, 186-192.

Khan, P.K., Sinha, S.P., 1996. Ameliorating effect of vitamin C on murine sperm toxicity induced by three pesticides (endosulfan, phosphamidon and mancozeb), Mutagenesis 11, 33-36.

Makkar HPS, Becker K. 1996. Nutritional value and antinutritional components of whole and ethanol extracted *Moringa oleifera* leaves. Anim Feed Sci Technol **63**: 211-2228.

Navaro, C.M., Montila, P.M., Martin, A., Jimenez, J., Utrilla, P.M., 1993. Free radicals scavenger and antihepatotoxic activity of Rosmarinus. Plant Med. 59, 312-314.

Rahman, M.F., Siddiqui, M.K., Jamil, K., 2000. Acid and alkaline phosphatase activities in a novel phosphorothionate (RPR-11) treated male and female rats. Evidence of dose and time-dependent respone. Drug Chem. Toxicol. 23, 497-509.

Ramachandran C, Peter KV, Gopalakrishnan PI. 1980. Drumstick (*Moringa oleifera*): a multipurpose Indian vegetable. Econ Bot34: 276-283.

Rivarola, V.A., Balegno, H.F., 1991. Effects of 2,4-dichlorophenoacetic acid on polyamine synthesis in Chinese hamster ovary cells. Toxicol. Lett. 56, 151-157.

Sahin, K., Kucuk, O., Sahin, N., Sari, M., 2002. Effects of vitamin C and Vitamin E on lipid peroxidation status, serum hormone, metabolite. And mineral concentrations of Japanese quails reared under heat stree (34 degreees C). Int. J. Vitam. Nutr. Res. 72, 91-100.

Silva, C.R., Oliveira, M.B., Melo, S.F., Dantas, F.J., de Mattos, J.C., Bezerra, R.J., Caldeira-de-Araujo, A., Duatti, A., Bernardo-Filho, M., 2002. Biological effects of stannous chloride, a substance that can produce stimulation or depression of the central nervous system. Brain Research Bulletin 59, 213-216.

Singh KK, Kumar K. 1999. Ethnotherapeutics of some medicinal plants used as antipyretic agent among the tribals of India. J Econ Taxon Bot 23: 135-141.

Yamaguchi, M., Sugii, K., Okada, S., 1981. Changes in mineral composition and its related enzyme activity in the femur of rats orally administered stannous chloride. J. Pharmacobiodyn. 4, 874-878.

Yousef, M.I., 2004. Aluminium induced changes in hemato biochemical parameters, lipid peroxidation and enzyme activities of male rabbits: Protective role of ascorbic acid. Toxicology 199 (1), 47-57.

2013, Environmental Biotechnology *Pages* ***217–223***
Editors: **D.R. Khanna, A.K. Chopra, Gagan Matta, Vikas Singh & Rakesh Bhutiani**
Published by: **BIOTECH BOOKS, NEW DELHI**

Chapter 24

Potential of Vermicompost from Sugarcane Trash and Cow-Dung (1:1) on Growth and Yield of *Vicia faba* L.

Anil Gupta and R.K.Rampal

Department Environmental Sciences,
University of Jammu, Jammu, J&K

In present studies, an attempt has been made to determine the impact of vermicompost produced from Sugarcane trash and cow-dung (1:1) on the growth and yield of *Vicia faba* L. Sugarcane trash is the waste generated after extracting juice from Sugarcane. The critical analysis of data after statistical evaluation of Pearson co-efficient of correlation (r) revealed that increase in vermicompost concentration exhibited significant ($p<0.05$) positive correlation (r) between vermicompost concentration and most of the growth parameters of *Vicia faba* L.

Keywords; Vermicompost, Sugarcane trash, cow-dung, Vicia faba

Introduction

Vermitechnology is the process by which biological degradation of organic wastes takes place in controlled conditions due to earthworms feeding on the materials.

Vermitechnology is comprised of Vermiculture- the science of breeding and rearing earthworms and vermicomposting- the bioconversion of organic waste materials into organic fertilizer through earthworm consumption (Gupta and Dwivedi, 2001). Moreover vermicompost contains higher level of C, N, P, K, Na, Mg and Ca than that in the surrounding soil. It also contains trace elements such as Manganese (Mn), Copper (Cu), Boron (B) and Molybdenum (Mo) which are essential for growth of crop plants. In addition, if the garbage contains cattle dung and human excreta, it may contribute to the presence of soluble sugars, amino-acids, nucleic acids and enzymes such as proteinase, lipase, urease, phosphatase and invertase which can improve the soil's nutrient status and dehydrogenase and ATPase which can improve the aeration in the soil and the energy status of the soil to which this compost is applied (Sinha and Sinha, 2000).

In this paper, an attempt has been made to determine the potential of vermicompost produced from Sugarcane trash and cow-dung (1:1) on the growth and yield of *Vicia faba* L. Sugarcane trash is the waste generated after extracting juice from Sugarcane. Sugarcane juice is taken as a soft drink particularly in summer. In Jammu large number of sellers are involved in Sugarcane juice selling and thus producing a lot of Sugarcane Trash which is either burnt or thrown into rivers or becomes breeding ground for flies, mites and microbes. If this is used as a raw material for production of Vermicompost, it not only abates pollution load of this waste but also leads to conservation of resources.

Materials and Methods

Vermicompost produced from Sugarcane trash and Cow-dung (1:1) by the three species of earthworm namely *Metaphire posthuma* (Vaillant), *M.houlleti*(Perrier) and *Amynthas morrisi*(Beddard) was mixed together to determine its potential on the quantitative morphological features of broad bean *i.e. Vicia faba* L. Certified seeds of *Vicia faba* L. were procured from the local market of Jammu city and after soaking for 24 hours seeds (@ 10 per polythene) were sown at a depth of 1cm – 2 cm in polythenes bags (each measuring 22 cm in diameter, 29 cm in depth) at an equal distance from one another. Five sets each comprising of 3 polythenes was installed. One set comprising of 3 polythenes was separated as *Control Set* with only 100 per cent soil. Four sets each comprising of 3 polythenes were designated as *Set I* with 25 per cent Vermicompost (Sugarcane trash and Cow-dung) + 75 per cent Soil, *Set II* with 50 per cent Vermicompost (Sugarcane trash and Cow-dung) + 50 per cent Soil, *Set III* with 75 per cent Vermicompost (Sugarcane trash and Cow-dung) + 25 per cent Soil and *Set IV* with 100 per cent Vermicompost (Sugarcane trash and Cowdung) only.

After seed germination thinning was done in each polythene bag of each set to avoid competition and only three seedlings in each bag were allowed to grow for further investigation. After harvesting all the pods, the uprooting of all the seedlings was done. During uprooting, polythene bags were opened and with the help of a fine jet of water soil were flushed out carefully to expose roots to calculate quantitative morphological features like the average root and shoot length with standard deviation; the average dry weight of the root and dry weight of the shoot with standard deviation using Precisa Balance XB 120A. Besides this cumulative number of flowers, fruits and seeds per set; the average dry weight of the fruits (including seedless fruits and excluding seedless fruits); average dry weight of the seeds was also determined. The percentage of pod formation per experimental set; the percentage of pods with no seed each experimental set; the percentage of one seeded pods per set; the percentage of two seeded pods per set; the percentage of three seeded pods per set; the average number of pods per plant, the average number of seeds per pod and per plant and yield in terms of total dry weight (mg) of the fruits per set and in terms of total dry weight (mg) of the seeds per set was also calculated.

Results and Discussion (Tables 24.1–24.4)

The percentage of seed germination in experimental sets *I, II, III and IV* exhibited values of 73.33 per cent, 86.67 per cent, 96.67 per cent and 100 per cent respectively whereas in *control set* the percentage of seed germination exhibited value of 70 per cent. The average shoot length and root length was observed to exhibit minimum value of 32.0 ±9.2 cm and 12.8±8.7 cm respectively in the experimental *Set I* but statistically insignificantly higher than that of *control set*. Whereas shoot length as well as root length exhibited statistically significantly higher values in *Set II, Set III* and *Set IV* with increase in vermicompost concentration except root length in *Set II* (Table 24.1). Edwards and Burrows (1988) also observed that most of the plants germinated earlier and grew better in vermicompost than in commercial plant growth media. Gunathilagaraj and Ravignanam (1996), Kalembasa (1996) and Garg and Bhardwaj (2000) also observed positive impact on quantitative morphology of different crops using different vermicomposts.

The experimental *Set I* exhibited minimum average dry weight of shoot as well as root in the range of 260-1000 mg and 50-1290 mg but statistically insignificantly higher than that of control set. The average dry weight of shoot as well as root in the experimental sets *i.e. Set II, Set III* and *Set IV* also exhibited statistically significantly higher values with increase in vermicompost concentration except dry shoot weight in *Set II* (Table 24.1). The cumulative number of flowers per set exhibited statistically significantly higher values of 99, 116, 105 and 94 in experimental sets I, II, III and IV respectively whereas the cumulative number of pods per set was also observed to exhibit statistically significantly higher values of 57, 69, 67 and 92 in the respective sets. The control set was observed to exhibit minimum values of 89 and 33 for the cumulative number of flowers per set and the cumulative number of pods per set (Table 24.2).

The percentage of pod formation in experimental sets *I, II, III* and *IV* exhibited values of 57.57 per cent, 59.48 per cent, 63.80 per cent and 97.87 per cent respectively whereas in control set the percentage of pod formation exhibited value of 37.07 per cent (Table 24.2).

The percentage of seedless pods in experimental sets exhibited higher values with increase in vermicompost concentration. Though the percentage of one, two and three seeded pods, average dry weight of the pod, average number of seeds per pod exhibited variable values in the control set as well as experimental sets yet the total number of seeds per set in experimental *sets I, II, III* and *IV* exhibited values of 70, 71, 76 and 83 respectively. The average dry weight of the seed was observed to be 117

Table 24.1: Impact of Sugarcane Trash and Cow-Dung (1:1) Vermicompost on Percentage Seed Germination and Vegetative Growth of *Vicia faba* L.

Sl.No.	*Parameter Studied*	*Control Set*	*Experimental Sets*			
			Set I	*Set II*	*Set III*	*Set IV*
1	Percentage of seed germination	70	73.3	86.67	96.67	100
2	Avg. Shoot Length (cm)	25.2±5.3 17.1-35.2	32.0±9.2 21.0-46.9*	33.3±8.07 18.5-40.4**	35.1±9.3 19.2-45.7**	38.7±10.5 21.2-50.4**
3	Avg. Root Length (cm)	11.1±3.1 8.0-18.1	12.8±8.7 7.2-35.7*	19.6±10.0 12.6-44.3*	20.6±7.9 6.3-30.4**	22.1±10.4 10.6-45.2**
4	Avg. dry wt. of shoot (mg)	270±100 140-530	490±200 260-1000*	500±300 280-1040*	590±200 200-930**	770±500 300-1560**
5	Avg. dry wt. of root (mg)	160±40 110-240	260±400 50-1290*	350±200 80-770**	640±400 200-1210**	660±500 200-1400**

*: Not significant at 0.05 (5 per cent) level; **: Signficance at 0.05 (5 per cent) level.

Set I: 25 per cent Vermicompost + 75 per cent soil; Set II: 50 per cent Vermicompost + 50 per cent soil; Set III : 75 per cent Vermicompost + 25 per cent soil; Set IV: 100 per cent Vermicompost only; Control Set: 100 per cent Soil only.

mg in *Set I*, 146 mg in *Set II*, 152 mg in *Set III* and 145 mg in *Set IV*. The total number of seeds and the average dry weight of the seed were observed to exhibit values of 22 and 92 mg in the *control set* (Table 24.3). The average number of pods per plant and average number of seed per plant exhibited increasingly higher values in experimental sets *I, II, III* and *IV* respectively.

Table 24.2: Impact of Sugarcane Trash and Cow-Dung (1:1) Vermicompost on Flowering and Fruiting of *Vicia faba* L.

Sl.No.	*Parameters Studied*	*Control Set*	*Experimental Sets*			
			Set I	*Set II*	*Set III*	*Set IV*
1.	Cumulative no. of Flowers/set	89	99*	116*	105*	94*
2.	Cumulative no. of Pods/set	33	57*	69*	67*	92*
3.	Percentage of Pod formation	37.07	57.57	59.48	63.80	97.87
4.	Percentage of pods with (0) seed	45.45	36.84	43.48	46.27	50.00
5.	Percentage of pods with (1) seed	45.45	14.03	24.64	8.96	15.22
6.	Percentage of pods with (2) seed	6.06	38.60	17.39	29.85	29.35
7.	Percentage of pods with (3) seed	3.03	10.53	14.49	14.92	5.43
8.	Avg. dry wt. of Pod (mg) including seedless Pods	90	188	188	231	190
9.	Avg. dry wt. of Pod (mg) excluding seedless Pods	166	299	333	430	380
10.	Avg. no. of Seeds/pod	1.22 (0-3)	1.94 (0-3)	1.82 (0-3)	2.11 (0-3)	1.80 (0-3)

*: Significant at 0.05 (5 per cent) level.

Set I: 25 per cent Vermicompost + 75 per cent soil; Set II: 50 per cent Vermicompost + 50 per cent soil; Set III: 75 per cent Vermicompost + 25 per cent soil; Set IV: 100 per cent Vermicompost only; Control Set: 100 per cent Soil only.

The overall yield in terms of fruits on dry weight basis was observed to be 10760 mg, 12990 mg, 15490 mg and 17480 mg in experimental sets *I, II, III* and *IV* respectively. The *control set* exhibited overall yield of 2980 mg in terms of fruits. INORA (1998) conducted studies on the effect of graded doses of vermicompost on Cotton, Sugarcane and Paddy. They observed chemical fertilizer was essential to achieve higher yield but application of vermicompost alongwith chemical fertilizers increased the cotton yield by 6.5 per cent over the recommended dose of chemical fertilizers whereas vermicompost alone increase the yield of seed cotton by 12.5 per cent over the set with no fertilizer. The application of worm casts or vermicompost have shown higher yield of wheat, paddy and green leaves ranging from 10 to 200 per cent (Govindan, 1998). The application of chemical fertilizer along with vermicompost increased the nutrient uptake and net production of Wheat and Sugarcane, respectively (Kale *et al.*, 1992).

Whereas overall yield in terms of seeds on dry weight basis was observed to be 8210 mg in *Set I*, 10360 mg in *Set II*, 11570 mg in *Set III* and 12040 mg in *Set IV*. Overall yield of 2040 mg in terms of seeds was observed in *control set* (Table 24.3).

Table 24.3: Impact of Sugarcane Trash and Cow-Dung (1:1) Vermicompost on Yield of *Vicia faba* L.

Sl.No.	Parameters Studied	Control Set	Experimental Sets			
			Set I	Set II	Set III	Set IV
1.	Cumulative no. of Seeds/set	22	70	71	76	83
2.	Avg. dry wt. of Seed (mg)	92	117	146	152	145
3.	Avg. no. of Pods/plant	3.67	6.33	7.67	7.44	10.22
4.	Avg. no. of Seeds/plant	2.44	7.78	7.89	8.44	9.22
5.	Yield in terms of Fruit (dry wt. mg)	2980	10760	12990	15490	17480
6.	Yield in terms of Seed (dry wt. mg)	2040	8210	10360	11570	12040

Set I: 25 per cent Vermicompost + 75 per cent Original soil sample; Set II: 50 per cent Vermicompost + 50 per cent Original soil sample; Set III: 75 per cent Vermicompost + 25 per cent Original soil sample; Set IV: 100 per cent Vermicompost only; Control Set: 100 per cent Soil only.

Table 24.4: Correlation Co-efficient (r) Value and Level of Significance at 5 per cent (p) Between Mixed Vermicompost Concentration and Growth Parameters of *Vicia faba* L.

Sl.No.	Parameters	(r)-Value	(p)-Value	Status at 0.05 (5 per cent) Level
1.	Vermicompost and Shoot length	+0.46	0.001	Significant
2.	Vermicompost and Root length	+0.46	0.000	Significant
3.	Vermicompost and Shoot weight	+0.48	0.000	Significant
4.	Vermicompost and Root weight	+0.52	0.000	Significant
5.	Vermicompost and cumulative no. of Pods/set	+0.90	0.447	Insignificant
6.	Vermicompost and no. of Seeds/set	+0.96	0.417	Insignificant
7.	Vermicompost and yield in terms of Fruits	+1.00	0.002	Significant
8.	Vermicompost and yield in terms of Seeds	+0.96	0.001	Significant

Conclusion

The study of various morphological characteristics of *Vicia faba* L. seedlings like percentage of seed germination, average root length, average shoot length, average dry weight of root, average dry weight of shoot, average dry weight of pod, average dry weight of seed, average number of seeds per pod, average number of pods and seeds per plant and number of flowers, pods and seeds/set and percentage of pod formation revealed that mixed vermicompost in different concentrations had positive impact on the growth of *Vicia faba* L. as compared with the seedlings of control set in which no vermicompost as well as fertilizer was added. The critical analysis of data after statistical evaluation of Pearson co-efficient of correlation (r) revealed that increase in vermicompost concentration exhibited positive correlation (r) between vermicompost concentration and growth parameters *i.e.* a significant positive correlation was observed between vermicompost and shoot length (r=0.46; p=0.001), vermicompost and root length (r=0.46; p=0.00), vermicompost and dried shoot weight (r=0.48; p=0.00), vermicompost and dried root weight (r=0.52; p=0.00), vermicompost and yield in terms of fruits (r=1; p=0.002) and vermicompost and yield in terms of seeds (r=0.96; p=0.001) and a positive but insignificant correlation was observed between vermicompost and cumulative number of Pods/set (r=0.90; p=0.447), vermicompost and number of Seeds/set (r=0.96; p=0.417) (Table 24.4).

The plants of *Vicia faba* L. growing in experimental *Set IV* with 100 per cent Vermicompost exhibited higher value of dry weight of root, shoot as well as yield as compared with dry weight and yield of plants growing in experimental *Sets III, II* and *I* with 75 per cent 50 per cent and 25 per cent vermicompost respectively.

References

Edwards, C.A. and I. Burrows (1988) The potential of earthworm composts as plant growth media. In: *Earthworms in Waste and Environmental Management* (Eds. Edwards, C.A. and E.F. Neuhauser). SPB Academic, The Hauge: 211-219.

Garg, K. and N. Bhardwaj (2000). Effect of vermicompost of *Parthenium* on two cultivars of wheat. *Indian J. Ecol.* 27 (2): 177-180.

Govindan, V.S. (1998) Vermiculture and vermicomposting (In: Ecotechnology for pollution control and Environmental Management Eds. R.K. Trivedy and Arvind Kumar). *Enviro Media*. 49-57.

Gunathilagaraj, K. and T. Ravignanam (1996) Effect of vermicompost on mulberry sapling establishment. *Madras Agric. J.* 83 (7): 476-477.

Gupta, P. and C.P. Dwivedi (2001) Vermitechnology: An emerging technology for waste management. *Indian Farmers Digest*: 36-40.

INORA (1998) Newsletter of Institute of Natural Organic Agriculture, April, 2 (2): 4.

Kale, R.D., B.C. Mallesh, K. Bano and D.J. Bagyaraj (1992) Influence of vermicompost application on the available macronutrients and selected microbial populations in a paddy field. *Soil Biol. Biochem.* 24 (12): 1317-1320.

Kalembasa, D. (1996) The effects of vermicomposts on the yield and chemical composition of tomato. *Zeszyty-Problemowe-Postepow-Nauk-Rolniczych*. 437: 249-252.

Sinha, R.K. and A.K. Sinha (2000) Waste Management. INA Shree Publishers Jaipur: 143-157.

2013, Environmental Biotechnology *Pages 225–227*
Editors: **D.R. Khanna, A.K. Chopra, Gagan Matta, Vikas Singh & Rakesh Bhutiani**
Published by: **BIOTECH BOOKS, NEW DELHI**

Chapter 25

Ethylmethane Sulphanate Induced Bushy Head Mutants in Jute (*Corchorus olitorius* L. Variety JRO-632)

P.K. Ghosh[1] and A. Chatterjee[2]

[1]CSB, CSR&TI, Berhampore – 742 101, West Bengal
[2]Centre of Advanced Study in Cell and Chromosome Research, Department of Botany, University of Calcutta, 35, Ballygunge Circular Road, Kolkata – 700 047

Presoaked seeds of jute (*Corchorus olitorius* L. Variety JRO-632) were treated with 0.5 per cent Ethylmethane Sulphonate (EMS) for 24 hours. Bushy head mutants were screened in M3 in contrast to the normal plants. Bushy head mutants otherwise looked normal excepting the nature of early flowering habit. A number of yield component growth parameters were recorded like plant height, basal diameter, plant spread, root length, pod per plant, seeds per pod, pod length/ breadth ratio, number of primary branches per plant, number of secondary branches per plant, leaf angle, branching angle, first flowering date, 100 per cent flowering date, total duration, per cent of pollen sterility and weight of 100 seeds which were found to vary from the control plant. Chromosome analysis revealed a number of aberrations like stickiness, fragmentation, clumping, polyploidy, and laggard and bridge formation etc. at very low frequency. These early flowering mutant plants gives more fiber yield than the control plants with superior quality. Multiple cropping has been possible with the availability of irrigation water and a number of early maturing varieties have introduced in case of various other crops. There should be a suitable early maturing variety of jute also to be best fitted in the multiple cropping patterns. With this objective in view the work on induction of mutation with chemical mutagen Ethylmethane sulphonate (EMS) was initiated.

Keywords: *Ethylmethane sulphonate (EMS), Corchorus olitorius L., Bushy head Mutant, 24 hours, chromosome, Concentration 0.5 per cent.*

Introduction

Jute (*Corchorus olitorius* L. Variety JRO-632) is one of the very important fiber yielding cash crops with great demand in International market. A number of mutants in jute were reported through genetic manipulation by application of ionizing radiations (Kundu, 1944, Ghosh, 1969, Hossain, 1970 and Basu, 1967, Bandhyopadhyay, 1974 and Monti, 1968). However, chemical mutagenesis in jute is still lacking although considerable worked has been done on this line in other commercial crops, the present work was therefore, undertaken to investigate the potentiality of host chemical substances to induce mutation in jute.

Materials and Methods

Jute seeds (*Corchorus olitorius* L. Variety JRO-632) obtained from Jute Agriculture Research Institute, ICAR, Barrackpore, W.B. were presoaked in distilled water for 24 hours and then treated with 0.5 per cent Ethylmethane sulphonate for 24 hours. The seeds were thoroughly washed with distilled water and then sown in the field directly with equal spacing for raising M1 generation. The individual M1 plants were harvested separately for growing M2 generation in progeny rows. The bushy head mutants were screened and again harvested for raising M3 generation. The mutants were screened in M3 generation. A number of essential yield components were recorded. Cytological anomalies and pollen sterility were recorded as per schedule techniques.

Results and Discussion

The present investigation indicates that some bushy head mutants were screened after M3 generations. These bushy head mutants otherwise looked normal plants like. Variation of yield component growth parameters was recorded. The segregation behavior in M2 generations was fitted to a ratio 3:1.In M2 almost all the plants were 8-locular fruit mutants excepting one or two cases of normal plants. No much noticeable variation in chromosome anomalies was recorded. Segregation behavior indicates that this is due to a single gene of recessive nature and pollen sterility was also recorded. Multiple cropping has been possible with the availability of irrigation water and a number of bushy head varieties have introduced in case of various other crops. There should be a suitable early maturing variety of jute also to be best fitted in the multiple cropping patterns. With this objective in view the work on induction of mutation with chemical mutagen Ethylmethane sulphonate (EMS) was initiated.

References

Bandyopadhyay, B. (1974).Studies on induced mutation in jute (*Corchorus capsularies* L. and *C.olitorius* L.) by single and combined treatments with X-rays and chemicals.Ph.D.Thesis, University of Calcutta, Calcutta.

Basu, R. K. (1967).Induction of high yielding mutants in jute.Mutat.Res.4:163-167.

Ghosh, N. (1969).Studies of variability in jute followed by irradiation and hybridization. Ph.D. Thesis, BCKV.

Hussein, M.M. (1970).Cytogenetic studies of mutants and colchiploids of some cultivated varieties of jute.Ph.D.Thesis, BCKV.

Kundu, B.C. (1944).A note on the preliminary studies on effect of X-irradiation upon growth and development of jute plant. Sci. and Cult.9:559-560.

Monti, L.M. (1968).Mutations in peas induced by Diethyl Sulphate and X-rays.Mutat.Res.5:187-191.

2013, Environmental Biotechnology *Pages* ***229–230***
Editors: **D.R. Khanna, A.K. Chopra, Gagan Matta, Vikas Singh & Rakesh Bhutiani**
Published by: **BIOTECH BOOKS, NEW DELHI**

Chapte 26

Induced Crumple Leaf Mutant in Jute (*Corchorus olitorius* L. Variety-JRO-632)

P.K. Ghosh[1] *and A. Chatterjee*[2]
[1]*CSB, CSR&TI, Berhampore – 742 101, West Bengal*
[2]*Centre of Advanced Study in Cell and Chromosome Research, Department of Botany, University of Calcutta, 35, Ballygunge Circular Road, Kolkata – 700 047*

Presoaked seeds of jute (*Corchorus olitorius* L.Variety JRO-632) were treated with 2 per cent Ethylamine (EA) for 24 hours. Crumple leaf mutants were screened in M3 in contrast to the normal looking fruit plants. A number of yield component parameters were recorded including plant height, basal diameter, plant spread, root length, pod per plant, seeds per pod, pod length/breadth ratio, number of primary branches, number of secondary branches, leaf angle, branching angle, first flowering date, 100 per cent flowering date, total duration, percentage of pollen sterility, and weight of 100 seeds which were found to vary from the control plant. Chromosome analysis revealed aberrations like stickiness, fragmentation, polyploidy, clumping, laggard and bridge formation etc.Multiple cropping has been possible with the availability of irrigation water and a number of early maturing varieties have introduced in case of various other crops. There should be a suitable crumple leaf mutants of jute also to be best fitted in the multiple cropping patterns. With this objective in view the work on induction of mutation with chemical mutagen Ethylamine (EA) was initiated.

Keywords: *Corchorus olitorius L., Chromosome, Crumple leaf mutants, Ethylamine, 24 hours, Concentration 2 per cent.*

Introduction

Jute (*Corchorus olitorius* L. Variety JRO-632) is one of the very important fiber yielding cash crops with great demand in International market. A number of mutants in jute was reported through genetic manipulation by application of ionizing radiations (Kundu, 1944, Ghosh, 1969, Hossain, 1970 and Basu, 1967,Chattrerjee and Jana,1974,Nayar,1979).However. Chemical mutagenesis in jute is still lacking although considerable worked has been done on this line in other commercial crops, the present work was therefore, undertaken to investigate the potentiality of host chemical substances to induce mutation in jute.

Material and Methods

Jute seeds (*Corchorus olitorius* L. Variety JRO-632) obtained from Jute Agriculture Research Institute, ICAR, Barrackpore, W.B. were presoaked in distilled water for 24 hours and then treated with 2 per cent Ethylamine for 24 hours. The seeds were thoroughly washed with distilled water and then sown in the field directly with equal spacing for raising M1 generation. The individual M1 plants were harvested separately for growing M2 generation in progeny rows. The crumple leaf mutants were screened and again harvested for raising M3 generation. The mutants were screened in M3 generation. A number of essential yield components were recorded. Cytological anomalies and pollen sterility were recorded as per schedule techniques.

Results and Discussions

The present investigation indicates that some crumple leaf mutants were screened after M3 generations. The rosette leaf mutants otherwise looked normal plants like. Variation of yield component growth parameters was recorded. The segregation behavior in M2 generations was fitted to a ratio 3:1.In M2 almost all the plants were crumpleleaf mutants excepting one or two cases of normal plants. No much noticeable variation in chromosome anomalies was recorded. Segregation behavior indicates that this is due to a single gene of recessive nature and pollen sterility was also recorded.

References

Basu, R. K. (1967). Induction of high yielding mutants in jute.Mutat.Res.4:163-167

Chatterjee, A and Jana, M.K. (1974).A new early maturing type of jute mutant.Sci. and Cult.40:319-320.

Ghosh, N. (1969). Studies of variability in jute followed by irradiation and hybridization. Ph.D. Thesis, BCKV.

Hussein, M.M. (1970). Cytogenetic studies of mutants and colchiploids of some cultivated varieties of jute. Ph.D.Thesis, BCKV.

Kundu, B.C. (1944).A note on the preliminary studies on effect of X-irradiation upon growth and development of jute plant. Sci. and Cult. 9:559-560.

Nayar, G.G. (1979).EMS induced high yielding early mutant in linseed (*Linum usitatissimum* L.). Curr.Sci.48:214-216.

2013, Environmental Biotechnology *Pages 231–233*
Editors: **D.R. Khanna, A.K. Chopra, Gagan Matta, Vikas Singh & Rakesh Bhutiani**
Published by: **BIOTECH BOOKS, NEW DELHI**

Chapter 27

Chromosomal Changes in Root Tip Explants during *In vitro* Growth in Jute (*Corchorus olitorius* L. Variety JRO-632)

P.K. Ghosh[1] *and A. Chatterjee*[2]
[1]*CSB, CSR&TI, Berhampore – 742 101, West Bengal*
[2]*Centre of Advanced Study in Cell and Chromosome Research, Department of Botany, University of Calcutta, 35, Ballygunge Circular Road, Kolkata – 700 047*

The root tips of jute (*Corchorus olitorius* L. variety JRO-632) were grown in in vitro cultured aseptically on 20 ml solid nutrient medium Murashige and Skoog's(1962)(MS) and Schenk and Hildebrandt's(1972)(SH) media were tried with various combinations and concentrations of different auxins(NAA,IAA,IBA and 2,4-D used separately o,o170 mg/L to 0.3500 mg/L) and cytokines (BAP,Kn used separately 0.1400 mg/L to 3.3000 mg/L,coconut milk 10-35 per cent (V/V). It has been observed that the callus growing in medium containing NAA revealed different degrees of ploidy and mitotic abnormalities such as stickiness, clumping, diplo-chromatids and spindle disturbances from the early stage and the mitotic index was also recorded. The frequency of chromosomal abnormalities gradually increases with the age of the callus tissues increases. A comparison of the cytology of the callus growing in NAA and the regenerating callus growing in IBA revealed differences in the rate of division and mitotic abnormalitiesd. Callus growing in NAA showed a comparatively lower rate of division and higher rate of mitotic abnormalities, while that growing in IBA exhibited just the opposite effcct.It appears that NAA has a distinct role to play in influencing Karyological instabilities and mitotic rate of cells and also observed the role of different hormones in inducing karyological changes during in vitro growth. The object of the present investigation was that to find out the role of different constituents of the medium in controlling in vitro growth and the karyological changes which they induce during organogenesis and callus formation of jute explants (*Corchorus olitorius* L.Variety JRO-632).

Keywords: *Root tip, Chromosome, Corchorus olitorius, Auxins, Cytokinins.*

Introduction

Morphogenesis in multicellular plants through in vitro induction depends upon the integration and mutual interaction of the various organs, tissues and cells. Together they form a complex system which can be analyzed by studying the origin and movement of the substances primarily involved in morphogenesis *e.g.* hormones and correlations between the differential capabilities of the separate parts of the plant to synthesize and transport primary nutrients and metabolities.Multicellular systems are highly organized cells and rapidly changed in callus tissues in in vitro culture. Induction of callus and their growth nature were discussed by Yamada, 1977, Torry, 1967, Digby and Wareing, 1966a, 1966b).Report of jute plant is still lacking. The object of the present investigation was that to find out the role of different constituents of the medium in controlling in vitro growth and the karyological changes which they induce during organogenesis and callus formation of jute explants (*Corchorus olitorius* L.Variety JRO-632).

Materials and Methods

Jute seeds (*Corchorus olitorius* L. Variety JRO-632) obtained from Jute Agriculture Research Institute (ICAR), Barrackpore, West Bengal were washed with 0.5 per cent aqueous teepol solution for 15 minutes, surface sterilized by 0.1 per cent mercuric chloride solution for 5 minutes and washed thoroughly with autoclaved sterile double distilled water (at ib/psi for 15-20 minutes at 121°C) in a chamber sterile by ultraviolet rays for one hour. The seeds were then aseptically inoculated in 250 ml culture flasks on 30 ml nutrient agar medium on White (1963) in which no hormone was added.Ph being mentioned at 5.6-5.8.The cultures were kept in the culture room having the temperature22± °C degree centrigrade under photoperiod at 16 hours light and 8 hours dark period at 55-60 per cent relative humidity. Root tip explants of were obtained from in vitro growing plants.MS and SH basal medium were prepared as per suggestion of Murashige and Skoog (1962) and Schenk and Hildebrandt (1972) and used a number of different set of cytokinins and auxins were used for this set of experiments. Cytological analysis was done as per schedule techniques.

Results and Discussions

The present investigation indicate that MS basal medium were much better than SH basal medium. When basal medium supplemented with different combinations of auxins and cytokinins induction of callus tissues of leaf segments were recorded.). It has been observed that the callus growing in medium containing NAA revealed different degrees of ploidy and mitotic abnormalities such as stickiness, clumping, diplo-chromatids and spindle disturbances from the early stage and the mitotic index was also recorded. The frequency of chromosomal abnormalities gradually increases with the age of the callus tissues increases. A comparison of the cytology of the callus growing in NAA and the regenerating callus growing in IBA revealed differences in the rate of division and mitotic abnormalitiesd.Callus growing in NAA showed a comparatively lower rate of division and higher rate of mitotic abnormalities, while that growing in IBA exhibited just the opposite effcct.It appears that NAA has a distinct role to play in influencing Karyological instabilities and mitotic rate of cells and also observed the role of different hormones in inducing karyological changes during in vitro growth. The object of the present investigation was that to find out the role of different constituents of the medium in controlling in vitro growth on root tip explants and the karyological changes which they induce during organogenesis and callus formation of jute explants (*Corchorus olitorius* L.Variety JRO-632).

References

Digby J, Wareing PF. (1966a).The effect of applied growth hormones on cambial division and the differentiation of cambial derivatives. *Ann Bot*, 30:539-548.

Digby J, Wareing PF. (1966b).The effect of growth hormones on cell division and expansion in liquid suspension cultures of Acer pseudoplatanus.*J Exp Bot*, 17:718-725.

Murashige T, Skoog F. (1962).A revised medium for rapid growth and bioassays with tobacco tissue cultures. *Physiol Plant*,15:473-497.

Schenk RU, Hildebrandt AC.(1972).Medium and techniques for induction and growth of monocotyledonous and dicotyledonous plant cell cultures. *Canadian J Bot*, 50:199-204.

Torrey JG.(1967).Morphogenesis in relation to chromosome constitution in long term plant tissue cultures.*Physiol Plant*,20:265-276.

Yamada Y.(1977).Tissue cultures of cereals. In: Applied and Fundamental Aspects of Plant Cell, Tissue and Organ Culture.(Reinert J,Bajaj YPS Eds.)pp 144-159, Springer-Verlag, Berlin.

2013, Environmental Biotechnology *Pages* ***235–240***
Editors: **D.R. Khanna, A.K. Chopra, Gagan Matta, Vikas Singh & Rakesh Bhutiani**
Published by: **BIOTECH BOOKS, NEW DELHI**

Chapter 28
Histological Changes in the Gills of the Freshwater Fish *Channa punctatus* Exposed to Cypermethrin

V.T. Tantarpale, P.H. Rohankar and Santosh Borkar
P.G. Department of Zoology,
Vidhya Bharati Mahavidhya Amravati – 444 602, M.S.

Introduction

The indiscriminate use of synthetic pyrethoids for control of various types of pests is regularly increasing by farmer,and other people in garden and domestics purpose. These pesticides drained into rivers, tanks,ponds and other water bodies and pollute aquatic environment. These pesticides and other chemicals are highly toxic to non-target organism and obstructed human too. Different types of pollutants pollute the aquatic ecosystem and affecting very greatly the metabolism and physiology of the fishes and sometimes showed heavy mortality.

The synthetic pyrethroids,a new class of agricultural insecticides has emerged as a complement to organophosphates and other pesticides.In recent times the importance of synthetic pyrethroids in growing which is attributed to its low mammalian toxicity, Parker *et al.*, 1984; Haya, 1989; high biodegradability Leahey,1979; high efficacy and non-phyto-toxicity.These are well popular for their toxicity to wide range to insects including resistant strains as well.

Among pyrethroids cypermethrin widely used for control pest,Agnihothrudu and Gaur, 1982; Patel *et al.*, 1986; Meister, 1996 and indirectly caused aquatic organism the non target animals fishes which caused severe metabolic disorders Coats *et al.,*; 1989; Reddy *et al.,* 1991; Reddy and Philip, 1994; Thakur and Bais 2000. It has been extensively popular because of its high insecticidal potency,low mammalian toxicity and very short persistence. However cypermethrin is extremely toxic to fishes it is due to gill uptake inefficient and detoxification and elimination and sensitivity at the site of action Bradbury *et al.,* 1987; Alazemi *et al.,* 1996; Das and Mukherjee, 2000; Tripathi *et al.,* 2001.

Materials and Methods

The experimental freshwater fish *Channa punctatus* collected from Wadali Lake and from local Friday market of Amravati, Maharashtra state.Fishes were brought in the laboratory in plastic container.The fishes were acclimatized in glass aquarium in the laboratory condition for a week.During this period fish were fed on locally available artificial food. The experimental purpose the fishes length were 11 to 16 cm and 15 to 45gm of weighed.They were disinfected with 0.1 per cent KMNO4 solution to avoid fungal and any other infection.

The experimental purpose,freshwater fish *Channa punctatus* divided into two groups. One for normal or control and another group for experimental.

Group I (control): As control and sacrificed at the end of each experimental period.

Group II: The freshwater fish *Channa punctatus* exposed to sublethal concentration of synthetic pyrethroid a Cypermethrin.The Lc50 values were calculated 0.00078 µ/ lit at 96 hrs. exposure period.

Results and Discussion

In the present work synthetic pyrethroid,a cypermethrin was treated with freshwater fish *Channa punctatus* for toxicity evaluation. The freshwater fish *Channa punctatus* exposed to sublethal concentration of cypermethrin showed histological alterations in the structure of the gill and showed hyperplasia in the respiratory surface of the folds in gills. The bulging of tip of primary gill filaments with distortion of the shape of secondary filaments, secondary gill lamellae breaks. The pillar cells showed Oedema, pillar cell nucleus necrosis. In the portion of secondary gill epithelium showed vacuolation, degenerated secondary lamella, Hyperplasia in gill lamella observed in the treated fish gill of *Channa punctatus* (Plate 28.2) due to pesticide stress.

The gills, being delicate structure, get affected easily if the surrounding media is contaminated (Roy and Munshi 1991). In the present study the gill of treated fish showed hyperplasia in the respiratory surface of the folds in the gills, oedema and accumulation of the cell cytoplasm at a specific region similar findings by Erkman *et al.,* 2000; Cengiz and Unlu 2002; Cengiz and Unlu 2003 showed similar effects of pesticides in the gill of fishes. Another investigator Ziynet Yildirim *et al.,* 2006 studies deltamethrin a synthetic Pyrithroids contaminating aquatic ecosystem as a potential toxic pollutant on *Tilapia oreochromis.* Also the treatment with ammonia showed histopathological lesions as lamellar fusion and hyperplasia in the respiratory surface

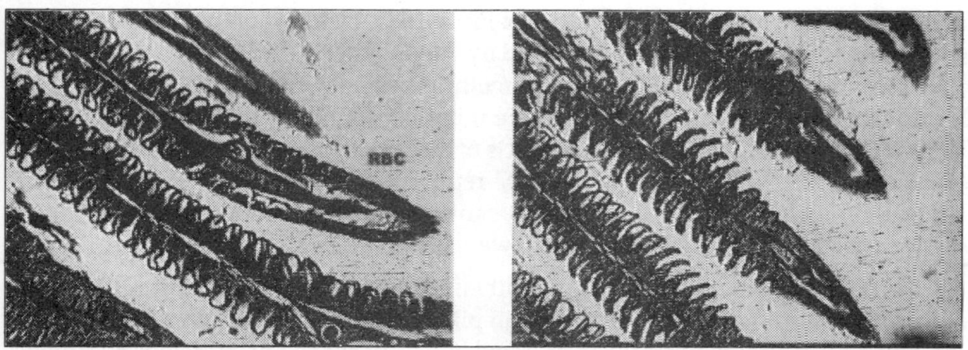

Plate 28.1: Gills of Control *Channa punctatus*

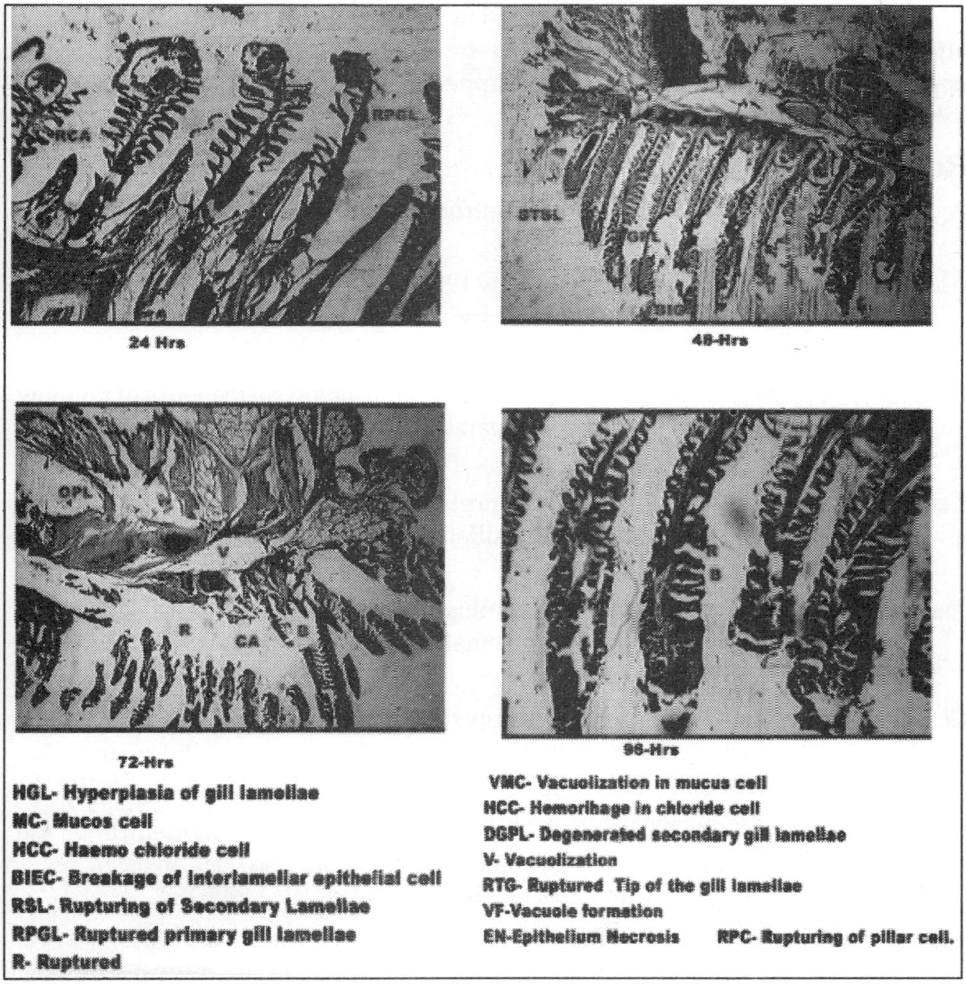

24 Hrs

48-Hrs

72-Hrs

96-Hrs

HGL- Hyperplasia of gill lamellae
MC- Mucos cell
HCC- Haemo chloride cell
BIEC- Breakage of interlamellar epithelial cell
RSL- Rupturing of Secondary Lamellae
RPGL- Ruptured primary gill lamellae
R- Ruptured

VMC- Vacuolization in mucus cell
HCC- Hemorihage in chloride cell
DGPL- Degenerated secondary gill lamellae
V- Vacuolization
RTG- Ruptured Tip of the gill lamellae
VF-Vacuole formation
EN-Epithelium Necrosis RPC- Rupturing of pillar cell.

(45X) Labomed Research Microscope

Plate 28.2: Effect of Sublethal Concentration of Cypermethrin Exposed to Freshwater Fish *Channa punctatus* at Different Hours

of fold in the gill observed by Smith and Piper 1975. Detoriation of lamellar ridges surface areas similar observation reported by Fukunga *et al.*, 1992; Gupta and Rajbansi 1995. Also the synthetic pyrithroids deltamethrin treated with *Cyprinus carpio* showed histopathological effects on the gill by Cengiz; 2006. Similar finding by Cengiz and Unlu 2006 were observed sublethal effects of deltamethrin on the structure of gill of *Gambusia affinis*, Velmurugan *et al.*, 2007 reported that sublethal concentration of lambda cyhalothrin exposed to Cirrhinus mrigala in which oedema, shortening of secondary lamellar and lamellar fusion were observed in gill tissue.

Hyperplasia and fusion of gill epithelium due to separation of epithelium, necrosis of gill epithelium, degeneration of pillar cells and vacuoles were observed in the epithelium these changes were observed in fishes exposed to pesticides by Haines and Schofield 1980; Love 1980; Roy and Dutta mushy 1991; Sunitha and Sahai 1993.

Through this present study concluded that very little concentration of pesticide affected the delicate organ Gills of the freshwater fish *channa punctatus*. The stress appeared on the fish and severe change appears in their histological structure and there is no recovery spontaneous.

References

Agnihotthrudu, V. and T.B. Gour 1982; control of cotton bollworms with fenvalerate in India crop protection 1(2) 231-234.

Alazemi, B.M; J.W.Lewis and E.B. Andrew 1996; Gill damage in the freshwater fish Gnathonemus ptersii (Family:Mormyridae) exposed to selected pollutants: an ultra structural study Environment Technology 17. 225-238.

Bradbury, S.P; James M.Mekim R. Joel Coats 1987: physiological response of rainbow trout (salmo gairdueri) to acute fenvalerate in toxication J. Pest Biochem. Physiol 27,275-288.

Cengiz, E. I E. Unlu 2002: Histopathological changes in the gills of mosquito fish Gambusia affinis exposed to endosulfan Bull. Environ. Contam. Toxicol; 68: 290-296.

Cengiz, E.I. and E. Unlu 2003: Histopathology of gills in mosquito fish Gambusia affinis after long term exposed to sublethal concentration of Malathion J. Environ. Sci. Health, B. 38 (5): 581-589.

Cengiz Elif Ipek 2006: Gill and kidney histopathology in the freshwater fish Cyprinus Carpio after acute exposure to deltamethrin Environ.Toxiocol. Pharmacol. Vol. 22, 2, pp. 200-204.

Cengiz, E. I. and E.Unlu 2008: Sublethal effects of commercial deltamethrin on the structure of the gill, liver and gut tissues of mosquito fish Gambusia affinis A microscopic studý Environ, Toxicol. Pharmacol. Vol. 21, 3,246-253.

Coats, J. R., D. M. Symonik, S. P. Bradbury, S. D.Dyer, L. K. Timson and G. J. Atchison 1989. Toxicology of synthetic pyrethroids in aquatic organisms: An overview Environ. Toxicol. Chem. 8,671-679.

Das, B. K. and S. C. Mukherjee 2000: Histopathological study of carp (Labeo rohita) exposed to hexachlorocyclohexane Veterinarski Arhiv. 70 (4): 169-180.

Erkman, B; M. Caliskan and S.Y.Yerli 2000: Histopathological effects of Cyphenothrin on gills of the Lepistes reticulates vet. Hum. Toxicol 42:5-7.

Fukunge, k. Suzuki, T. Arita, M. Suzukis Hora A Yamauchi, K. Shineiki N; Ishizaki k. and K. Takma 1992:Acute toxicity of Ozone against morphology of gill and erythrocytes of Japanese chars. (Salvelinus Leucomaenis) comp. Biochem. Phyriol 101 C: 331-336.

Gupta, A.K. and V.K. Rajbanshi 1995 Mercury poisoning Architectural changes in the gill of Rasbora daniconius (Hamilton) J. Environ. Biol. 16: 33-36.

Haya, K. 1989: Toxicity of pyrethroid insecticides to fish Environ. Toxicol. Chem. 8, 381-391.

Hainies, T.A and C.I. Schofield 1980 Resonses of fishes to acidification of stream and lakes in Eastern North America, In: Restoration of lakes and inland waters International Symposium of inland waters and lake restoration Portland Maine, U.S. Environ. Port. Age. Washington DC EPA: 44051-81-010. 467-473.

Leahey, J. P. 1979. The metabolism and environmental degradation of the pyrethroid insecticides outlook. Agric 10: 135 -142.

Love, R.M. 1980: The chemical biology of fishes Vol. 2 Academic Press, New York.

Meister, R. T. 1996: Farm Chemicals Handbook 96 Vol.82, Meister publishing Company, Willoughloy OH 294-295.

Parker, C. M., D.R. Peterson. G.A. Vangeldes, E.B. Gordon, R.M. Parry, R.C. Chander and R.M. Shahani 1984. A rapid and sensitive assay of muramidase Proc-50 C. Rxp. Bio. Med. 199: 384-386.

Patel, B. K; N.B. Rote, N. P. Mehta and A.H. Shah 1986. Judicious use of fenvalerate based on economic level of bollworms in hybrid 4 cotton pesticide 20 (2): 31-34.

Reddy, P. M. and G. H. Philip (1994): In vivo inhibition of ACHE and ATPase activities in the tissues of freshwater fish Cyprinus Carpio exposed to technical grade cypermethrin Bull. Environ. Contam. Toxicol 52, 619-626.

Reddy, P. M, G. H. Philip and M. D. Bashamohideen 1991 Fenvalerate induced biochemical changes in the selected tissues of the freshwater fish *cyprinus carpio* Biochem. Int. 23, 1087-1096.

Roy, K. J. and J.S. Duttamunshi 1991 Malathion induced structural and morph metric changes of gills of a freshwater major carp *Cirrhinus mrigala* (Ham). J. Environ. Biol. 12,79-87.

Sunitha, S. and S. Sahai 1993 histopathological change in the gill of *Rasbora daniconius* induced by BHC. J. Environ. Biol. 5,65-69.

Thakur, P.B. and V.S.Bais 2000 Toxic effects of aldrin and fenvalerate on certain Haematological parameter of freshwater teleost *Heteropneustes fossilis* (Bloch) J. Environ. Biol. 21 (2); 161-163.

Tripati G.; G. P. Varma, G. C. Pandey and Y.C.Tripati 2001 Biological effects of fenvalerate In: Current trends in Environmental Sciences Eds: G.Tripati and G.C. Pandey \ ABD. Publishers, Jaipur India.

Velmurugan Babu Mariadoss, Sevanayagam Elif Ipeak Cengis and Erhan Unlu 2007 Histopathology of Lambda – Cyhalothrin on tissues (gill, kidney, liver and intestine) Cirrhinus mrigala Environ. Toxicol and Pharmcol. Vol. 24, (3) pp 286-291.

Ziynet Yidirim, M.A.Cauglan Karasu Benl; Mahmut Selvi; Ayhan Ozkul,Figen Erkoc and Oner Kocak 2006 Acute toxicity, bevavioral changes and histopatological effects of deltamethrin tissues (gills, liver,brain,spleen,kidney, muscle, skin) of Nile tilapia (Oreochris niloticus L.) fingerlings.Environ.Toxicol. 6, vol.21,614-620.

2013, Environmental Biotechnology *Pages* **241–250**
Editors: **D.R. Khanna, A.K. Chopra, Gagan Matta, Vikas Singh & Rakesh Bhutiani**
Published by: **BIOTECH BOOKS, NEW DELHI**

Chapter 29

Effect of Nutrient Types on the Growth and Yield of a Potential Medicinal Plant, *Ammi majus* L.

Amarjeet Bajaj[1], *K.J. Cherian*[1] *and Megha Bhambri*[2]
[1]*Professor and Head of Botany, MVM College, Bhopal, M.P.*
[2]*Hislop College, Civil Lines, Nagpur*

Majority population of developing countries still rely on plant based traditional drugs. Cultivation of medicinal plants for their products is in great demand especially due to high cost of their products. The present work is focused on improving the living standard of poor farmers by introducing most effective and profitable farming of medicinal herbs without discontinuing their traditional farming. *Ammi majus* L. a member of family Apiaceae has several medicinal properties. Its seeds have contraceptive and diuretic properties.It is mainly used in the treatment of vitiligo and psoriasis. It is also used as tonic or in the treatment of asthama and angina. The present demand of its seeds in the world market is worth for about 14 billion US dollars per year. Over exploitation of noncultivated medicinal plants has become a threat to biodiversity in the forest areas in India. The seeds of *Ammi majus* L. were obtained from Hamdard University, New Delhi and were cultivated in farm land as well as Hislop college experimental field with control and treatment of various nutrient types. The present work has proved that this plant can be cultivated from February to May in the Vidarbha region of Maharashtra, India. The organic farming techniques have proven to be better yielding.

Keywords: *Medicinal Plants, Ammi majus L.,Vitiligo, Psoriasis, Contraceptive and Diuretic.*

Introduction

The plant *Ammi majus* L. is a native of Nile Delta of Egypt. In India, *Ammi majus* L. was introduced in forest research institute, Dehradun in 1955 through the efforts of UNESCO for its medicinal and ornamental value (Bradu and Atal,1970; Singh,1963 and Umrao Singh *et al.*, 1982). Since then, its experimental cultivation has been tried in several parts of India including Jammu, Dehradun, Mumbai, Chennai, Delhi and Punjab.

The plant is used for the treatment of leucoderma and psoriasis (Anup Kumar, 1988 and Hansen,1979). It has been recommended as a diuretic, expectorant and useful in Jaundice (Khan and Rehman, 1985 and Lal, 1977). The fruits are used in vitiligo and also in the formation of suntan lotion (Anonymous, 1985, 1986). The essential oil from *Ammi majus* L. seed has been extracted by Ashraf *et al.*, in 1972 who also studied its quality compound for the first time.

The seed is contraceptive, diuretic and tonic. An infusion is used to calm the digestive system, whilst it is also used in the treatment of asthma and angina. Its decoction is also used as a gargle in the treatment of toothache. It was an Egyptian professor Abdel Monem. El Mofty, who observed plants used in Egyptian folk medicine (e.g *Ammi majus* L.) and began the development of modern photochemotherapy (PUVA) for vitiligo and psoriasis. Klaber (1942) introduced the term phytophotodermatitis to emphasize the necessity of plants and light to cause the reaction. Duke (1988) have reported that furocoumarins have bactericidal, fungicidal, insecticidal, larvicidal, moluscicidial, nematicidal, ovicidal, virincidal and herbicidal activities.

Ammi majus L. is an important medicinal perennial herb belonging to family – Apiaceae. Its common name is Aatrilal, bishops weed, bullwort, False queen anne's lace, lace flower or mayweed. It has a striated subglaucous stems, leaves acute, serrulate, alternate, bipinnate and lobes oblong. Inflorescence a compound umbel, flowers bisexual, polygamous, bracteate, calyx teeth small, petals obovate, stamens epigynous, ovary inferior, two locular, stigma capitate. The plant prefers bright light, sandy medium (Loamy) or heavy (Clay) soils. The plant prefers acid, neutral or basic (Alkaline) soils. It can grow in semi shade (Light woodland) or no shade. It requires moist soil.

Material and Methods

Preparation of Experiment Field

The experimental field of Hislop College and a local farm, about 18 km away from the city were selected for the study. The field was ploughed to clear off the weeds and also for solarisation of the underneath soil layer to get rid off the unwanted soil fungal flora. The soil was mixed with organic compost manure. In local farm the field was divided in to 5 plots, each plot 100 ft. long and 5 ft. wide (500 sq. ft.). Due to lack of sufficient area the experimental field of the college was divided in to 5 parts each having length 5 ft. and width1 ft. Sufficient gap between each plot was given to prevent the influence of a particular treatment to the neighbouring plots.

Seed Collection and Sowing

The seeds of *Ammi majus* L. were obtained from Hamdard University,Hamdard Nagar,New Delhi. They were cleaned with sieve and weight of 1000 seeds was obtained by digital single pan balance in the laboratory. On the basis of weight, packets of 1000 seeds were made for the convinence in sowing in the farm. 1000 seeds/plot were sown in the farm while 10 seeds/plot were sown in the college experimental field. The seeds are sown at a depth of about 1.5mm-2mm and at a distance of 1 foot between each other. After germination thining and transplanting of the seedling was carried out in a manner that 500 strong plants were left in each plot while 5 plants/plot remained in the college field. Regular watering was done as per the need to keep the soil sufficiently moistured. Deweeding and hoeing was done twice a month to avoid the problems created by weeds and also to make the soil soft with sufficient aeration.

Manures

Nutrient Solution

An organic nutrient solution was prepared by fermenting fresh cowdung with neem oil cake. 200 liters of water with 5kg. cowdung and 250 grams of neem oil cake were kept in a plastic drum. 2 gm of urea was added as a stimulant for bacterial fermentation. The fermentor drum was kept in optimum conditions of temperature *i.e.* 20-30 °C for 15 days. The medium was stirred regularly for better fermentation. The supernatant liquid was taken as stock solution which was further diluted to 10 times with water before giving to the plants. The stock solution was diluted to 20 times for the use of foliar spray to the plants.

Vermi-compost

Vermi-compost was prepared in the college composting unit. It is of 40 ft. by 20 ft. in size and is divided in to 4 compartments by brick partition with holes to connect each other. The degradable garden waste was added to one compartment daily and cowdung occasionally till the compartment became full. Earth worms were introduced into the compartment along with cowdung when the compartment was half full. A layer of about 1 ft. thick cowdung was added at the top and covered with a jute sheet. Required moisture was maintained by spraying water daily. Same process was repeated in each compartment one by one. As after about 70 to 80 days the composting process gets over, the water spraying was stopped so that Earth worms automatically move into the next compartment where moisture and semi degraded organic matter are available.

Inorganic Fertilizers

The inorganic fertilizers were purchased from the market. They are Urea, Sterameal and DAP. Each plot with 480 sq.ft. size was given 100 gm Urea,10 gm Sterameal and 50 gm DAP twice during the crop period. Each 5 sq. ft. plot in the college was given 10 gm Urea, 1 gm Sterameal and 5 gm DAP twice during the crop period.

Sowing of the seeds was done on 12th feb 2011.Thining, hoeing and transplanting was done after about two weeks of sowing (from 25th February to 1st March) in all experimental plots.Out of 5 plots 1st plot was kept as control. While other four plots were given the 1st dose of nutrients after about three weeks of sowing (on 5th March 2011) in the farm land. Different types of nutritional combinations used are: A) Control without any additional nutrition, B) Diluted nutrient solution, C) Vermicompost, D) Vermicompost and nutrient solution, E) Inorganic nutrient solution. Similar treatments were also applied in the college experimental field about three weeks after sowing of seed *i.e.* on 6th March 2011. Foliar spray of nutrient solution was given after four weeks i.e 10th March 2011 to 'B' compartment only.The second dose of nutrients was given from 2nd and 3rd april 2011 in the same manner after a period of eight weeks since sowing.

About 70 per cent germination was observed both in the farm land and the college field.About 500 plants were maintained in each plot of farm land while only 5 plants each were raised in college field. The rest were removed by thining and transplanting.Two doses of manures were given within a gap of four weeks. Flowering occurred after about eight weeks of sowing(7th April 2011) and fruits were matured after about twelve weeks of sowing(5th May 2011).The fruits were plucked and collected in 5 different polythene bags separately from farm land and college field from each plot.They were allowed to dry in the lab by spreading it over newspaper and covered with net cloth for 15 days.Majority of the fruits were splited and rest were broken by soft grinding by hand.They were cleaned by a sieve and then packed in separate transparent polythene bags.The weight of 1000 seeds and total weight of seeds was measured. The results are depicted in Table 29.1 for farmland and Table 29.2 for the experimental field.

Result and Discussion

In the present investigation the effectiveness of four different types of manure/ nutrient treatement on the seed yield and seed weight of *Ammi majus* L. was determined. The findings are depicted in Tables 29.1 and 29.2 and Figures 29.5 and 29.6. On perusal of data given in table:I and II, it becomes obvious that the nutrient type T4(Vermicompost + nutrient solution) has excelled over other treatment type both in terms of seed yield and seed weight. It has exhibited an accretion of about 55 per cent in the seed yield in farm field and about 58 per cent in the Hislop College experimental field. Similarly, an increase of about 53 per cent and 65 per cent is recorded in seed weight from the yield of farm field and Hislop College experimental field respectively.

The overall pattern of effectiveness of different nutrient/manuring combinations in descending order can be expressed as follows:

T4 > T2 > T3 > T5 >T1 where,

T1 stands for Control, T2 for Nutrient Solution, T3 for Vermicompost, T4 for Vermicompost + Nutrient and T5 for Inorganic solution.

It is worth mentioning that better seed yield and seed weight has comparatively been recorded from Hislop College experimental field. This could be attributed to soil quality(garden soil) and moisture availability at this experimental site.

Many scholars have worked on the cultivation techniques of *Ammi majus* L. (Bradu *et al.*, 1970; Kumar.A, 1988; Singh V.P,1963; Sobti *et al.*, 1978.) The present

Figure 29.1: Vegetatative Phase (Control)

Figure 29.2: Vegetatative Phase (Vermicompost+Nutrient)

Figure 29.3: Flowering Phase (Control)

Figure 29.4: Flowering Phase (Vermicompost+Nutrient)

Table 29.1: Effect of Nutrient Types on the Yield of *Ammi majus* L. (Farm Field)

Sl.No.	*Treatment*	*Average Yield Per Plant*	*Total Weight of Crop (Seeds)*	*Percent Increase in Seed Production*	*Weight of 1000 Seeds*	*Percent Accretion in Seed Weight*
1.	Control (T1)	8.140 gm	4070.10 gm	—	0.3650 gm	—
2.	Nutrient solution (T2)	11.822 gm	5911.20 gm	45.235 per cent	0.5220 gm	43.013 per cent
3.	Vermicompost (T3)	10.413 gm	5206.80 gm	27.98 per cent	0.4960 gm	35.89 per cent
4.	Vermicompost + Nutrient (T4)	12.625 gm	6312.60 gm	55.096 per cent	0.5600 gm	53.424 per cent
5.	Inorganic (T5)	8.520 gm	4260.20 gm	4.67 per cent	0.3950 gm	8.219 per cent

Table 29.2: Effect of Nutrient Types on the Yield of *Ammi majus* L. (Experimental Field)

Sl.No.	*Treatment*	*Average Yield Per Plant*	*Total Weight of Crop (Seeds)*	*Percent Increase in Seed Production*	*Weight of 1000 Seeds*	*Percent Accretion in Seed Weight*
1.	Control (T1)	0.776 gm	3.880 gm	—	0.340 gm	—
2.	Nutrient solution (T2)	1.16 gm	5.800 gm	49.484 per cent	0.520 gm	52.941 per cent
3.	Vermicompost (T3)	0.996 gm	4,980 gm	28.350 per cent	0.495 gm	45.58 per cent
4.	Vermicompost + Nutrient (T4)	1.226 gm	6.130 gm	57.989 per cent	0.560 gm	64.705 per cent
5.	Inorganic (T5)	0.818 gm	4.090 gm	5.412 per cent	0.380 gm	11.764 per cent

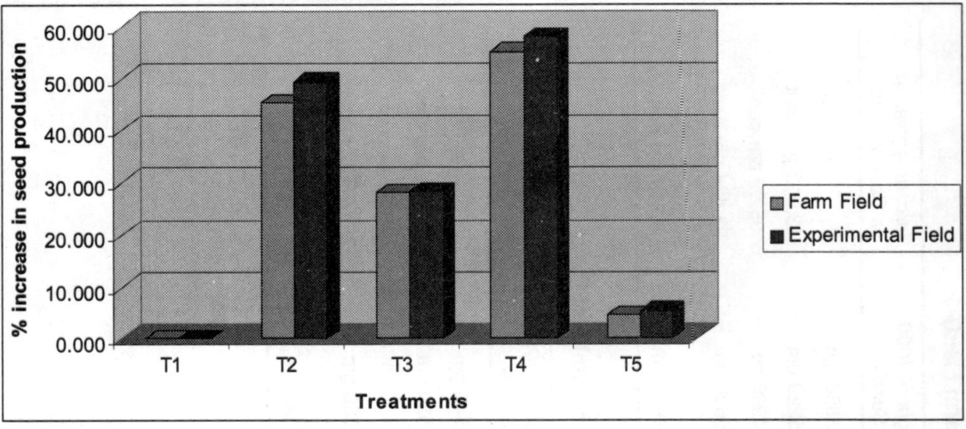

Figure 29.5

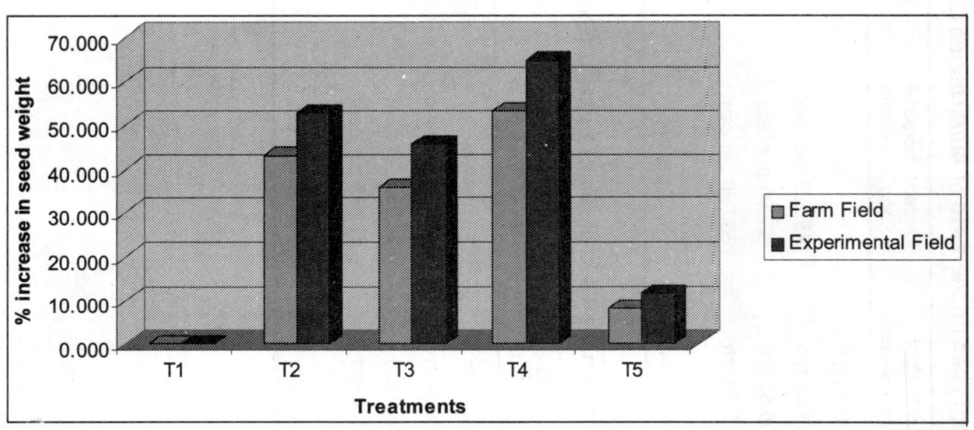

Figure 29.6

experiments were carried out during the month of February to May 2011. Many research workers have suggested different sowing timing for *Ammi majus* L. cultivation in India.

Panda (2002) has stated that an ideal time for direct sowing of *Ammi majus* L. in the field is September whereas Singh and co-workers 1963 had recommended 30th October as the most suitable date for sowing of *Ammi majus* L. at Chakroli (Jammu). They have observed that sowing beyond this date cause significant reduction in the seed yield. These findings are in concurrence with the timing of Duhan and co-workers cf. Panda (2004).

As per Panda (2002) the crop of *Ammi majus* L. on an average yield 12 q/ha of dry seeds. A yield of 1375 kg/ha has been obtained under experimental conditions and 900-1200 kgs/ha under large scale cultivation under Jammu Condition. In Palampur Baijnath area the yield of just 600 kg/ha has been obtained. In the present study a maximum yield of 600 kg/ha has been recorded using organic farming technique during off season months. A better yield could be expected during the optimum growing season between October-May as suggested by various worker (Duke 1988; Panda 2002, 2004; Singh 1963-1983).

As mentioned earlier, the combination treatment of nutritent solution and vermicompost together gives the best possible yield. Thus, one can very rightly go for organic farming in the cultivation of *Ammi majus* L. in Vidarbha region of Maharashtra state where maximum number of farmers suicides occurred due to failure of other crop and economic hardship.

The yield obtained from this research during February to May 2011 that too from the control indicates that the crop of *Ammi majus* L. can be used as a summer crop in the local climatic conditions after the harvesting of the regular crop. The organic farming methods tried can be used to increase the yield remarkably and at the same time maintaining the fertility of the soil. It will definitely boost the economic condition of our poor farmers in Vidarbha region.

Conclusion

The present research work on cultivation of *Ammi majus* L. indicates that this medicinal herb can be grown as an additional summer crop after the harvesting of regular crops which will be over by February. The present study has revealed the importance of nutrients to enhance the production of this crop. Thus, it can be concluded that the organic farming will help to increase the productivity of this crop without compromising quality. Total organic farming of *Ammi majus* L. can yield better result in production without affecting the health of the soil. The production of about 600 kg. per acre will fetch between Rs.10 lacs to 11 lacs in the open market. This will boost the economy, help to improve earning potential of the farmers and enhance their living conditions alongwith earning valuable foreign exchange for the country.

References

Abdel-Monelm EL-Mofty, M.D. (1952) cf. Botanical Dermatology, Phythophotodermatitis, http://telemedicine.org/botanica/bot5.htm

Ashraf,M.; Ahmad,R and Bhatty, M.K 1979 studies on the essential oil content of the Pakistani species of the family Umbelliferae. Part XXX *Ammi majus* seed oil. Pak J Sci Ind Res 22(5); 255-257

Anonymous 1985. The wealth of India – Raw Materials. Vol I council of scientific industrial research, New Delhi

Anonymous 1986. Useful plants of India. Council of Scientific Industrial Research, New Delhi.

Bradu, B.L and Atal, C.K. 1970. Cultivation of *Ammi majus* Linn. In Jammu Indian J. Pharma Sci 165-167

Duke (1988) cf. Botanical Dermatology, *Phythophotodermatitis*, http://telemedicine.org/botanica/bot5.htm

Duhan *et al.*, cf. Panda, H.2004. Cultivation and utilization of aromatic plants – NI 155.

Hansen N.E. 1979. Development of acute mycloid leukaemia in a patient with psoriasis treaed oral 8 methoxypsoralen and long wave ultraviolet light scand J Haematol 22:57-60.

James, A.Duke. 1988. Economic Botany Vol.42, No.3 pp 442-445.

Klaber (1942) cf. Botanical Dermatology, Phythophotodermatitis, http://telemedicine.org/botanica/bot5.htm

Khan,N.A and Rahman, S.Z., 1985. A.Study on Atrilal (*Ammi majus* Linn) a traditional drug for leucoderma Nagarjun 29(3); 4-7.

Kumar, A. 1988. Cultivation and Utilization of *Ammi majus*: A review. Current Res.Med Aromat P1. 10(1); 34-39

Lal, J.1977. Pharmacognostic Investigation of the fruit of *Ammi majus* Linn J. Res Indian Med Yoga Homoeop. 12(3); 68-75

Panda, H.2002. Medicinal Plants – Cultivation and their uses. Asia Pacific Business Press Inc. New Delhi. Pp 11-16.

Panda, H.2004. Cultivation and utilization of aromatic plants – NI 155.

Singh, V.P 1963. Cultivation of Egyptian Herb- *Ammi majus* and *Ammi visnaga.* Indian For 89(8):555-557

Sobti *et al.*, 1978 cf. Panda,H.2002. Cultivation and their uses. Asia Pacific Business Press Inc. New Delhi. Pp 11-16.

Umrao Singh Wadhwani, A.M and Johri B.M 1983 (Repr. Ed) Dictionary of Economic plants of India. Indian Council of Agricultural Research, New Delhi.

2013, Environmental Biotechnology
Editors: D.R. Khanna, A.K. Chopra, Gagan Matta, Vikas Singh &
Rakesh Bhutiani
Published by: BIOTECH BOOKS, NEW DELHI
Pages 251–256

Chapter 30

Response of Phule Jyoti (*Capsicum annuum* L.) Cultivators to Foliar Application of Blue Green Algal Extract, Bioforce, Amruta and Recommended Dose

A.R. Abhang[1] *and S.D. Pingle*[2]
[1]*Arts Science and Commerce College, Rahuri*
[2]*K.J.S. College, Kopergoan, M.S.*

Introduction

Chilli is one of the most important vegetable crop used in our daily diet. Fruit have pungency due to volatile alkaloid "oleoresin capsaicin" which is widely used for medicinal purpose. Chilli as a vegetable is used for its pungency, spicy taste beside the appealing colour it adds to the food. It also used in the different forms,such as spices, condiments, sauces and pickles.

Considering the importance of chilli in daily diet and its demand for export as a condiment and for medical purpose, increase the production by bringing more area under cultivation and increasing yield per unit area of the cultivation.

In India, chilli is grown over an area of 917.6 thousand hectors with an annual production of 779.6 thousand metric tones of dry chillies while in Maharashatra the total area under this crop is about 132.7 thousand hectors with the production of 68.40 thousand metric tones (Anonymous 1996).

Contemporary agriculture has managed to increase the productivity of Horticultural crop with the minimum use of plant nutrients. However, the intensive application of chemical fertilizers deteriorated the soil productiveness and distributed the biological balance of biodiversity (Hansra, 1993). Hence the current effort is to explore the possibility of an alternative for chemical fertilizers.

There are many evidences for the presence of growth hormones in many algal members (Thiman *et al.*, 1942; Knight, 1947; Williams, 1947; Burrows, 1956; Weber, 1958 and Bentley, 1958), but their effect on the growth of crops has not been investigated. As the green revolution is started in agriculture the concept of the use of fertilizers for enhancing growth and yield of crop is changed. Instead of application of fertilizer to the soil various kind of fertilizers in the form of liquid are sprayed on plant. Such types of liquid fertilizers are found to be more beneficial than the soil application.

Therefore, an attempt was made to assess potentiality of some blue-green algal extract on economically significant crop plant like Chilli. The aim of the work has been to determine the nature and extent of any effect on growth and yield of vegetable crop like Chilli. due to spraying of blue-green algal extract as a liquid biofertilizer in the form of foliar spray.

Materials and Methods

The present investigation on Chilli. (*Capsicum annuum* L) var. Phule Jyoti was carried out during 2006-07. Materials used and methods followed during the investigation are described in the succeeding paragraph:

Materials

1. The seeds of Chilli. (*Capsicum annuum* L) var. Phule Jyoti were obtained from All India Co-ordinated Vegetable Improvement Project on MPKV., Rahuri.
2. BGA cultures were developed from the soil samples by the serial dilution.
3. After isolation and identification unialgal cultures were used as a starter for mass cultivation.

Methods

1. Mass cultivation: Simple pit method (Venkatraman, 1969).
2. The algal mass dried under shade for week and then powdered.
3. Algal extract: Bhosale *et al.* (1975).
4. Algal concentration: 1 per cent, 5 per cent, 10 per cent,15 per cent and 20 per cent
5. Plan of layout: Factorial Randomised Block Design (FRBD) with seven treatments, which were replicated thrice.
6. Plot size: Season: Kharif

   Gross plot size: 3.4 x 2.8 m^2

   Net plot size: 2.4 x 2.7 m^2

Spacing: 60 x 45 cm

Crop variety: Phule Jyoti

Total plants per plot: 24

Source of irrigation: Well and canal Irrigation.

7. Concentration percentage of algal extract, Bioforce, Amruta (19:19:19) and recommended dose applied on Tomato.

Table 30.1

Sl.No.	Treatment	Symbol	Concentration Percentage	Remark	
				Spraying	Time of Spraying (DAS)
1.	*Nostoc calcicola* extract	T_1	5-20	1st spraying	Pre flowering
2.	*Lyngbya majuscula* extract	T_2	5-15		
3.	*Scytonema millei* extract	T_3	15-20		
4.	*Oscillatoria subbrevis* extract	T_4	5-15	2nd spraying	At flowering
5.	Bioforce	T_5	2 ml/lit		
6.	Amruta (19:19:19)	T_6	0.5 g/lit		
7.	Recommended dose (NPK) kg/ha	T_7	150:120:125	3rd spraying	Post flowering
8.	Control	T_8	Distilled water		

9. Growth and yield parameters such as plant height, number of branches, number of leaves, plant spread (cm), days for flower initiation, days required for 50 per cent flowering,length of fruits, diameter of fruit, number of fruits per plant, weight of fruit per plant and yield of crop (q/ha) were recorded by using standard methods. Then the yield per hectare was calculated.

Result and Discussion

Foliar application of various compounds (Algal extract, Bioforce and Amruta) had significant influence on plant heigh, number of leaves and branches, plant spread and yield characters like days for flower initiation, 50 per cent flowering, length and diameter of fruit, number of fruit, fruit weight and yield of crop (Table 30.2). Foliar application of BGA extract and two commercial preparation, Bioforce (an organic liquid vitaliser) and Amruta (100 per cent water soluble fertilizer 19:19:19) performed better than control in improving the growth, yield attribute and yield of chilli.

Between the two commercial preparations tried, Bioforce performed better than Amruta, but blue green algal extract like *Nostoc calcicola* extract recorded the tallest plant height [75.14 cm], highest number of leaves[440], number of branches[11.80], plant spread[58.13 cm], dimeter of fruit[1.26 cm], number of fruit per plant[238.73], weight of fruit per plant[678.33g] and fruit yield[171.00 q/ha].However maximum days required for flower initiation [41.66], and 50 per cent flowering [66.33] recorded in Recommended dose. On the contrary, unspread control [T8] registered the shortest plant height [43.64 cm], lowest number of leaves [266], number of branches[7.40],

Table 30.2: Response of Phule Jyoti (*Capsicum annuum* L) Cultivers to Foliar Application Blue Green Algal Extract Bioforce Amruta and Recommended Dose

Treatments	*Symbol*	*Plant Height (cm)*	*Nos. of Expanded Leaves*	*Nos. of Branches*	*Plant Spread (cm)*	*Days for Flower Initiation*	*Days Required for 50 per cent Flowering*	*Length of Fruits (cm)*	*Diameter of Fruits (cm)*	*Nos. of Fruits per Plant*	*Weight of Fruit per Plant (g)*	*Yield of Crops q/ha*
Nastoc extract	T_1	75.14	440	11.80	58.13	34.00	59.00	10.20	1.26	238.73	678.33	171.00
Lyngbya extract	T_2	70.47	337	10.00	56.06	38.66	61.00	09.80	1.20	217.53	620.80	156.50
Scytonema extract	T_3	75.55	407	11.13	57.43	34.33	60.00	10.10	1.16	227.00	624.96	157.55
Oscillatoria extract	T_4	70.20	320	09.80	55.90	38.66	60.66	9.70	1.13	211.13	568.66	143.35
Bioforce	T_5	65.25	310	09.54	54.57	40.33	61.00	9.50	1.10	209.40	565.20	142.48
Amruta (19:19:19)	T_6	58.65	313	09.46	52.67	41.00	63.00	9.20	1.04	208.40	561.20	141.47
Recommended dose	T_7	55.15	290	09.13	52.19	41.66	66.33	9.20	1.00	207.80	550.93	138.88
Control	T_8	43.64	266	07.40	46.65	42.33	68.33	8.50	0.94	197.26	463.73	116.90
Mean		64.231	335.78	9.783	54.200	38.876	62.04	9.525	1.10	214.66	579.22	146.02
S.E		0.165	3.279	0.508	0.153	0.552	0.857	0.130	0.019	0.546	0.254	45.17
CD 5 per cent		0.490	9.738	1.510	0.454	1.640	2.547	0.387	0.058	1.624	0.757	134.14
CD 1 per cent		0.672	13.355	2.070	0.623	2.249	3.493	0.531	0.079	2.227	1.038	183.97

plant spread[52.19 cm], fruit length[8.50 cm], fruit diameter[0.94cm], number of fruit per plant [197.26], fruit weight per plant[463.73g] and yield of crop[116.90 q/ha.] These results are in consonance with the results of Khemnar[2001], Dandawate[2006], Mohite[2007] and Renukabai et.al[2008].. The blue green algal extract proved its superiority over the commercial available formulation in influencing the growth and yield of chilli. It is more vivid that algal extract application in crops promotes the proliferation of root and root hair formation [Mohan *et al.*, 1994, Gencer and Ay.1997, Stephan *et al.*, 1985.] Further the low molecular weight blue green algal extract reported to be directly absorbed by plant, when it is applied on foliage Khemnar 2001, Mohite 2007, Venkatraman *et al.*, 1997].

It has been speculated that the treatment comparison of blue green algal extract application with Bioforce and Amruta had given significantly better results. This might be due to the stimulatory action of blue green algal extract that contain growth hormones (Knight 1947, Burrows 1955, weber 1958, and Bently 1958) which increased uptake from soil.

Effective utilization of foliar applied nutrients promotes of photosynthesis and respiration contributed by the protein quinine groups respectively of accumulated blue green algal extracts.

Conclusion

The present investigation clearly conclude that, by using the low cost technology *i.e.* use of blue green algal liquid biofertiliser has exhibited its assorted influence on growth and yield with great boost over control as well as other commercial liquid fertilizer treatments. It was also observed that different blue green algal extract proved to be superior in increase in height, leaves, branches, plant spread, length diameter of fruit, no. of fruits, fruits weight, yield of crops etc. so the inorganic fertilizers are known to be away from this aspects and long term, algal organic fertilizer are more useful.

Acknowledgement

Authors wishes to express sincere thanks to principal of P.V.P. College, Pravaranagar. And Arts, Science and Commerce College, Rahuri. for the providing laboratory facilities during the investigation.

References

Anonymous (19) Production Year Book. Food and Agricultural Organization of United Ntions.

Bentley, J. A. 1958. Role plant harmones in algal metabolism and ecology. *Nature*, **181:** 1499-1501.

Bhosale, N. B., Untawale, A. G. and Dhargalkar, V. K. 1975. Effect of seaweed extract on the growth of *Phaseolus vulgaris. Indian J. Mar. Sci.* **4:** 209-210.

Burrows, E. M. 1955. Growth control in Fucaceae. Abstract Int. Symp. Seaweed Symposium.

Dandwate, S. C. (2006). Studies on the impact of lift irrigation and agrobased industrial effluents on blue green algae and soils in Sangamner Taluka, Dist. Ahmednagar in Maharashtra. Ph. D. Thesis, University of Pune.

Gencer,O and H. Ay 1997. Effect of seaweed *Ascophyllum nodosum* extract, seaweed *Durvillea potatorum* suspension on the morphological and technological properties of cotton, *Gossypium hisutum*, L.Proc. FAO IRCRNC, *Cotton nutrition and growth regulators, Cairo*, Egypt pp.177-182.

Hansra, B. S. 1993. Transfer of agricultural technology on irrigated agriculture. Fer. News, **38**(4):31-33.

Khemnar, A. S. 2001. Screening of seaweeds from the coast of Maharashtra for their potential as liquid fertilizer. Ph. D. Thesis, University of Pune.

Knight, M. 1947. Biological activity of *Fucus vesiculosus* and *Fucus cerratus*. Proc. Linn. Soc. London. **159:** 87-90.

Mohan, V. R., Venkatraman Kumar, Murugeswari, R. and Muthuswamy, S. (1994). Effect of crude and commercial seaweed extracts on seed germination and seedling growth in *Cajanus cajan* L. *Phykos*, **33** (1-2): 47-51.

Mohite, A.K. (2007) Biochemical studies of freshwater algae and their screening for various potential. Ph.D. Thesis, University of Pune, Pune.

Renuka Bai, N., Laila banu, N. R., Jaquilin Goldi, S. and Prakash J. W. (2008). Effect of seaweed extracts (SLF) on the growth and yield of *Phaseolus aureus* L. *Indian Hydrobiology*, **11** (1): 113-119.

Stephen, A.B., J.K. Macleod, L.S. PALNI and D.S. Lotham 1985. Detection of cytokinins in a seaweed extract. *Phytochemistry* **24**:2611-2614.

Thimann, K.V.,Skoog, F. and Byer, A.C. (1942). The extraction of auxin from plant tissues. II *Amer.J. Bot.* **27**: 598-660

Venkatraman, G. S. 1969. The cultivation of Algae. Pub. Indian Council of Agricultural Res., New Delhi, pp. 319.

Venkatraman Kumar and Mohan, V. R. 1997. Effect of seaweed liquid fertilizer on Black gram. Phykos, **36** (1 and 2): 43-47.

Weber, W. 1958. Zur Polaritat Von Vaucheria, Z. bot. **46**: 161-198.

2013, Environmental Biotechnology *Pages* **257–267**
Editors: **D.R. Khanna, A.K. Chopra, Gagan Matta, Vikas Singh & Rakesh Bhutiani**
Published by: **BIOTECH BOOKS, NEW DELHI**

Chapter 31

Successive Extraction, Phytochemical Investigation and Antimicrobial Screening of *Litchi chinensis* Leaves from Dehradun Region, India

Rishi Kumar Shukla, Abha Shukla, Deepak Painuly and Anirudh Porwal
Department of Chemistry, Gurukul Kangri Vishwavidyalaya, Haridwar – 249 404

Litchi chinensis belongs to family Sapindaceae. In India it is wildly cultivated. The aim of the present study is to establish a correlation between litchi leaves extracts and their activity against human pathogenic bacteria. For this fresh litchi leaves were collected from Dehradun region, (India) and dried in shade; powdered and extracted exhaustively and successively with solvents of different polarity. Phytochemical investigation was performed using conventional natural products identification tests and antimicrobial screening was performed for two Gram negative and three Gram positive human pathogenic bacterial strains. Antimicrobial test was performed by disc diffusion method. All the tests were performed in a triplicate. Presence of majority of phytoconstituents in acetone extract may be responsible for its prominent activity against all the pathogens.

Keywords: *Litchi chinensis, Successive extraction, Phytochemical investigation, and Antimicrobial screening.*

Introduction

The wealth of India is stored in the enormous natural flora which has been gifted to her, endowed with a wide diversity of agro-climatic conditions. India possesses all type of climatic conditions varying from temperate in the Himalayas to tropical in South India, dry in Central India to humid and wet in Assam and Kerala, thus providing conditions favorable for the growth of variety of medicinal and aromatic plants. Due to climatic diversity the chemical constituents in a same plant vary from region to region. In north India Uttarakhand is known for its climatic diversity, Dehradun is popular for its litchi production. The litchi (*Litchi chinensis* sonn.) is an evergreen tree belonging to the family sapindaceae. It is the plant native to South china but now exotic to other parts of the world like Bangladesh, Indonesia, Australia, USA, New Zealand, South Africa and India. In India it is wildly cultivated for its high nutritive value. The fruit of litchi is preferred for its characteristic sweet-acidic taste, excellent aroma, high nutritive value, and bright red colour of its peel [(1)]. It is a rich source of vitamin C[(2)]. The chemical composition of litchi reveals that it had the edible portion 74.5 per cent, moisture 78.5 per cent, citric acid 1.2 per cent, ash 0.69 per cent and sugar 13.57 per cent [(3)]. Medicinally the fruit of litchi is tonic to heart, brain, and liver; allays thirst; very wholesome to the body (yunani) [(4)]. In the last few years, a number of studies have been conducted in different countries to prove antimicrobial efficacy of botanicals [(5),(6)]. The potential of cultivated plants as a source for new anti-microbial drug and botanical pesticides is still largely unexplored. This is also true in India and only a small percentage of plants of this region have been evaluated for antibacterial activity [(7)]. The present study is designed to explore the preliminary phytochemical and anti-microbial screening of leaves extract of *Litchi chinensis* of Dehradun region and to establish a correlation between the phytoconstituents and their pharmacological activity.

Materials and Methods

Collection of plant material

Fresh leaves of *Litchi Chinensis* were collected from "Bombay bagh" Dehradun in the month of April. Two set of plant herbarium were provided for authentication at *Botanical Survey of India, (BSI), 192 Kaulagarh road, Dehradun*. One set of the sample is deposited in the herbarium of Botanical Survey of India, Northern regional centre, Dehradun (BSD). One set of authenticated voucher specimen with *Acc. No. - 113638* of the plant was received and deposited in Department of Chemistry, Gurukul Kangri University, Haridwar, Uttarakhand. The collected leaves of litchi were dried in shade, ground in to powder and stored in polythene bags before use.

Figure 31.1

Extraction of Plant Material

50 gm Shade dried, grinded leaves of *Litchi chinensis* were extracted in a soxhlet exhaustively and sequentially in 700 ml of Petroleum ether (40-60°C) and Acetone. A minimum of 75 cycles of siphoning was done for each solvent. The obtained extracts were concentrated at reduced pressure in a rotary vacuum evaporator. After concentration solvent free extracts were kept in refrigerator for phytochemical and anti-microbial screening.

Preliminary Phytochemical Investigation

Phytochemical analysis for major phytoconstituents of the obtained extracts was undertaken using standard qualitative methods [(8), (9)]. The plant extracts were screened for the presence of biologically active compounds like Alkaloids, Carbohydrates, Steroids, Terpenoids, Protein, Tannin and Phenolic Group, Flavonoids and Amino Acids. Following qualitative test were carried out on each extract separately.

Alkaloids

Solvent free extract (50 mg) is stirred with 2-3 ml of dilute hydrochloric acid and filtered. The filtrate is subjected to following test

Wagner's Test

To a 2-3 ml of filtrate, few drops of Wagner's reagent are added by the side of the tube. A reddish-brown precipitate confirms the test as positive.

Hager's Test

to a 2-3 ml of filtrate, 1 or 2 ml of Hager's reagent (saturated aqueous solution of picric acid) are added. A prominent yellow precipitate indicates the test as positive.

Carbohydrates

Molisch's Test

The extract solution was treated with 2-3 drops of alcoholic α-naphthol. 0.2 ml of concentrated H_2SO_4 is added slowly through the sides of the test tube, purple to violet colour ring appears at the junction.

Fehling's Test

Equal volume of Fehling's A and Fehling's B reagents are mixed along with 2-3 drops of extract solution, boiled, a brick red precipitate of cuprous oxide forms, which indicate the presence of reducing sugars.

Amino Acids

Two drop of Ninhydrin solution is added to 2 ml of aqueous filtrate. A characteristic purple colour indicates the presence of amino acids.

Proteins

Millon's Test

To 2 ml of filtrate, 2-3 drops of Millon's regent is added. White precipitates indicate the test as positive.

Biuret Test

An aliquot of 2 mi of filtrate is treated with one drop of 2 per cent copper sulphate solution. To this, 1 ml of ethanol (95 per cent) is added, followed by excess of potassium hydroxide pellets. Pink colour in the ethanolic layer indicates the presence of protein

Tannins

Ferric Chloride Test

The extract (50 mg) is dissolved in 5 ml of distilled water. To this 2-3 drops of neutral 5 per cent ferric chloride solution are added. A dark green colour indicates the presence of phenolic compounds.

Alkaline Reagent Test

Aqueous solution of the extract when treated with 10 per cent sodium hydroxide solution gives *yellow* to *red* precipitate within short time.

Steroids and Terpenoids

Libermann-Burchard Test

Extract is treated with 2-3 drops of acetic anhydride, boiled and cooled, and then concentrated sulphuric acid is added from the side of the test tube, if it shows brown ring at the junction of two layers and the upper layer turns green then steroids are present, formation of deep red colour indicates the presence of Terpenoids.

Salkowski's Test

Extract in chloroform is treated with few drops of concentrated Sulphuric acid, shaken well and allow to stand for some time, red colour appears in the lower layer indicates the presence of Steroids and formation of yellow colour in lower layer indicates the presence of Terpenoids.

Flavonoids

Zinc-Hydrochloride Reduction Test

To the extract solution a mixture of Zinc dust and conc. Hydrochloric acid is added. It gives red colour after few minutes, which indicates the presence of flavonoids.

Alkaline Reagent Test

To the extract solution 3-4 drops of Sodium hydroxide solution is added, formation of an intense yellow colour, which turns to colourless on addition of 2-3 drops of dilute acetic acid, indicates the presence of flavonoids.

Microbiological Study

Microorganism Used

A total of five out of which two Gram negative and three Gram positive human pathogenic bacterial strains were used in the present study. The bacterial species used for the test are *Escherichia coli* (*ATCC 433*), *Bacillus cereus* (*ATCC 11778*), *Bacillus licheniformis* (*ATCC 1483*), *Salmonella typhi* (*ATCC 733*), *and Staphylococcus aureus*. All the stock cultures in pure form were collected from S.G.R.R.I.T.S department of Life Sciences, Dehradun. All the clinical isolates were biochemically and serologically characterized by standard methods.

Culture Media and Inoculum

Muller Hinton Agar and Agar Nutrient broth (Hi-Media Pvt. Ltd., Bombay, India) are used as culture for the test microorganism. Microbial cultures, freshly grown at 37°C were appropriately diluted in sterile normal saline solution and the turbidity of the suspension is adjusted equivalent to a 0.5 McFarland standard by adding more organisms, so as to obtain the cell suspension between 10^5 to10^8 CFU/ml.

Preparation of Test Extract

Solvent free extracts obtained were dissolved in sterilized and filtered DMSO (filtered with whatman filter of pore size 0.45 micron) to prepare a test solution of extract of desired concentration.

Antibacterial Screening

The anti-bacterial activity of the extract prepared from leaves of *Litchi Chinensis* was performed by standard protocol of Kirby- Bauer disc diffusion susceptibility methods [10]. In this method, 6 mm sterilized filter paper discs (Whatman No.1) were saturated with sterilized extracts of the test solution. The discs absorb approximately a 10 µl of the test solution. The discs were stored and dried in the cold. The impregnated discs are then placed on to the surface of MHA medium. The MHA media has pre-inoculated with test bacteria (inoculum size between 10^5-10^8 CFU/ml).The disc devoid of extract and presence of DMSO served as control. Tetracycline (30 µg/disc) was used as standard. The plates were kept at 4 °C for 1 hour for diffusion of extract, thereafter the plates were incubated at 37 °C for 24 hours. After incubation, zone of inhibition if any around the disc was measured in mm (millimeter). All the test processes were performed in a triplicate in laminar chamber.

Results

Extractive Yield

Shade dried grinded leaves of *Litchi chinensis* when subjected to exhaustive and sequential soxhlet extraction with Petroleum ether and Acetone and further their concentration yields the extracts with different yield and consistency. Table 31.1 represents the per cent yield w/w of the obtained extract. Figure 31.2 represent the per cent yield on pie chart.

Table 31.1: Percentage Extractive Values and Physical Characteristics of Different Extract of *Litchi chinensis* Leaves

Extracts	*Weight of Sample (gm)*	*Weight of Extract (gm)*	*w/w per cent Yield*	*Colour*	*Consistency*
Petroleum ether	50	1.526	3.052	Yellowish	Waxy
Acetone	50	2.726	5.452	Yellowish	Sticky

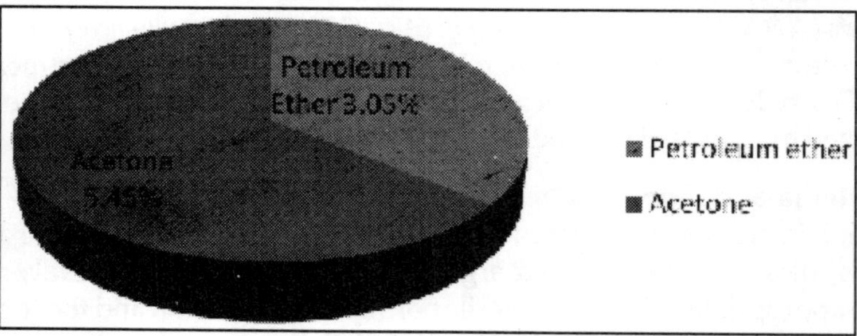

Figure 31.2: w/w per cent Yield of the Extract on Pie Chart

Preliminary Phytochemical Investigation

The preliminary phytochemical screening results of leaf extract of *Litchi chinensis* showed the presence of bioactive secondary metabolite constituents such as alkaloids, flavonoids, steroid, Terpenoids, tannin and phenolic components. Table 31.2 represents the presence of various phytoconstituents in different extracts.

Table 31.2: Phytoconstituents Present in Various Concentrations in Different Extract of *Litchi chinensis* Leaves

Phytoconstituents	*Extracts*	
	Petroleum ether	*Acetone*
Carbohydrates	–	–
Alkaloids	–	+++
Amino Acids	–	–
Steroids	+++	+
Terpenoids	–	++
Protein	–	–
Tannin and Phenols	++	++
Flavonoids	–	+++

+++: High concentration; ++: Moderate concentration; +: :ow concentration; –: Absent.

Acetone extract show the presence of majority of phytoconstituents. Alkaloids, Steroids, Terpenoids, flavonoids, tannins and phenolic compounds are present in an appreciable. Petroleum ether shows the presence of steroids in high concentration. Tannins and phenolic components are present in both extracts. Carbohydrate, proteins and amino acids are absent in all the extracts.

Antimicrobial Screening

The results of antimicrobial screening for both extracts of *Litchi chinensis* leaves against the selected microorganism were investigated by disc diffusion method. The antimicrobial activity showed significant reduction in bacterial growth in terms of zone of inhibition, indicating that the plant exhibited antimicrobial activity against the selected microorganisms and the zone of inhibition was recorded and tabulated in Table 31.3. The Acetone extract exhibit the maximum inhibitory effect against all Gram positive and Gram negative bacterial strain. Petroleum ether shows zone of inhibition aginst *Escherichia coli, Bacillus licheniformis,* and *Salmonella typhi* only. Figure 31.3 shows graphical representation of zone of inhibition of both extracts against the selected pathogens.

Discussion

Emergence of multi-drug resistance in human and animal pathogenic bacteria as well as undesirable side effects of certain antibiotics has triggered immense interest in the search for new antimicrobial drugs of plant origin. In the present study extracts from leaves of *Litchi chinensis* were tested against drug-resistant Gram negative and Gram positive bacteria. Table 31.1 of the results shows that the acetone fraction contains a greater proportion by mass of the component compound. Phytochemical screening shows the presence of various bioactive secondary metabolites which are

Table 31.3: Antimicrobial Screening of Different Extracts of *Litchi chinensis* Leaf. All the values are mean zone of inhibition ± SD

Bacterial Strain	Zone of Inhibition in mm. (Mean ± SD)		
	Ref Drug (T)	Pet. Ether	Acetone
Escherichia coli	14 ± 1.732	11.66 ± 2.08	14.0 ± 2.0
Bacillus cereus	29.0 ± 1.00	–	23.66 ± 1.52
Bacillus licheniformis	24.33 ± 2.08	+	26.33 ± 2.08
Salmonella typhi	19.33 ± 7.09	12.0 ± 2.0	15.33 ± 3.05
Staphylococcus aureus	25.33 ± 1.52	–	24.33 ± 3.78

+: Light zone of inhibition; –: No zone of inhibition.

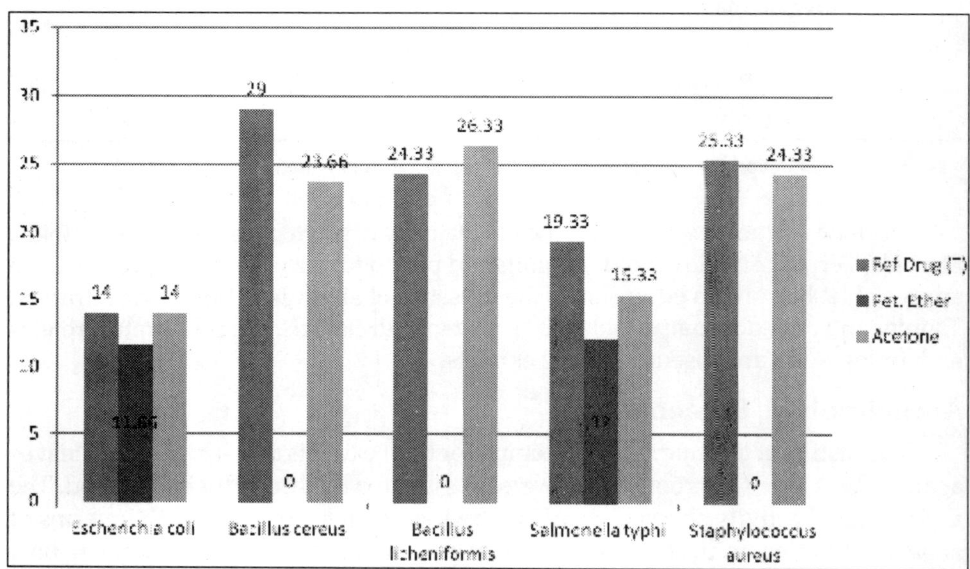

Figure 31.3: Graphical Representation of Zone of Inhibition of Different Extract Against the Selected Pathogens

well known to have curative activity against several human pathogenic microorganisms and therefore could suggest the use traditionally for the treatment of various diseases [(11), (12)]. The literature revealed that the presence of polyphenolic compounds including condensed tannin and flavonoids in *Litchi chinensis* are responsible for its potential anti-cancer, anti-oxidant, [(13), (14)] and cardio-protective activity [(15)]. Acetone extract is evidenced for the presence of Terpenoids which shows cytotoxic, [(16)] anti-tumor, anti-inflammatory, anti-viral and anti- bacterial activity [(17)]. The anti-microbial activity of the extracts could be as a result of alkaloids, flavonoids, steroid, Terpenoids, tannin and phenolic components. On correlating the results an inference can be drawn that presence of majority of phytoconstituents in acetone extract may be responsible for its prominent activity against all the pathogens.

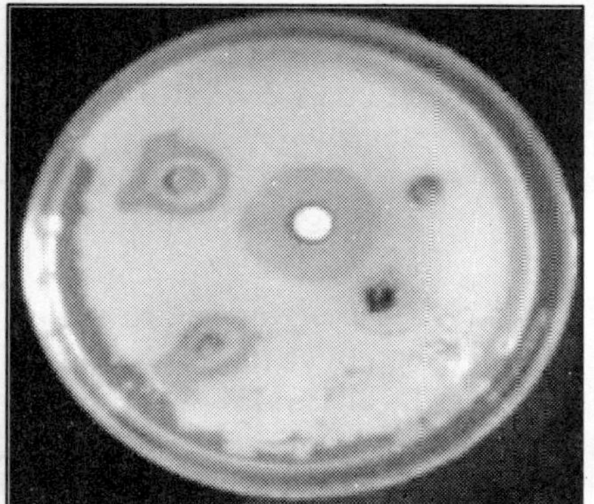

Bacillus licheniformis

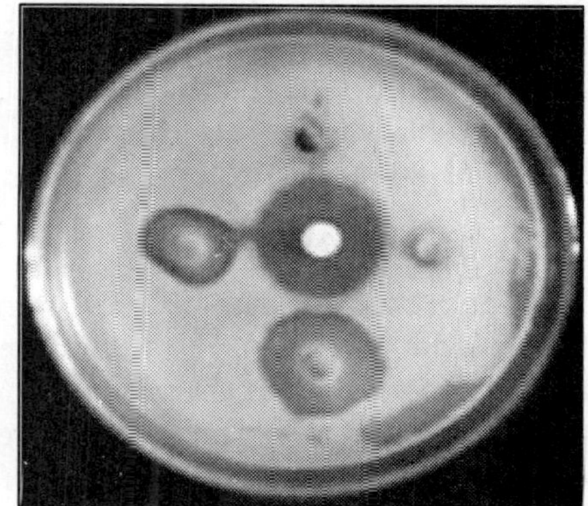

Bacillus cerus

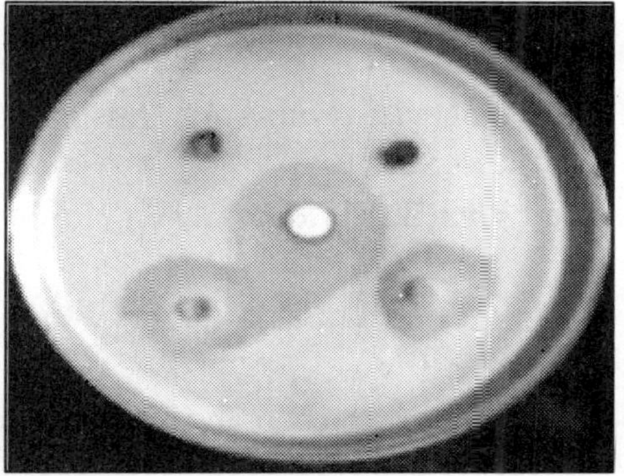

Staphylococcus aureus

Figure 31.4: Antimicrobial Activity of *Litchi chinensis* Leaves Extract (Photographic View)

The active components usually interfere with growth and metabolism of microorganisms in a negative manner [(18)]. Antimicrobial properties of several plant extract have been attributed to these secondary metabolites [(19), (20), (21)]. The reason for the difference in sensitivity between the gram-positive and gram-negative bacteria could be ascribed to the morphological differences between these microorganisms. Gram- negative pathogens having an outer phospholipidic membrane carrying the structural lipopolysccharide components. This makes the cell wall impermeable to lipophilic solutes, while porins constitute a selective barrier to hydrophilic solutes. The gram positive bacteria should be more susceptible having only an outer peptidoglycone layer which is not an effective permeability barrier [(22)]. Phenolic content of plant extract possess antimicrobial activity [(23)] and highly oxidized phenols are more inhibitory because of phenolic toxicity to microorganisms [(24)]. In addition, leaf extracts also possess antimicrobial potential against all pathogens which may be due to the presence of steroids [(25)].

The above mechanism also confirms that due to the presence of appreciable amount of Phenolic compounds in acetone extract, it shows potent antimicrobial activity but it is also evidenced that more hydrophilic phytoconstituents are present in acetone extract; due to which it shows more prominent antimicrobial activity against Gram-negative bacteria. Lastly further explorations on plant derived antimicrobials are needed, to determine the identity of that particular compound in this plant and also to determine their full spectrum.

Acknowledgement

The authors are thankful to Department of Chemistry, Gurukul Kangri University for providing facilities. The authors are grateful to thank Dr. Divya Juyal, Dr. Kunal and Dr. Keerti Katiyar (Department of Life Sciences, S.G.R.R.I.T.S) for their help in proving pure bacterial strains. We are also thankful to Dr. Vishal Kumar Deshwal and Mr. Suryabhan Singh (Department of Microbiology D.P.M.C and Hospital) for their constant help during the work.

References

1. Holcroft, D.M., Mitcham, E.J. (1996). *Post Bio Tech.* 9, pp. 265–281.
2. Ahmad, S. (1956). *The Litchi Leaflet.* Ayub Agric. Res. Inst. Faisalabad–Pakistan, p. 122.
3. Cabin, M. (1954). The lychee in Florida. Bulliten No. 546. Floride Fruit Res. Institute, U.S.A.
4. Kritkar, K.R., Basu, B.D. (1998). *Indian Medicinal Plants.* Lalit Mohan Basu Publication. Vol. 1, pp. 636-637.
5. Sharma, B., Kumar, P. (2009). Extraction and pharmacological evaluation of some extracts of Tridax procumbens and Capparis deciduas. *International Journal of Applied Research in Natural Products,* 1(4). Pp. 5-12.
6. Thenmozhi, M., Rajeshwari, S. (2010). Phytochemical analysis and antimicrobial activity of polyalthia longifolia. *International Journal of Pharma and Bio Sciences,* 1(3) pp. 1-7.
7. Varma, J., Dubey, N.K. (1999). Prospectives of botanical and microbial products as pesticides of tomorrow. *Current Science,* 76 pp. 172-179.
8. Kokate, C.K., Purohit, A.P., Gokhale, S.B. (2006). *Pharmacognosy.* Nirali prakashan, edt. 35, pp. 593-597.
9. Harborne, J.B. (1984). *A Guide to Modern Techniques of Plant Analysis.* Chapman and Hall, London, pp. 4–80.
10. Baur, A.W., Kirby, W.M.M., Sherris, J.C., Turch, M. (1966). Antibiotic susceptibility testing by a standardized single disc method. *American Journal of Clinical Pathology,* 45, pp. 494–496
11. Hassan, M.M., Oyewal, A.O., Amupitan, J.O., Abdullahi, M.S., and Okonkwo, E.M. (2004). Preliminary phytochemical and antibacterial investigation of crude extract of the root bark of Detariaummicrocarpum. *Journal of chemical society Nigeria,* 29, pp. 26-29.

12. Usman, H., and Osuji, J.C. (2007). Phytochemical and *in vitro* antimicrobial assay of the leaf extract of Newbouldia laevis. *African journal of traditional complementary and alternative medicine*, 4, pp. 476-480.
13. Wang,X., Yuan, S., Wang, J., Lin, P., Liu, G., Lu, Y., Zhang, J., Wang, w., Wei, Y. (2006). Anticancer activity of litchi fruit pericarp extract against human breast cancer in vitro and in vivo. *Toxicology and applied pharmacology*, 216, pp. 168-178.
14. Li, J., and Jiang, Y. (2007). Litchi Flavonoids: Isolation, Identification and Biological Activity. *Molecules, 12, pp. 745-758.*
15. Deng, J.Y., Yuan, Y.C., Chin, L.H., Cheng, W.L., Yu, W. (2010). Protective effect of a litchi (*Litchi chinensis*)-flower-water-extract on cardiovascular health in a high-fat/cholesterol-dietary hamsters. *Journal of Food Chemistry*. 119, pp. 1457-1464.
16. Xinya,X., Haihui, X., Jing, H.,Yueming, J., Xiaoyi, W. (2010). Eudesmane sesquiterpene glucosides from lychee seed and their cytotoxic activity. *Journal of Food Chemistry*. 123, pp. 1123-1126.
17. Sashi, B.M., and Sucharitra, S. (1997). *Phytochemistry*, 44, pp. 1185-1236.
18. sAboba, O.O., Etuwape, B.M. (2001). Antibacterial properties of some Nigerian species. *Bio resources community*.13, pp. 183-188.
19. Sharma, D.K. (2006). Pharmacological properties of flavonoids including flavonolignans- integration of petro crops with drug development from plant. *Journal of industrial research*, 65, pp. 477-484.
20. Kosalec, I., Pepeljnjak, S., Bakmaz, M., Vladimir-Kenezevic, S. (2005). Flavonoids analysis and antimicrobial activity of commercially available propolis products. *Acta pharma*, 55, pp. 423-430.
21. Tim Cushine, T.P., Lamb, A.J. (2005). Antimicrobial activity of flavonoids. *International journal of antimicrobial agents*, 26, pp. 343-356.
22. Nikaido, H., Vaara, M. (1985).molecular basis of bacterial outer membrane persmeability. *Microbiology review*, 49 (1), pp. 1-32.
23. Acar, G., Dogan, N.M., Duru, M.E., Kivrak,I. (2010). Phenolic profiles, antimicrobial and antioxidant activity of various extract of crocus species in Anatolia. *African journal of microbiological research*, 4 (11), pp. 1154-1161.
24. Ciocan, I.D., Bara, I.I. (2007). Plant products as antimicrobial agents. *Genetic si biologie molecular*, 8 (1), pp. 151-156.
25. Cowan, M.M. (1999). Plant products as antimicrobial agents. *Clinical microbiology review*, 12 (4), pp. 564-582.

2013, Environmental Biotechnology *Pages* ***269–273***
Editors: **D.R. Khanna, A.K. Chopra, Gagan Matta, Vikas Singh & Rakesh Bhutiani**
Published by: **BIOTECH BOOKS, NEW DELHI**

Chapter 32

Nutraceutical Evaluation of *Brachystelma edulis* (Coll. and Helmsl.)

Varsha D. Jadhav, Swati R. Deshmakh, Sujata R. Valvi and Seema R. More

Department of Botany, Shivaji University, Kolhapur – 416 004, M.S.

Many wild edible plants provide minerals, vitamins, fibers, fatty acid etc. The present paper deals with the analysis of mineral composition, proximate values and antioxidant capacity of leaves and tubers of *Brachystelma edulis*. Among the entire minerals, calcium was found in large quantity in leaves and tuber. Where as leaves are rich in polyphenols and ascorbic acid.

Keywords: *Mineral analysis, Antioxidant capacity, Proximate value, Brachystelma edulis.*

Introduction

The wild edible plants play a vital role in fulfilling nutritional requirements of the rural people in remote areas of our country. A wide variety of wild edible plants such as leaves, fruits, flowers, roots, barks, nuts, tubers etc. are consumed by rural masses. The edible tubers not only enrich the diet of the people but also possess medicinal properties. Also many of tropical tuber species are used in the preparation of stimulants, tonics, carminatives and expectorants (Edison *et al.*, 2006). The wild edible tuberous plants are important but fast disappearing element of diet so far have been largely neglected in scientific studies. Nevertheless, a careful examination of the literature reveals that there are still a large number of wild edible tuberous species which are inexpensive and commonly used by local people, whose nutritional potential have not yet been adequately studied.

Material and Methods

Sample Preparation

Selected wild edible tuberous plants were collected from various localities of Satara, *Brachystelma edulis.* Efforts made to collect these plants in flowering and fruiting conditions for the correct botanical identification. Healthy and disease free edible plant part/s selected and dried them under shade so as to prevent the decomposition of chemical compounds present in them. All the dried material powdered in blander for further study.

Proximate Composition

Dry matter and moisture of the material were determined by following the method by AOAC (1990).Ash value was determined by following the method of AOAC (1990). Carbohydrates were estimated according to the method described by Nelson (1944). The Crude fat and crude fiber content was determined by following the method of Sadasivam and Manikam (1992). Total nitrogen was estimated according to the method of Hawk *et al.* (1948). The quantity of protein was calculated as 6.25×N (AOAC, 1990).

Mineral Content

The acid digestion method of Toth *et al.* (1948) has been followed for the analysis of inorganic constituents. Five hundred mg oven dried powdered was transferred to 150 ml clean borosil beaker and to that 10 ml concentrated HNO_3 were added. It was covered with watch glass and kept for an hour till the primary reactions subsided. Then, it was then heated on hot plate till all the material was completely dissolved. It was allowed to cool to room temperature and then 10 ml of Perchloric acid (60 per cent) were added to it and mixed thoroughly. Then, it was then heated strongly on the hot plate until the solution became colourless and reduced to about 2-3 ml. While heating, the solution was not allowed to dry. After cooling, it was transferred quantitatively to 100 ml capacity volumetric flask, diluted to 100 ml with distilled water and kept overnight. Next day the extract was filtered through Whatman No. 44 (Ashless) filter paper. The filtrate was stored properly and used for analysis of inorganic constituents.

Antioxidant Analysis

The method of Folin and Denis (1985) was employed for determination of the total polyphenols content in plant material. A titrimetric method described by Sadasivam and Manikam (1992) was followed to determine the ascorbic acid content. To study the enzyme peroxidase activity the method of Maehly (1954) was followed.

Catalase activity was assayed by following the method of Luck (1974) as described by Sadasivam and Manikam (1992). Superoxide dismutase was determined by following the method described by Giannopolitis and Ries (1977), with slight modifications. The carotenoids were estimated following the method of Kirk and Allen, 1965.

Results and Discussion

Proximate Nutrient Analysis

In these experiments moisture, dry matter, ash, crude fiber, crude fat, crude protein, reducing sugar, total sugar, starch and energy content of the wild edible tuberous plant were analyzed. The results are shown in Figure 32.1.

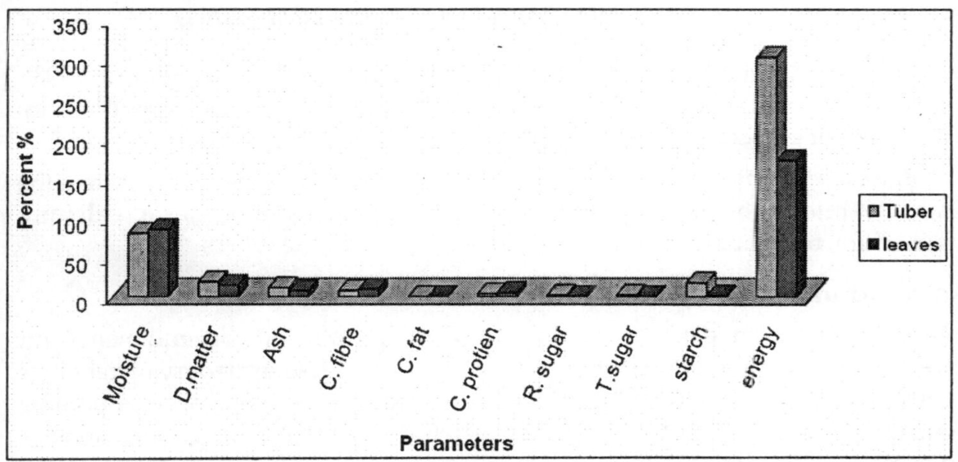

Figure 32.1: Proximate Composition of *Brachystelma edulis*

In tubers of *Brachystelma edulis* dry matter, ash, crude fat, reducing sugar, total sugar, starch and energy is higher as compare to that leaves. In leaves moisture, crude fiber, crude protein is more than the tuber.

Mineral Analysis

Living organism requires a continuous supply of large number of substances from food to complete their life cycle. This supply is called as nutrition. The mineral nutrition is an important aspects and it play pivoted role in human life for healthy growth. Such type of mineral is easily available in wild edible plants. Thus it is through worth to study the mineral nutrition of *Brachystelma edulis*. Minerals may be broadly classified as macro and micro elements. The macro-minerals include calcium, phosphorus, sodium and chloride, while the micro-elements include iron, copper, cobalt, potassium, magnesium, iodine, zinc, manganese, molybdenum, fluoride, chromium, selenium and sulfur (Eruvbetine, 2003).Along with several organic compounds, it is now well established that many trace elements play a vital role in general well-being as well as in the cure of diseases. Minerals are chemical constituents used by the body in many ways. Minerals are not provided energy, but

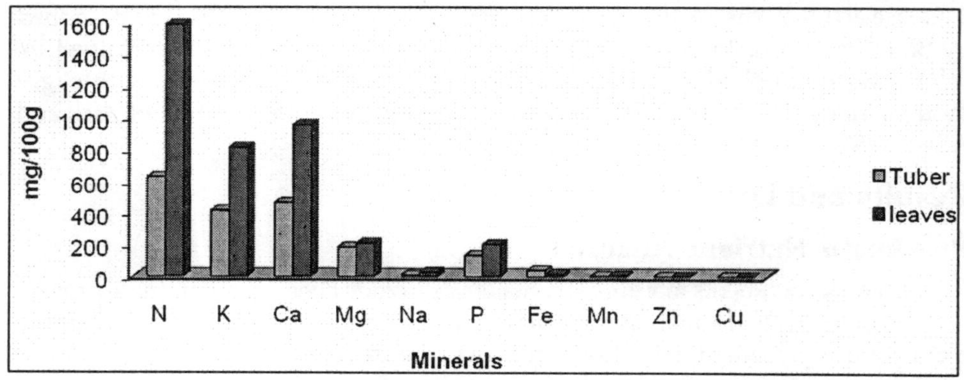

Figure 32.2: Mineral Composition of *Brachystelma edulis*

they played important roles in many activities in human body (Malhotra, 1998). About 14 elements are essential to human health. Their deficiency causes diseases; where as their presence in excess may result in toxicity.

Brachystelma edulis showed higher amount of iron and copper in tuber than the leaves. While higher amount of nitrogen, phosphorus, potassium, calcium, magnesium, manganese, sodium and zinc in leaves than the tubers.

Antioxidant Activity

Antioxidants have become synonymous with good health. Several plants and vegetables used in traditional medicine can provide diverse secondary metabolites with antioxidant potentials in that most of which are isolated phenolic compounds (Ramarathnam *et al.*, 1997). The continued search among plant secondary metabolites for natural antioxidants has gained importance in recent years because of the increasing awareness of herbal remedies as potential sources of antioxidants.

Polyphenols, ascorbic acid, enzyme peroxidase, catalase, superoxide dismutase, and carotenoid content of tubers and leaves of the wild edible tuberous plants examined in this study are shown in Figure 32.3.

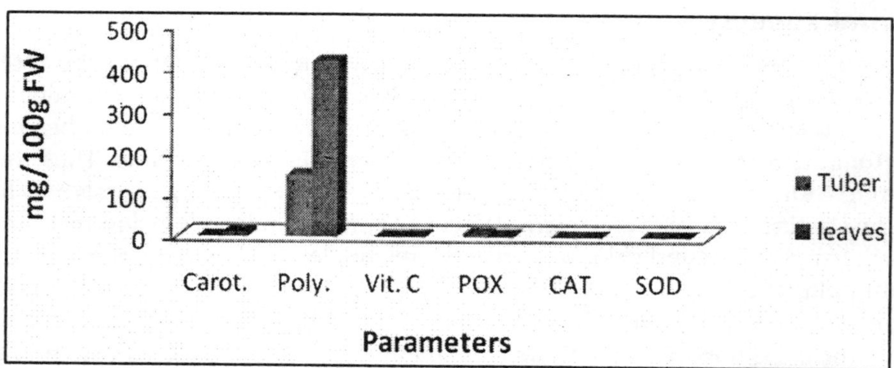

Figure 32.3: Antioxidant Activity of *Brachystelma edulis*

Tuber of *Brachystelma edulis* showed higher amount of peroxidase while in leaves, polyphenol, ascorbic acid, catalase, superoxide dismutase and caratenoids are more in leaves. Leaves and tuber are rich in polyphenols.

References

AOAC, (1990). Official Methods of Analysis. *Association of Official Analytical Chemists*, Washington DC.

Edison, S., Unnikrishnan, M., Vimala, B., Pillai, S.V., Sheela, M.N., Sreekumari, M.T. and Abraham, K. (2006). Biodiversity of tropical tuber crops in India; NBA Scientific Bulletin 7. National Biodiversity Authority of India, Chennai, Tamil Nadu, India: 60.

Eruvbetine, D. (2003). Canine nutrition and health. A paper presented at the seminar organized by kensington pharmaceuticals Nig. Ltd.

Folin, O. and Denis, W. (1985). A colorimetric estimation of phenols (and phenolic derivatives) in urine. *J. Biol. Chem.*, 22: 305-308.

Ginnopolitis, C. N. and Ries, S.K. (1977). Superoxide dismutase: Occurance in Higher Plants. *Plant physiol.*,59: 309-314.

Hawk, P.B., Oser, B.L. and Summerson, W.H. (1948). Practical physiological chemistry (Publ.). The lockiston Co. USA

Kirk, J.O.T. and Allen, R.L. (1965). Dependance of chloroplast pigment on actidione Arch. *Biochem. Biophys.Res.Commun.*, 21: 523-530.

Luck, H. (1974). In: Methods in Enzymatic Analysis II (ed.) Bergmeyer. (Publ.) Academic Press, New York: 885.

Maehly, A.C. (1954). Methods in biochemical analysis (Ed.) Glick D. (Publ.) Interscience publishers Inc. New York: 385-386.

Malhotra, V.K. (1998). Biochemistry for students 10th Edition Jaypee Brothers medical publishers (P) Ltd. New Delhi India.

Nelson, N. (1944). A Photochemical adaptation of the Somogyi method for the etermination of glucose. *J. Biol. Chem.*, **153**: 375-380.

Ramarathnam, N., Ochi, H., Takeuchi, M. (1997). Antioxidants Defense system in Vegetable Extracts. In: Shahidi F (ed). Natural antioxidants: Chemistry, Health Effects and Applications. Aoca Press: Champaign, IL: 76-87.

Sadashivam, S. and Manikam, A. (1992). Biochemical method for agricultural sciences, Willey, Eastern Ltd.: 105.

Toth, S.J., Prince, A.L., Wallace, A. and Mikkenlsen, D.S. (1948). Rapid quantitative determination of eight mineral elements in plant tissue vy systematic procedure involving use of a flame photometer. *Soil Sci.*,66: 459-466.

2013, Environmental Biotechnology *Pages* **275–281**
Editors: **D.R. Khanna, A.K. Chopra, Gagan Matta, Vikas Singh & Rakesh Bhutiani**
Published by: **BIOTECH BOOKS, NEW DELHI**

Chapter 33

Toxicological Effects of Inhaled Mosquito Coil Smoke on the Rat Trachea and Lung: A Histological Study

S.R. Akarte and Y.D. Akhare
Department of Zoology, VidyaBharati Mahavidyalaya, C.K. Naidu Road, Camp, Amravati, Maharashtra

The allethrin containing mosquito coils are extensively used as an insect repellent specially advertised as mosquito repellent to escape from mosquito-bite. The availability and relative cheapness of this insecticide have made mosquito coil very popular. The inhalation of the smoke of these burning mosquito coils may cause the adverse effect on respiratory tract. Eighteen adult rats of wild strain were randomly divided in 3 groups of 6 rat each, where group I served as control, second group was exposed for 15 days and third group was exposed for 30 days with smoke of one burned coil every day. Few studied have demonstrated the pathologic reaction yielded by smoke inhalation on the airways in rats. The experiment was conducted on wild strain rat to know the effect of excessive smoke inhalation to observe the histological changes in lungs and trachea. The histomorphological observations after 15 and 30 days inhalation by rats revealed the focal decilation of tracheal epithelium and metaplasia of epithelial cells. The alveolar distortion was also observed in 30 day-exposed rats. Even after feeding the same quantity of food to both groups *i.e.* experimental and control, the rats exposed to smoke revealed the reduction in body weight as compared to control.

Keywords: *Mosquito coil smoke, Trachea, Lung, Rats.*

Introduction

The interaction between humans and the materials that make up for environment occur continuously with each having an influence on the other. With chemicals and physical agents, the response of the living organisms comes after contact with the agent. There are three main results by which a person can come into contact with these agents. A material can land on one's body surface (dermal exposure), it can be swallowed after contact with the mouth and oral cavity (oral exposure) and it can be breathed through the nose or mouth into the lungs (inhalation exposure). Marketing of repellents in India is well organized, so that many brands can be found throughout the country. The current Indian market for various repellents is in the range of Rs 500-600crores with annual growth of 7 to 10 per cent. This increase in growth rate indicates the day by day increase in use of these repellents, resulting of constant addition of such latent environmental pollutants, affecting the human health.

The mosquito coils are burned to generate smoke which effectively drives mosquitoes out of the room or may kill it. This practice is currently used in numerous households in Asia, Africa and South America. The major active ingredients of the mosquito coil are allethrins, the active substance, which is a synthetic analogue of natural pyrethrum. (Lukwa and Chandiwana, 1998). Pyrethrins have low chronic toxicity to humans and low reproductive toxicity in animals. Although headache, nausea and dizziness was observed in human being by spray of 0.01-1.98 mg/m^3 pyrethrins for 0.5 – 5 hr (Zhang *et al.*, 1991).

The experiment conducted on female albino rats exposed to mosquito coil smoke indicated that, there were sign of toxicity to liver and spleen (Liu and Wong, 1987). It has been suggested that the toxic effect of mosquito coil smoke is caused by its combustion products such as sub micron particles coated with heavy metals, allethrin and a wide range of organic vapours like phenol, O-cresol, benzene and toluene (Liu and sun, 1988). Chronic exposure of high doses of allethrin (50-200mg/kg/day for two years) produced the signs such as increased liver and kidney weights with adverse morphological changes in liver tissue in female rats (WHO, 1989). There was however, no observation of carcinogenic or teratogenic and/or developmental effects in exposed rats (US Environmental Protection Agency, 1988).

The human exposure to the vapors emanating from mosquito coil is on a daily basis in most residential areas. The main active ingredient of the mosquito coil is the d-trans allethrin. The d-trans allethrin is effectively used for the control of adult mosquitoes in the house. This study was therefore designed to examine the effect of inhaling mosquito coil smoke on the histomorphological integrity of the trachea and lung in wild black rats.

Materials and Methods

Adult male and female wild black rats weighing 200-230 g. were kept in rearing cages. They were provided with the pellet diet and water *ad libitum* for 7 days to observe the normal health or any sign of lethargic activity. The rats were found in good active form throughout this period. After this, the rats were exposed for inhalation of the fumes of mosquito coil for 15 and 30 days in two groups with one coil burned every day. For exposure the cages containing rats were kept in gas chamber of 2×2×3 feet. Glass chamber was having a ventilation to avoid the suffocation of the experimental animals.

The experimental animals were exposed with the fumes of mosquito coil daily in the glass chamber up to the completion of one coil for 15 and 30 days and sacrificed at the end of each experimental period. The changes in the body weight of both control and experimental rats were recorded at every alternate day for 30 days. Rats were dissected to take out the trachea and lung which were blotted and weighed on the electronic balance. The tissues like lung and trachea were used for the histological study.

Results and Discussion

Inhaled gases and vapors may be absorbed throughout the respiratory tract. The net driving force for absorption of inhaled gases or vapors is diffusion which is the net movement of molecules from a region of high concentration to a lower concentration. Vapors absorption continues until the vapor concentration in the air and tissue are the same. At this point the system is at equilibrium and no further transport of vapor occurs.

Table 33.1: Effect of Mosquito Coil Smoke Inhalation after 15 and 30 Days on Body Weight (g) of Rats

Duration	*Group*	*Initial Body wt. (g.)*		*Final Body wt. (g.)*	
		Range	*Mean±SD*	*Range*	*Mean±SD*
15 Days	Control	198.24–222.34	211.24±9.6	203.55–253.23	219.87±18.76
	Experimental	198.84–240.33	210.73±15.79	195.14–212.87	204.14±7.69
30 Days	Control	198.24–222.34	211.24±9.6	205.84–240.14	217.14±15.30
	Experimental	199.74–220.14	209.53±6.63	192.74–210.14	202.07±6.39

Values are the mean ±SD of 6 animals.

Inhaled mosquito coil smoke may be absorbed directly with in the nasal cavity, thereby sparing the lung from direct exposure. The smoke may be either physically absorbed within the nasal tissues, a reversible process driven in part by the relative solubility of the smoke in air or tissue or irreversibly bound to tissue components.

In the present study the mean body weight was slightly increased in the control rats and slightly reduced in all the groups exposed to the mosquito coil smoke though the slight increase and reduction in body weight were not statistically significant (Table 33.1). The body weight loss observed in this study though not significant agrees with other works of the same nature (Parker *et al.*, 1984, Ishmael and Litchfield, 1988).

The lung tissue of control rats in the study, showed numerous alveoli and it communicate with one another by apertures in their walls. Respiratory bronchioles are seen which having ciliated epidermis (Figure 33.2a). While the lung tissue of smoke-inhaled rats in the study, showed thickening of the bronchiolar epithelial wall, thickening of alveoli septa and consolidation in alveolar areas after 15 days of exposure which are indicative of toxic lung insult (Figure 33.2b). After 30 days mosquito coil exposure to rat showing disrupted and sloughed epithelial lining of the bronchioles. Many surface lining cells appeared enlarged with cytoplasmic vacuolation (Figure 33.2c). These findings are consistent with observations made when mice were exposed to pneumotoxicants like naphthalene, butylated hydroxytolune and 1,1-dichloroethylene (DCS), which caused time-dependant damage to Clara and alveolar cells of the lung with concomitant reductions in some

Figure 33.1a: Sagital Section of Trachea of Control Rat Showing the Epithelial Cells, Membranous Part and Cartilaginous Part. H and E Mag. ×100.

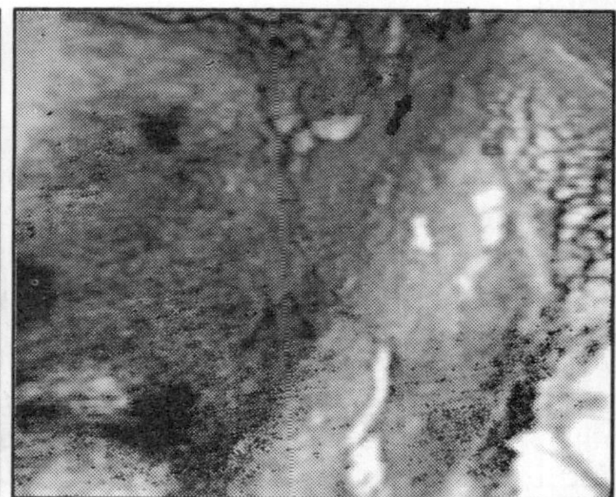

Figure 33.1b: Sagital Section of Trachea of Rat Exposed to 15 Days Mosquito Coil Smoke Showing Epithelial Sheding. H and E Mag. ×100.

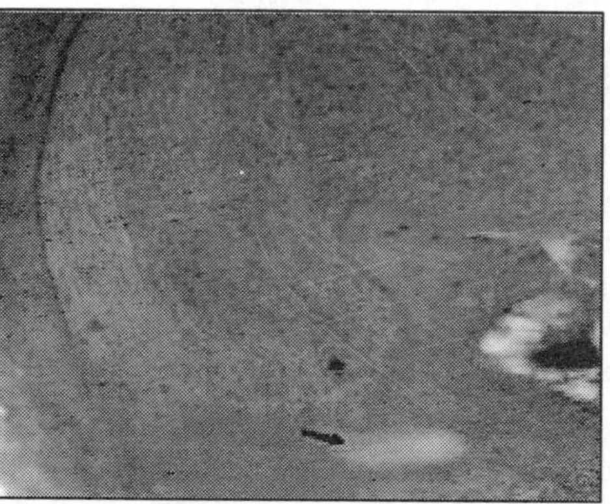

Figure 33.1c: Sagital Section of Trachea of Rat Exposed to 30 Days Mosquito Coil Smoke Showing Local Oedema in Epithelial Cells. H and E Mag. ×100.

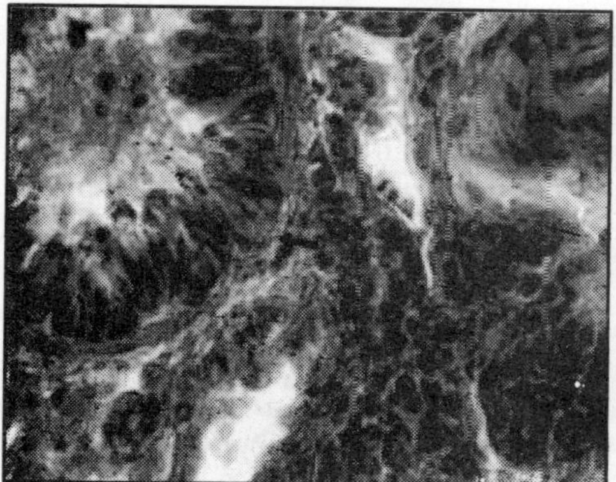

Figure 33.2a: Microphotograph of Lung of Control Rat Showing Numerous Alveoli and Respiratory Bronchioles having Ciliated Epidermis. H and E Mag. ×400.

Figure 33.2b: Microphotograph of Lung of Rat Exposed to 15 Days Mosquito Coil Smoke Showing Thickening of Bronchiolar Epithelial Wall and Alveolar Septa. H and E Mag. ×400.

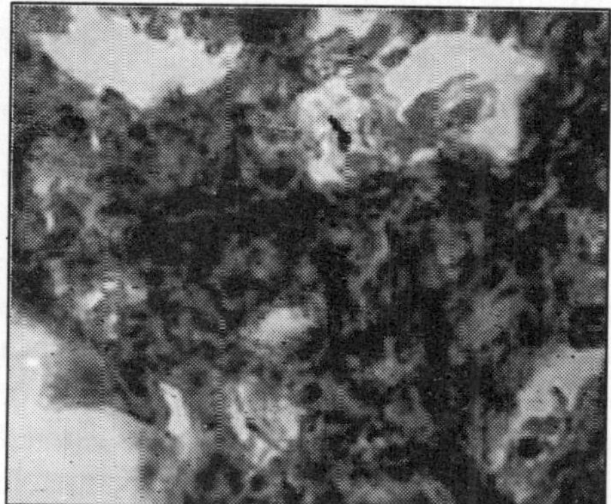

Figure 33.2c: Microphotograph of Lung of Rat Exposed to 30 Days Mosquito Coil Smoke Showing Vacuolated and Disrupted Lining of Bronchioles. H and E ×400.

mono-oxygenase activities and increase in lung wet weight (Okine *et al.*, 1985, Tong *et al.*, 1981).

In the present study, histologically, the trachea of rat was clearly divided into the membranous and cartilaginous portions. Respiratory epithelium was composed of pseudostratified ciliated columnar cells on a basal lamina which contained the basal cells. Below the basal lamina was the lamina propria containing the trachealis muscle which was closely attached to the external surface of the cartilaginous rings and separated the cartilages from the epithelial layer (Figure 32.1a). The tracheal epithelium of rat from the excessive mosquito coil smoke inhaled groups, however, appeared uniformly flatted. This aspect was constant and was due to a change in the epithelial structure, which appeared monostratified. No cell lysis, local oedema or epithelial sheding was observed and the usual cell subtypes, such as ciliated cells were present. The basement membrane appeared normal and inflammatory cells occasionally infiltrated the lamina propria.

The branching pattern of the tracheobronchial tree influences air flow dynamics and the amount and location of particle deposition is lined with a mucus-covered ciliated epithelium. Insoluble particles that deposit in this are cleared by mucocilliary transport toward the oral cavity for removal by swallowing or expectoration. Clearance in this region is generally foster for the larger rather than the smaller airways. Particles may enter the peribronchiolar regions by crossing bronchial epithelium or be absorbed into the circulation after dissolution in the mucus layer.

Epidemicologic studies have shown that long-term exposure to mosquito coil smoke can induce asthma and persistant wheeze in children (Azizi and Henry, 1991; Fagbule and Ekanem, 1994; Koo and Ho, 1994). When a group of 30 female albino rats were exposed to mosquito-coil smoke in a 22.5-(m-sup.3) chamber for 8 hr per day, 6 days per week, for 6 months, these rats lost typical ruffed membranes of their alveolar macrophages. In addition, the levels of total protein and lecithin and the activities of lactate dehydrogenase, acid phosphatase, and (beta)-glucuronidase in the lung- lavage fluid of the rats were significantly higher than those in a control group that was exposed to air for the same exposure duration (Liu *et al.*, 1989). The toxic effects of allethrin result from its action on the nervous system. After intravenous injection with a lethal dose of bioallethrin (4mg/kg body weight), initial tremors were followed by death within 20 min. Hyperexcitation and tremors usually developed a few minutes after application (Verschoyle and Barness, 1972; Carlton, 1977).

The remaining components of mosquito coil are organic fillers, binders, dyes, and other additives capable of smoldering well. The combustion of the remaining materials generates large amounts of sub micrometer particles can reach the lower respiratory tract and may be coated with a wide range of organic compounds, some of which are carcinogens or suspected carcinogens, such as polycyclic aromatic hydrocarbons (PAHs) generated through incomplete combustion of biomass (mosquito coil base materials). Because coil consumers usually use mosquito coils for at least several months every year, cumulative effects from long-term exposure to the coil smoke may also be a concern.

References

Azizi, B. H. O. and R. L. Henry, 1991. The effects of indoor environmental factors on respiratory illness in primary school children in Kuala Lumpur. Int. J. Epidemiol., 20: 144-149.

Carlton, M., 1977. Some effects of cismethrin on the rabbit nervous system. Pestic. Sci., 8: 700-712.

Fagbule, D. and E. E. Ekanem, 1994. Some environmental risk factors for childhood asthma; A case-control study. Ann. Trop. Paediatr., 14: 15-19.

Ishmael, I. and M. H. Litchfield, 1988. Chronic toxicity and carcinogenic evaluation of permethrin in rats and mice. Fundam. Applied Toxicol., 11: 308-322.

Koo, L. C. L. and J. H. C. Ho, 1994. Mosquito coil smoke and respiratory health among Hong Kong Chinese epidemiological studies. Indoor Environ., 3: 304-310.

Liu, W. K. and M. H. Wong, 1987. Toxic effects of mosquito coil (a: mosquito repellent) smoke on rat. Toxicol. Lett., 39: 223-239.

Liu, W. K. and S. E. Sun, 1988. Ultrastructural changes if tracheal epithelium and alveolar macrophages of rats exposed to mosquito coil smoke. Toxicol. Lett., 41: 145-157.

Liu, L., He, F. and S. Wong, 1989. Clinical manifestations and diagnosis of acute pyrethroid poisoning. Arch. Toxicol., 63: 54-58.

Lukwa N., S. K. Chandiwana, 1988. Efficacy of mosquito coil containing 0.3 per cent and 0.4 per cent pyrethrins against *An gambiae sensu lato mosquitoes*. Cent. Afr. J. Med. 44(4): 104-107.

Okine L. K., J. M. Goochee and J. E. Gram, 1985. Studies on the distribution and covalent binding *in vivo*. Biochem. Pharmacol. 34(2): 4051-4057.

Parker, C. M., D. R. Patterson and Van G. A. Gelder, 1984a. Chronic toxicity and carcinogenicity evaluation of fenvalerate in rats. J. Toxicol. Environ. Health, 13: 83-97.

Tang, S. S., V. T. Hirokata, M. A. Trush, E. G. Mimnaugh, E. Ginsburg, M. C. Lowe and T. E. Gram, 1981. Clara cell damage and inhibition of pulmonary mixed-function oxidase activity by naphthalene. Biochem. Biophy. Res. Comm., 100: 944-950.

US Environmental Protection Agency, 1988. Pesticide fact sheet number 158 Allethrin stereoisomers. Office of pesticides and toxic substances, Washington, 2-8.

Verschoyle, R. D. and Barness, J. M., 1972. Toxicity of natural and synthetic pyrethrins to rats. Pestic. Biochem. Physiol., 2: 308-311.

World Health Organisation, 1989. Allethrins; Allethrin, D-allethrin, Bioallethrin, S-biollethrin- Environmental Health Criteria 87. International Programme on chemical safety, Geneva, Switzerland: 2-20.

Zang, W., J. Sun, S. Chen, Y. Wu, F. He, 1991. Levels of exposure and biological monitoring of pyrethroids in spray men. Br. J. Ind. Med. 48: 82-86.

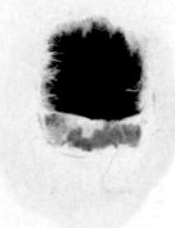

2013, Environmental Biotechnology *Pages* ***283–287***
Editors: **D.R. Khanna, A.K. Chopra, Gagan Matta, Vikas Singh & Rakesh Bhutiani**
Published by: **BIOTECH BOOKS, NEW DELHI**

Chapter 34

Potentiality of Eugenol as a Source of New Contraceptive Principle in Male Albino Rat

D.S. Kulkarni

Department of Zoology,
Bharatiya Mahavidyalaya, Amravati

Eugenol mixed with olive oil as vehicle was used to study its effect on male reproductive function (sperm count and reproductive hormones) in male albino rats. Animals of test group received 0.2ml eugenol up to 30 days, while control group received 0.2ml olive oil only for the same duration. Sperm count and hormonal estimation (testosterone, FSH and LH) were carried out in serum samples of both the groups and compared. A significant decrease was noted in the sperm count in the experimental rats. Serum levels of the reproductive hormones (testosterone, FSH and LH) tested showed low levels in eugenol treated animals as compared to control. The results suggest the potential use of eugenol as an effective male contraceptive agent.

Keywords: Sperm count, Testosterone, FSH and LH, Eugenol, Male rats.

Introduction

Ocimum sanctum (OS) commonly known as tulsi or holy basil is one of the medicinal plants having a versatile role in traditional medicine in Indian subcontinent

(Ghosh, 1995). The whole plant of OS has medicinal value, although mostly leaves and sometimes the seeds are used (Gupta, 2002). Studies had shown that the leaf extract of OS contains potent bioactive components (essential oils) made up of urosilic acid, apigenin, luteolin and eugenol. It also reported that leaves and flowering tops of OS contain alkaloids, glycosides, saponin, tannin (Kokate, *et al.*, 2000).Leaves of OS had been reported to have antifertility effect in male albino rats (Kantak and Gogate, 1992) and benzene extract of OS decreased the sperm count and sperm motility (Singh V, *et al.*, 2010).

One of the major constituent of the tulsi leaves is urosilic acid and it has been reported that it possesses antifertility effect. This effect had been attributed to its anti-estrogenic activity, responsible for arrest of spermatogenesis in male. Eugenol (4-allyl-2 methoxy phenol) is also the potent bioactive components of OS and since no study is available to document effect of eugenol on the levels of reproductive hormones mainly LH and FSH the present study was undertaken to analyze the effect of eugenol on sperm count and reproductive hormones in male albino rats.

Materials and Methods

Animals

Healthy adult albino rats weighing 170 ± 10gm were procured from animal house of the Dr. Panjabrao Deshmukh Memorial Medical College, Amravati (MS, India). The animals were housed under controlled condition (28±2° C; 12:12hr L: D). Animals were fed with standard pellet diet and water *ad libitum* for a period of 30 days.

Drug

Eugenol (Loba Chem,India).

Experimental Setup

Animals were divided into two groups of 6 each. Group I (control) rats were injected intramuscularly (im) 0.2ml olive oil/day/rat up to 30 days as vehicle. Group II experimental rats were administered (im) with eugenol + olive oil, 2ml/day/rat up to 30 days (0.2ml contains 200mg eugenol).

After 30 days blood sample was taken from the eye of the rats for assessment of hormone levels (testosterone, FSH and LH) from both groups. The hormones were assayed by RIA using the method given by Shille, *et al.*, 1983.The animals were sacrificed by cervical dislocation on 31st day. Sperm count was assessed in all the animals. The cauda epididymis from both sides were removed and washed repeatedly in 10ml of normal physiological saline to remove traces of blood. Spermatozoa were counted by using 1ml aliquots of sperm specimen with the help of a haemocytometer (Saalu, 2007).

Body Weight and Organ Weight

During the experiment the rats were weighed daily and sacrificed on 31st day by cervical dislocation. Testis and cauda epididymus were isolated carefully and weighed.

Statistical Analysis

Students "t" test was used, P< 0.05 was regarded as moderately significant and P< 0.01 as significant (Fischer,1950).

Results

In the present study, a significant decrease (41.65 per cent) in sperm count was observed in rats administered with eugenol. A marked decrease in serum testosterone level was noted as compared to control. The FSH and LH levels in the experimental group decreased by 66.40 per cent and 74.76 per cent respectively as compared to that in control group (Table 34.1).

Table 34.1: Effect of Eugenol on Sperm Count and Sex Hormones of Male Albino Rats

Group	*Sperm Count (10^7)*	*Serum Testosterone (ng/ml)*	*Serum FSH (MIU/ml)*	*Serum LH (MIU/ml)*
Control	6.34±0.64	2.74±0.30	6.25±0.03	4.20±0.06
Test	3.70±0.52(-41.66)	0.68*±0.06(-75.18)	2.10*±0.65(-66.40)	1.06±0.20(-74.76)

Values are mean ±SE of six animals per group and refer to the average combined weight of right and left organs.

*: Significant, p<0.05 compared to control, NS: Not significant.

Figures in parenthesis indicate percent change over control.

The body weight of the test group rats showed 4.92 per cent increase but the testis and epididymus weights were decreased when compared with control readings (Table 34.2).

Table 34.2: Effect of Eugenol on Body Weight, Testis and Caput Epididymus Weight of Male Albino Rats

Group	*Body Weight (gm)*		*Testis Weight (g)*		*Cauda Epididymus Weight (mg)*	
	Initial	*Final*	*Initial*	*Final*	*Initial*	*Final*
Control	170.10±12.00	175.08±11.09 (+2.92)	1.126±0.02	1.217±0.08 (+8.08)	196.01±11.40	234.46±12.27 (+19.61)
Test	173.60±12.11	182.15*±12.01 (+4.92)	1.076±0.01	0.624±0.07 (-42.00)	182.10±11.08	142.12*±11.76 (-21.95)

Values are mean ±SE of six animals per group and refer to the average combined weight of right and left organs.

*: Significant, p<0.05 compared to control; NS: Not significant.

Figures in parenthesis indicate percent change over control.

Male reproductive functioning is regulated by intricately balanced mechanism involving the hypothalamus-pituitary-testis axis and accessory sex organs. Sperm production occurs in the seminiferous tubules of the testis, which is controlled by testosterone, produced by Leydig (interstitial) cells of the testis (Dolores and Cheng, 2004). Testosterone production is directly dependent on the concentration of Luteinizing hormone (LH), in the milieu secreted by the anterior pituitary gland (Ganong, 2001). Follicle stimulating hormone (FSH), released also by the anterior pituitary stimulates the sertoli cells of the testis which give support and nourishment to developing spermatozoa (Christensen, 1975; Huang, *et al.*, 1991; Gangong, 2001). Testosterone has profound influence on germ cell development and differentiation.

Results in the present study clearly showed that eugenol administration brings about a reduction in sperm count which in agreement with earlier reports of OS leaves extract (Singh V, 2010). Khanna, *et al.* (1984) had reported significant decrease in sperm count and motility as well as decrease in weight of testis, epididymis, seminal vesicle and ventral prostate after long term feeding of OS leaves (Khanna, *et al.*1986). In the present study 30 days treatment with eugenol had shown to increase in the body weight, while decrease in the weight of testis and epidydimus. The results suggest that low sperm count in male rats seems to be due to impairment of spermatogenesis. Significant decrease in interstitial cell number and LH hormone concentration in experimental group could lead to decreased testosterone secretion from interstitial cell.

The study reveals that the intramuscular administration of eugenol has anti-spermatogenic effect and it is brought about by modulation of levels of reproductive hormones. Thus, from this preliminary study it is suggested that eugenol has potential to be used as an effective herbal male antifertility agent.

References

Gupta SK, Prakash J, Srivastava S (2002): Validation of traditional claim of Tulsi, Ocimum sanctum Linn., as a medicinal plant., Indian J. Exp.Bio; 40:765-73.

Kokate CK, Purohit AP, Gokhale SB (2000): Pharmacognosy, Published by Nirali Prakashan, 564.

Kantak NM, Gogate MG (1992): Effect of short term administration of Tulsi (*Ocimum sanctum* Linn.) on reproductive behavior of adult male rats. Ind. J. Physio. Pharmacol.36 (2):109-111.

Ghosh GR (1995): Tulasi (N. Tulsi (N.O. Labiatae, Genus-Ocimum), New approaches to Medicine and Health (NANAH); 3:23-9.

Singh V, Ambdekar S,Verma O (2010): *Ocimum sactum* (tulsi): Bio-pharmacological activities webmed central Pharmacology,1(10): WMC00, 1046.

Shille VM, Munra C, Walkerfarmer, Papkohh H, Stabenfeidt GH (1983): Ovarian and endocrine responses in the cat after coitus, J. Reprod. Fertile: 68, 29-39.

Saalu LC, Oliyami KA, Omotuyi IO (2007): Tocopherol attenuates testicular toxicity associated with experimental cryptorchidism in rats, African J. Biotechnology; 6:1373-7.

Dolores DM, Cheng CY (2004): Sertoli-Sertoli and Sertoli-Germ Cell interactions and their significance in germ cell movement in the seminiferous epithelium during spermatogenesis. Endocrine Rev.25: 747-806.

Ganong, WF (2001): Review of Medical Physiology 20th ed. Lange, New York: 239.

Cristensen AC (1975): Leydig cells in Greep PO, Astwood EB 9eds) Handbook of physiology (American Physiological Society, Washington, D.C):165.

Zitzmann M (2008): Effect of testes replacement and its pharmacogenetics in physical performance and metabolism. Asian J. Androl; 10 (3): 364-372.

Wade AP, Wilkinson GS, Davis JC, Jeffcoat TN (1968): The metabolism of testosterone, androstendione and estrogen by testes from a case of testicular ferminization, J. Endocrinol; 42(3):391-403.

Khanna S, Gupta SK, Grover JK (1986): Effect of long term feeding of Tulsi (*Ocimum sanctum* L) on reproductive performance of adult albino rats. Indian J. Exp Biol.; 24: 302-4.

2013, Environmental Biotechnology *Pages* **289–296**
Editors: **D.R. Khanna, A.K. Chopra, Gagan Matta, Vikas Singh & Rakesh Bhutiani**
Published by: **BIOTECH BOOKS, NEW DELHI**

Chapter 35

Effects of an Insecticide Endosulfan on Protein Content of Edible Freshwater Fish *Nemacheilus botia*

Anita Parashram Patil, Sunil Dagdu Patil and Kailas Haribhau Kapadnis

Department of Zoology, Mahatma Gandhi Vidyamandir's L.V.H. College, Panchvati, Nashik, Maharashtra

The average protein content of freshwater Fish *Nemacheilus botia* significantly depleted after acute treatment by endosulfan. The protein content in liver depleted from 21.82 to 20.67, 21.82 to 19.53, 21.85 to 18.37, 20.67 to 17.22 mg/gm wet tissue for 24, 48, and 72 hours respectively at 2.024 milli ug/lit. and alimentary canal decreased from 15.51 to 14.35, 15.51 to 13.21, 14.93 to 12.06, 14.35 to 11.48 mg/gm of wet tissue for 24, 48, 72, and 96 hours respectively at 2.024 milli ug/lit. Protein content in gills decreased from 12.64 to 11.48, 12.06 to 10.33, 11.48 to 09.18, 10.33 to 8.61 mg/gm of wet tissue for 24, 48, 72 and 96 hours respectively at 2.024 milli ug/lit. In the present investigation the protein level of *N. botia* after actual treatment depleted. The decrease in total protein content of tissue after treatment suggest enhancement of proteolysis to meet the high energy demand under endosulfan pollution stress condition.

Keywords: Protein, Gill, Alimentary canal, Endosulfan.

Chemical Name

6,7,8,9,10,10-Hexachloro-1,5,5a,6,9,9a-hexahydro-6,9-methano-2,4,3-benzodioxathiepin-3-oxide.

Alternative Names

CAS 115-29-7, Chlorothiepin, Cyclodan, ENT-23979, FMC-5462, HOE-2671, Insectophene, Kop-Thiodan, Malix, Thifor, Thimul, Thiodan, Thionex.

Introduction

Fish being a highly portentous food available at cheaper rate has become a necessary ingredient or component of the national economy of most of the third world countries, particularly countries with acute food shortage including India. Fishes are particularly sensitive to wide variety of pesticides, chemicals and conditions may arise as there will be extinction of fishes from the freshwater rivers. Brown (1976) listed 12 basic types of investigations of toxicity these are for preliminary screening of chemicals, for monitoring influence to determine the extent of risk to aquatic organisms and to determine the component causing death. The work of Brown (1976) has stimulated number of ecologist and opened the field of experimental areas concerned in evaluating the degree of damage to biota and suggesting the necessary protective measures to be employed. These pesticides enter into the environment through spillage or drift at time of pesticide application or through the industrial effluents. About 77420 tones of pesticides and insecticides are used for agriculture purposes in India every year. Endosulfan has been detected from food samples by Araujo A.C.(1999), around the world in Australian beef at 0.36 mg/kg, Panups (1996)) (2 times the Australian limit and 4 times the international limit), in cow's milk from tobacco farming areas in USA and food samples in USA and Canada reported Anon (1984). Fishes have been the most widely used as organisms to determine the toxicity of water and other pollutant. Kori Siakper and Ovie (2008). The wide use of fishes is probably due to their adaptability to laboratory conditions as well as their availability and their varying degree of sensitivity to the toxic substance (Verma *et al.*, 1980). The higher concentration of toxicant bring about the adverse effect on the aquatic Fauna and Flora having great capacity to adjust to their new circumstances by changing their metabolic activity and also behavior up to certain limit. *Nemachielus botia* is a freshwater fish. An attempt has been made to study the changes in protein content of tissues like liver, gills and intestine of fish; when it is exposed to different concentrations of endosulfan. Therefore the changes in protein level were studied in the present investigation. *Nemacheilus botia* is freshwater fish generally eaten by many poor people living along the banks of rivers in Maharashtra, India and in some areas it is also used as food for poultry bird, so that knowledge of its chemical constituent is extremely necessary since the nutritive value of that organism is reflected in its chemical composition.

Material and Methods

The edible fish *Nemachielus botia* is locally available freshwater fish. The tested fish *Nemachielus botia* were collected regularly in live conditions from river Godavari at the place Nandur Madhmeshwar a famous bird sanctuary from Nashik district,

M.S., India. Fish are acclimatized to laboratory conditions for a week avoiding overcrowding. The fish were fed with standard fish food on an alternate day. In this biostatic bioassay 10 mature and healthy fishes showing no sign of disease and wounds, of same size were selected and used. The important environmental parameters like pH, temperature, dissolved oxygen, total hardness were determined for the water as per standard method (APHA 1981). The cyclodane insecticide endosulfan(6,7,8,9,10,10-hexachloro-1,5,5a,6,9a-hexahydro-6,9-methano-2,4,3-benzo-dioxyanthiepin-3-oxide) is among the most toxic pesticides In this acute static bioassay, mature and healthy fishes used were of average weight is 2 ± 0.3gm. and sized is 4 ± 1 cm.

The effect of endosulfan on biochemical constituents of *Nemachelus botia*, were evaluated by exposing fish to 1/2, 1/3 ... 1/10th to sublethal concentration of endosulfan at 24, 48, 72 and 96 hours.. The control and treated fishes were used for analysis of biochemical constituents such as proteins and glycogen in liver, intestine and gills. The fishes were sacrificed, dissected and their intestine, gills and liver is isolated. The isolated tissues were freshly used for estimation of protein (Lowery 1951). The results were statistically analyzed by applying standard deviation and Chi-square test. Protein estimation was carried with the help of spectrophotometer, absorbance measured at 650 nm in 1 cm glass curettes (Lowry *et al.*, 1951, Stoschck CM. 1990 and Hartree 1972). Bovine serum albumin used for standard known concentration.

Results and Discussion

Biochemical changes in the body of a intoxicated fish gives an indication and help to understand the mode of action and type of pollutant. The analysis of biochemical components of freshwater form in India has been done for their nutritive value as these fish serve as a protein rich food source for most human beings (Anon 1962).

The average protein content of freshwater Fist *N. botia* significantly depleted after acute treatment by endosulfan. (Table 35.1) After acute treatment of endosulfan the protein content in liver depleted from 21.82 to 20.67, 21.82 to 19.53, 21.85 to 18.37, 20.67 to 17.22 mg/gm wet tissue for 24, 48, and 72 hours respectively at 2.024 milli ug/lit.Protein content in alimentary canal decreased from 15.51 to 14.35, 15.51 to 13.21, 14.93 to 12.06, 14.35 to 11.48 mg/gm of wet tissue for 24, 48, 72, and 96 hours respectively at 2.024 milli ug/lit. Protein content in gills decreased from 12.64 to 11.48, 12.06 to 10.33, 11.48 to 09.18, 10.33 to 8.61 mg/gm of wet tissue for 24, 48, 72 and 96 hours respectively at 2.024 milli ug/lit. The protein content in liver depleted from 21.82 to 20.10, 21.82 to 18.37, 21.85 to 17.22, 20.67 to 16.65 mg/gm of wet tissue for 24, 48, 72, and 96 hours respectively at 3.039 milli ug/lit. The protein content in alimentary canal depleted from the 15.51 to 13.78, 15.51 to 12.63, 14.93 to 63, 14.65 to 10.91 ml/gm of wet tissue for 24, 48, 72, and 96 hours respectively at 3.039 milli ug/lit. The protein content in gills depleted from the 12.64 to 10.91, 12.06 to 9.76, 11.48 to 9.18, 10.33 to 8.04 mg/gm of wet tissue for 24, 48 and 72 hours respectively at 3.039 milli ug/lit.

Table 35.1: Protein Content in mg/gm of Wet Tissue in Tissue Liver, Intestine, Gill of *N. botia* in Control and Treated Condition of Endosulfan Exposed

Concentration of Endosulfan	*Organ*	*24 hours*	*48 hours*	*72 hours*	*96 hours*
Control	Liver Intestine	21.821 (± 0.237)	21.821 (± 0.249)	21.25 (± 0.256)	20.67 (± 0.268)
	Gill	15.506 (± 0.312)	15.506 (± 0.321)	14.93 (± 0.333)	14.35 (± 0.351)
		12.635 (± 0.398)	12.06 (± 0.408)	11.48 (± 0.472)	10.33 (± 0.450)
Endosulfan Conc. 2.024 milli ug/lit.	Liver Intestine	20.675 (± 0.312)	19.527 (± 0.331)	18.37 (± 0.352)	17.22 (± 0.450)
	Gill	14.351 (± 0.356)	13.209 (± 0.367)	12.06 (± 0.383)	11.48 (± 0.348)
		11.486 (± 0.376)	10.331 (± 0.382)	9.18 (± 0.391)	8.62 (± 0.402)
Endosulfan Conc. 3.039 milli ug/lit.	Liver Intestine	20.1 (± 0.212)	18.37 (± 0.232)	17.22 (± 0.251)	16.65 (± 0.278)
	Gill	13.78 (± 0.312)	12.63 (± 0.323)	12.63 (± 0.341)	10.91 (± 0.372)
		10.91 (± 0.421)	9.76 (± 0.432)	9.18 (± 0.451)	8.04 (± 0.476)

The protein content of gill was depleted more as compared to liver and alimentary canal. The alterations in biochemical content in different tissue of fish due to toxic effect of different pesticide have been reported by numerous workers. James *et al.*

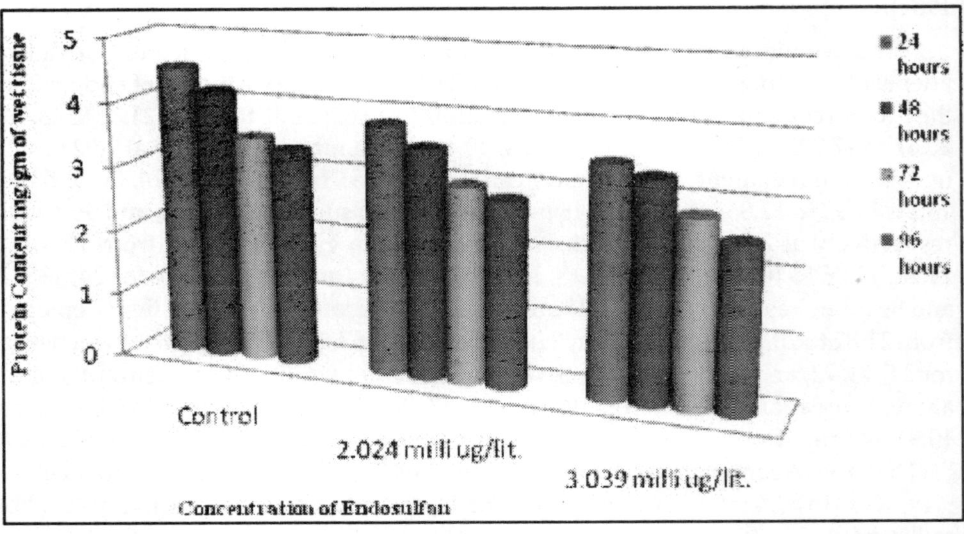

Figure 35.1: Protein Content in Liver of *N. botia* Exposed to Endosulfan for 24, 48, 72, and 96 hrs. at 2.024 and 3.039 milli µg/lit. Concentration

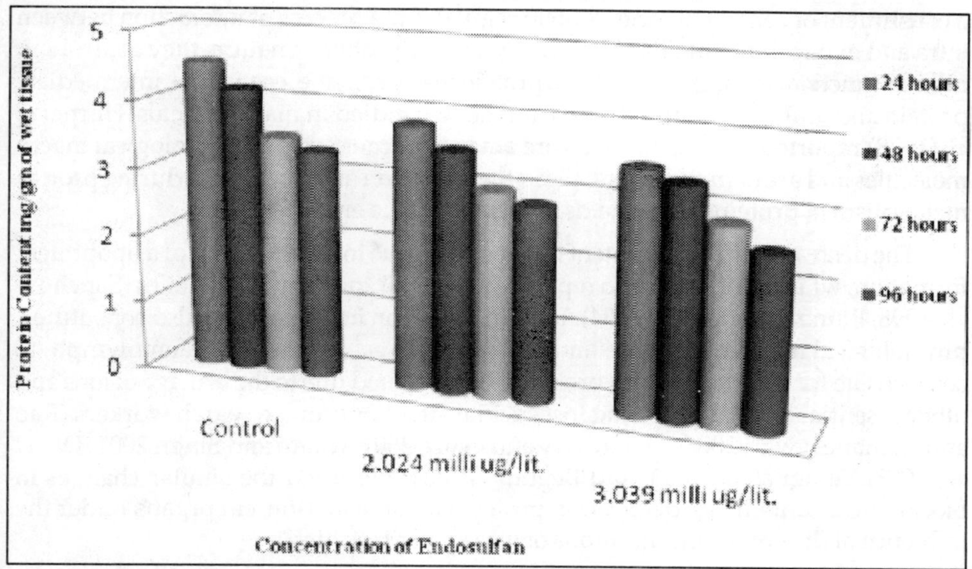

Figure 35.2: Protein Content in Intestine of *N. botia* Exposed to Endosulfan for 24, 48, 72, and 96 hrs. at 2.024 and 3.039 milli µg/lit. Concentration

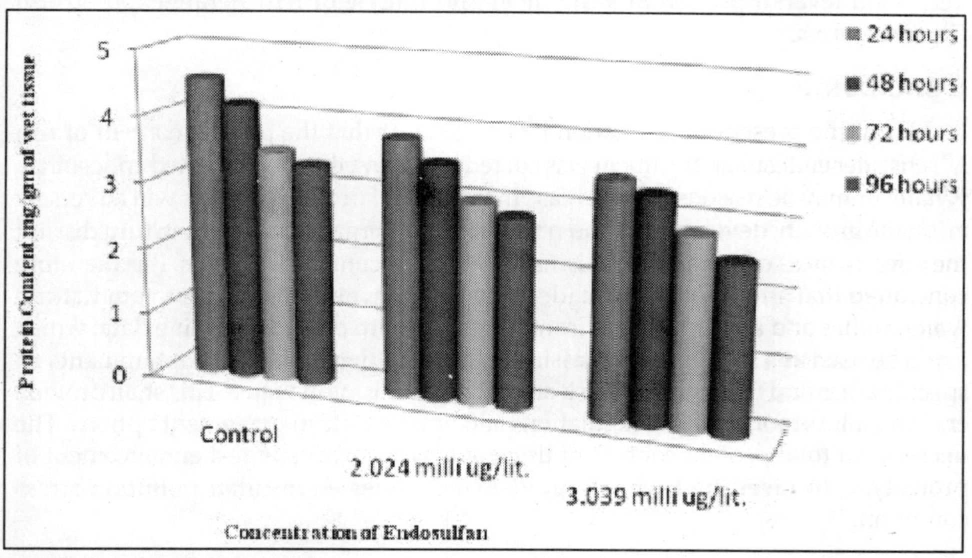

Figure 35.3: Protein Content in Gills of *N. botia* Exposed to Endosulfan for 24, 48, 72, and 96 hrs. at 2.024 and 3.039 milli µg/lit. Concentration

(1995), Sivakumari *et al.* (1997) Nanda *et al.* (2000), Tilak and Yacobu 2002, Yeragi *et al.* (2003). A change in glycogen content in various organism after exposure to various pollutant have been studied by many worker Mayer (1971), Koundinya and Rammurthi (1979), and Gill and Pant (1981). Protein is an important organic constituent in animal tissue and it plays a significant role in cellular metabolism. As a constituent of cell membrane, protein regulates the process of interaction between intra and extracellular media. All the enzymes are protein in nature, they control sub cellular function. In the metabolism of protein many enzyme, coenzyme, intermediate protein and amino acid are involved and hence studied in many animals. Harper *et al.* (1978) reported that the proteins are among the most abundant biological macro molecules and are extremely versatile in their function and interaction during protein metabolism in protein, amino acids, enzymes and co enzymes.

The decrease in protein content might also be due to a mechanism of a lipoprotein formation, which will be used to repair damaged all and tissue organelles (Sancho *et al.*, 1998, Rambabu and Rao, 1994). Protein depletion in tissues may also constitute a physiological mechanism by retaining free amino acid content in haemolymph, to compensate for osmoregulatory problems encountered due to the leakage of ions and other essential molecules during insecticidal stress. Various research workers (Rao and Ramaneshwari 2000, Chaudhary and Gaur, 2001, Khare and Singh 2001, Das *et al.*, 2003, Yeragi *et al.*, 2003, and Begum G.2004) observed the similar changes in biochemical contents *i.e.* decrease in protein contents in different organs under the influence of different concentrations of variety of pesticides.

The different in protein content in the fish at the experimental diet was small while lipid content was more variable Maria *et al.* (1986). Reddy *et al.* (1991) observed decreased level of protein in brain, liver and muscle of fenvalerate-exposed fish *Cyprinus carpio.*

Conclusion

From the present investigation it is quite clear that the protein content of fish *N. botia* after endosulfan treatment was altered indicating the effect of tested endosulfan. Whatever may be reasons, the decrease in the level of protein contents will adversely affect the growth, development and reproduction of organisms, which in turn disrupt the effectiveness of aquatic organisms in biological control programs. It is therefore concluded that effort should be made to study the levels of pesticides in our natural water bodies and adjourning environment in order to provide baseline data, which could be used as a strategy for assessing the distribution, impact of contaminants on specific areas and the ecological risk on native organisms in water. This shall provide critical tools for formulating regulations and appropriate management options. The decrease in total protein content of tissue after treatment suggest enhancement of proteolysis to meet the high energy demand under endosulfan pollution stress condition.

References

Anon,(1984), 'Environment Health Criteria 40- Endosulfan IPCS (International Programme on Chemical Safety),' *WHO* Geneva.

Anon, (1962): 'The wealth of India Raw Material of Vol. IV supplement fish and fisheries council of scientific and industrial research,' *New Delhi*, pp: 132.

APHA(1981): *American Public Health Association: Standard methods for the Examination of wastewaters* 17th edition, Washington, DC.

Araujo A. C., Telks D. L., Gorni R. and Lima L. L. (June 1999), 'Endosulfan residues in Brazilian Tomatoes and their impact on public health and the environment,' *Bulletin Environ. Contam. Toxicol* 62(6): 671-6.

Begum G, (2004), 'Carbofuran insecticide induced biochemical alternations in liver and muscle tissue of the fish *Clarias batrachus* (linn) and recovery response.' *Aquat Toxicol.* 7, 66(1): 83-92.

Brown V.M. (1976), 'Advances in testing the toxicity of substance to fish,' *chemy Ind.* 21, 143-149.

Choudhary A. and Gaur S. (2001), 'Effect of sodium fluoride on the muscle and liver of a freshwater fish *Cyprinus carpio*'. *J. Aqua. Biol.* 16(2): 67-68.

Das B.L.and Mukherjee S.C. (2003), 'Toxicity of cypermethrin in *Labeo rohita* fingerlings biochemical, and enzymatic and hematology cal consequences.' *Comp. Biochem. Physiol C. Toxicol pharmacol. Jan.* 134(1): 109-121.

Gill T.S. and Pant J.C. (1981), 'Effect of sublethal concentration of mercury in teleost, *putius conchonias* Biochemical and haematological response,' *Ind. J Exp. Biol.* 19: 571 – 573.

Harper H.A., Rodwell V.W. and Mayers P. A. (1978), 'A Review of Physiological chemistry, Long Medical Publications, California'.

Hartree E.F.,(1972), ' Determination of protein: a modification of the Lowry method that gives a linear photometric response,' *Anal Biochem.,*48; 422-427.

James R., Sampath K., Sivkumar V., Babu S. and Shanmuganandan P. (1995), ' Toxic effects of copper and mercury on food intake growth and proximate chemical composition in *Heteropneustes fossilis,*'. *J. Environ. Biol.*, 16(1): 1-6.

Khare A. and Singh S. (2002), 'Impact of mlathion on protein content in the freshwater fish *Clarias batrachus*'. *J. Ecotoxicol. Environ. Monit*, 12(2): 129-132.

Kondinya P.R. and Ramamurty R.,(1979), 'Hematological studies in sarotherodon *Tillapia mossombica(peters)* exposed to lethal conc. of Sumition and Seven, *Curr. Sci.*48(9):877-879.

Kori-Siakper, Ovie(2008): Acute toxicity of potassium permanganate to fingerlings of the African catfish, *Clarias gariepinus* (Burchell,1822). *African J. Biotech.* Vol.7 (14), pp. 2514-2520.

Lowry O.H., Rosenbrought N.J., Farr A.L and. Randall R.J.,(1951), 'Protein measurement with folin phenol reagent,'*J. Biol. Chem.* 196: 265 – 275.

Maria N.A., Vassiliki T., Elli P. and Papou L. (1986), ' Effect of Diet composition and protein level on growth, body composition hematological characteristics and cost of production of rainbow trout, (*Salmo gairdneri*)' *Aquiculture.*, 75-78.

Mayer F.L.and Ellersieck M.R. (1986), 'Manual of acute toxicity: interpretation and data base for 410 chemicals and 66 species of freshwater animals,' *US Dept Inter Fish Wildl Serv. Res. Publ.* 160.

Nanda P., Pande B.N. and Behare M.K. (2000), ' Nickel induced alterations in protein level of *Heteropneustes fossilis*,' *K. Environ. Biol.* 21(2): 117-129.

PANUPS "Endosulfan Residue in Australian Beef (Agro: World Crop Protection News, Aug 28- 1998, Jan 15 1998); PANUPS (Pesticide Action Network- Update Service- May 20 1996)"

Rambabu J.P. and Rao M.B. (1994), 'Effect of organochlorine and three organophosphate pesticides on glucose, glycogen, lipid and protein contents in tissues of the freshwater snail *Bellamya dissimilis* (Muller).' *Bull. Environ. Contam. Toxicol.* 53: 142-14

Rao L.M. and Ramaneswari K. (2000), 'Variation in acute toxicity of endosulphan and monocrotophos to *Labeo rohita, Mystus vittatus and Channa punctata.*' *Pollution Res* 19.

Reddy P.M., Philip G.H.and Bashamohideen M. (1991), 'Fenvalerate induced ... freshwater fish, *Cyprinus carpio Biochem.*' *Int.* 23(6) 1087-1096.

Sancho E. M.D., Ferrando C., Fernandez and Andreu E. (1998), 'Liver energy metabolism of *Anguilla Anguilla* after exposure to fenitrothion. *Ecotoxicol.' Environ. Saf.* 41: 168-175.

Shivkumari K., Manavalaramanujam R., Ramesh M. and Laxami P. (1997), ' Cypermethrin toxicity sublethal effects on enzyme activities in the freshwater fish *Cyprinus caripo*,', *J. Environ. Biol.* 18(2): 121-125.

Stoscheck C.M.,(1990), ' Quantitation of protein methods in enzymology,' J. Enzymol. 182; 50.

Tilak and Yacobu 2002, Tilak K.S. and Yacobu K. (2000), 'Toxicity and effect of fenvalerate on fish *Ctenopharyngodon fdellus*'. *J. Ecotoxicol. Environ. Monit*, 12(1): 9-15.

Verma *et al.*, 1980). Verma S.R., Rani S., Bansal S.K and Delala R.C.,(1980), ' Effect of pesticide thiotox, dichlorovous, and chloro furun on the fish *Mystus vittatus*,' *Water or Soil. Pollu.* 13 (2): 229 – 234.

Yeragi *et al.*, 2003, Yeragi S. G., Rana A.M. and Koli. V.A. (2003), 'Effect of pesticides on protein metabolism of mudskipper *Bielophthalmus dussumieri*,' *J. Ecotoxicol. Environ, Monit.*, 13(3): 211-214.

2013, Environmental Biotechnology *Pages* ***297–301***
Editors: **D.R. Khanna, A.K. Chopra, Gagan Matta, Vikas Singh & Rakesh Bhutiani**
Published by: **BIOTECH BOOKS, NEW DELHI**

Chapter 36

Impact of Flacherie Disease on the Total and Differential Haemocyte Count of *Philosamia ricini*

Ulka Yadav, Shobha Shouche and Yogita Sharma
Government Madhav Vigyan Mahavidyalaya, Ujjain, M.P

Philosamia ricini is an economically important insect and it is reared for procuring Eri silk. The rearing suffers by Flacherie (Bacterial disease) which affect the production of silk. Insect have effective defense mechanism. Cellular defense is mediated by haemocytes by different processes (depending on the size of invader and duration). Different haemocytes play their role in these processes. As per need, during infectious condition the THC and DHC fluctuates. In present study the THC and DHC of healthy and diseased worms were made (during different developmental stages), in respect of different types of haemocytes and compare with control (statistically).

Keywords: Philosamia ricini, Flacherie, Defense, Haemocytes.

Introduction

Silk is a fibrous protein of animal origin. Commercial silk are classified in to two categories 1.) Mulberry silk produce by *Bombyx mori.* 2.) Non mulberry silk known as Vanya silk (Tasar, Muga, Eri) produced by different insects. Eri silk known as nonviolence and poor person silk produced by Eri silk warm *Philosamia ricini*. Due to centuries of domestic life, silk worm has no natural resistance and also shows neither morphological nor behavioral adaptation to escape parasites and pets, so they are very delicate, have week résistance and they are very susceptible. They cannot tolerate diurnal and seasonal fluctuations in the environmental conditions. Proper disinfection, providing optimum conditions and hygienic rearing required to produce healthy worms, instead of these certain diseases like Pebrine (Protozon disease), Muscardine (fungal disease), Grasserie (viral disease) and Flacherie (bacterial disease) are affect the production of silk.

Like other animals, insect have effective defense mechanism which is capable of resisting containing and eliminating the foreign invader. Insect's defense include humoral and cellular defense mechanism. Cellular defense is mainly mediated by haemocytes by the three processes that are phagocytocis, Nodule formation and Encapsulation. Different haemocytes play their role in defense mechanism and as per need the percentage of their population fluctuates.

Material and Methods

The *Bacillus megaterium* used in this experiment was obtained from Dept. of Microbiology Govt. M.V.M College Ujjain. *Bacillus megaterium* is a short, rod shaped, 0.4-0.7μm in width and 3-5μm in length. The organisms are gram negative and stained uniformly the by usual amiline dyes.

The insect *Philosamia ricini* used in this investigation were obtained from Govt. Sericulture Dept, Indore(in the eggs form). The eggs where disinfected by the method described by Steinhaus E.A. (1949) and then reared. In this experiment the 3rd, 4th and 5th instar larvae were feed with castor leaves which had been dipped in a suspension of *Bacillus megaterium* contain 4×10^9 bacterial cells per ml of sterile distilled water. No motility happened of this treatment. The THC and DHC were calculated by the method suggested by Rosenburger, C.R. and J.C.Jones.1960; the numbers of haemocytes were calculated by the formula suggested by Jones, (1962), and compare with control statistically.

Observation

In the present study the total haemocytes count and differential haemocytes counts of Healthy and diseased worm were made (during different developmental stages) in respect of different type of haemocytes statistically.

Result

This study indicates that:

1. The total number of haemocytes increases significantly (Table 36.1) with the development during larval stages. THC was highest in the 5th instar mature larvae in normal condition (THC studies).
2. The Diseased Larvae have significantly low population of haemocytes (Table 36.2) they usually die after 2nd day of spinning in (THC studies).

Table 36.1: Heat Fixed THC of Normal Eri-silkworm *Philosamia ricini* during Development

Heat Fixed THC	*Range*	*Mean*	*Standard Derivation*	*Standard Error*
3rd Instar	4070-7250	5763/mm³	1446.52	+-457.76
4th Instar	6350-8930	7950/mm³	714.90	+-226.23
5th Instar	6690-10390	8748/mm³	1278.60	+-404.32

Table 36.2: Heat Fixed THC of Diseased Eri-silkworm *Philosamia ricini* during development

Heat Fixed THC	*Range*	*Mean*	*Standard Derivation*	*Standard Error*
3rd Instar	2410-4320	3450/mm³	1738.68	+-55470
4th Instar	3070-4850	3925/mm³	1742.38	+-554.89
5th Instar	2600-4075	2950/mm³	1812.55	+-577.24

Table 36.3: The Differential Haemocytes Count of Normal and Diseased (Flacherie) Worms of *Philosamia ricini* in Percentage

Haemocytes	*3rd Instar*		*4th Instar*		*5th Instar*	
	Normal	*Diseased*	*Normal*	*Diseased*	*Normal*	*Diseased*
Prehaemocytes	5.00	4.00	2.00	2.00	0.00	0.00
Plasmocytes	50.00	70.00	50.00	68.00	50.00	62.00
Granuler	58.00	45.00	55.00	48.00	53.00	45.00
Spherules cells	6.00	0.00	9.00	0.00	13.00	0.00

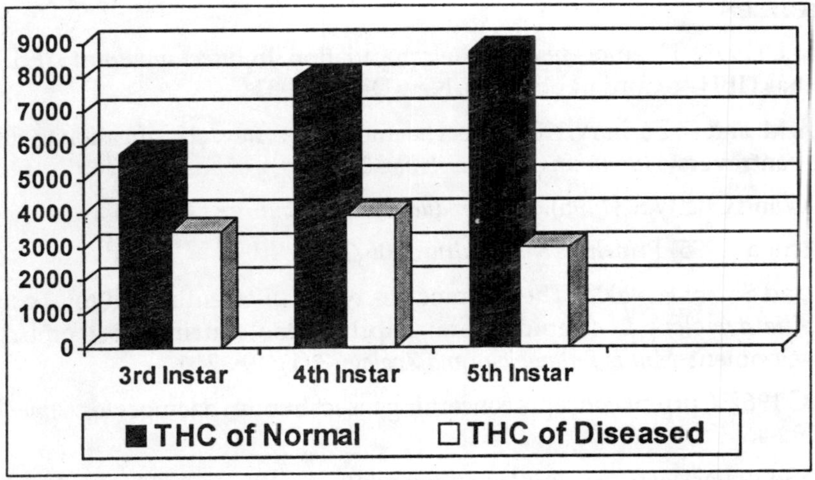

Figure 36.1: Comparative Analysis of Heat Fixed THC of Normal and Diseased Eri-Silkworm *Philosamia ricini* during Development

3. The Worm suffering by flacherie disease (Table 36.3) show significant increase in the population in the per cent of Plasmatocytes and decrease in the per cent of granular cells and spherule cells (DHC studies).

Discussion

Sericulture has been practiced in India for over 2000 years (Ramakrishana *et al.*, 1994) Bacterial Flacherie is one of the important and earliest Known Disease of silk worm and played a major role in decreasing the production of cocoons Vaidy (1960) Samson (1995) Nanavaty(1990), Patil (1990), Pathak (1995), Sharma J and Pathak (2002), Jalali J.and Salahi R.(2008), Bahadur J.(1993), Sharma P.R., Sharma O.P. and Saxena B.P.(2003) quoting data of loss due to disease.The Present study mainly emphasis on Bacterial disease which is very common and inflicts the maximum damage to industry. Diseases manifested only during certain physiological stress included by poor leaf quality, unhygienic rearing, excessive humidity, wide fluctuation in temperature, over feeding, over crowding etc resulting week and poorly developed larvae. Different bacterial diseases are; clear head, shrinking disease, satto disease, faecol chain disease (name according to symptoms) Ganga. G and Chetty.S. (1995), Hisao Aruga (1995) Therefore it is concluded that THC increases with the development in each larval state and decrease during infectious conditions and these variations are statistically significant. The fluctuation in DHC clearly indicates the role of plasmatocytes in the formation of nodule and phagocytosis because the size of invador is microscopic.

Acknowledgment

Author thankful to the Principal Govt. M.V.M Ujjain, M.P, India for providing the necessary lab facilities. Author also grateful to University Grant Commission for financial assistance in the form of sanction a minor research project (F.NO.MS 100/107037/09-10/CRO) Date- 31st March 2010.

References

Bahadur J. (1993). Haemocytes and their population. In *Insect immunity*. (Ed. J.P.N Pathak) IBH –oxford publication. New Delhi. 15-32

Brehelin, M. and D. Zachary.1986. Insect haemocytes: A new classification to rule out the controversy. *Immunity in invertebrates* 35-48.

Ganga. G and Chetty.S. (1995). *An introduction* to *Sericulture* 120-123.

Hisao Aruga (1995) *Principles of sericulture* 206-231.

Jalali J. and Salahi R. (2008) The Haemocytes types, differential and total count in *Papilio demoleus* L. (Lepidoptera: Papilionidae) during post embryonic development. *Munis Entomology and Zoology* 3 (1): 199-216

Jones, J.C.1962. Current concepts concerning insect humun haemocytes. *Amer. Zool.* 2: 209-46.

Karl M and Robert E (1975). *Invertebrate Immunity* 181-196.

Nanavaty, M., 1990, Silk production processing and marketing, Wiley Eastern Ltd. New Delhi: pp.176.

Pathak, J.P.N. (1995) N.S.P report. Studies on the pathological problems of silkworm *Bombyx mori* L rearing in Madhya Pradesh special reference to Bacterial flacherie.

Patil, C.S, 1990, Silkworm Disease and their management in Japan, *Indian Silk* 29 (5)31-34.

Ramakrishna, T.R., Gupta, M. and Kathavate, A.Y., 1994, Corporate Sericulture: A techno commercial review, *Indian Silk*, 33 (6):25-28

Rosenburger, C.R. and J.C.Jones. 1960. Studies on total blood counts of Southern army worm larva, *Prodenia eridania* (Lepedoptera). *Ann. Entomol. Soc.Amer.* 53:551-5.

Samson, M.V. (1995) Flacherie in *Bombaybyx mori*, L., *Indian Silk*, March 1995:31-34.

Sharma, J. (2002) A Study on the Bacterial Flacherie of *Bombyx mori* In Malwa Plateau (Thesis).

Sharma P.R.,Sharma O.P. and Saxena B.P. (2003) Effect of Neem gold on haemocytes of the tobacco armyworm, *Spodoptera litura* (Fabricius) (Lepidoptera; Noctuidae) *Current Science*,84(5),690-695

Steinlaus E, A 1949, *Principles of Insect Pathology*.

Vaidya, M.J. (1960), *Indian Sericulture*– Problem of mordinisation. Central Silk Board, Bombay, pp. 64.

2013, Environmental Biotechnology *Pages* **303–313**
Editors: **D.R. Khanna, A.K. Chopra, Gagan Matta, Vikas Singh & Rakesh Bhutiani**
Published by: **BIOTECH BOOKS, NEW DELHI**

Chapter 37

Comparative Evaluation of Activated Carbon with Rice Husk for Decolourisation of Textile Dye

Rekha Singh[1] *and J.L. Tarar*[2]
[2]*G4S Asia Middle East, Gurgaon, Haryana*
[2]*Department of Environmental Sciences, Institute of Science, Nagpur, M.S.*

In this paper comparative study of adsorption of textile dye forosol blue onto rice husk and PAC has been presented. It is found that the adsorption potential varies as a function of contact time, concentration, pH, and adsorbent doses. By using 500 mg/lt of processed rice husk, 100 per cent colour removal was achieved upto 10ppm dye solution, within 60 min at the pH 2 which was further achieved by only 200 mg/lt of PAC under similar conditions. Batch kinetics and isotherm studies were undertaken, and the data evaluated for compliance with the Freundlich isotherm model and Lagergren model.

Keywords: Rice husk, PAC, Forosol blue, XRF scan, Freundlich model, Lagergren model.

Introduction

The textile industry is in no way different than other chemical industries, which cause pollution of one or the other type. Effluent being discharged from the textile industries into the sewage system or neighbouring water receiving bodies is presently a cause of major environmental and health concern. Wastewater discharged from textile industry is solution of complex chemical with high colour value. The colour in the effluent is mainly due to unfixed dye. The concentration of unused dyes in the effluent depends upon the nature of dyes and dyeing process underway at the type. The wastewater from the textile processing industry differs greatly in composition and no single method of treatment can be suggested that will suit all effluent (Edward, 2003a; Edward, 2003b).

Colour removal is generally considered more important than the removal of the colourless chemical contaminants because it is an aesthetic problem and also detrimental to microbial life. The presence of dye in textile wastewater is responsible for raising the COD level of the effluent.

The colour present in the water also interferes in the photosynthetic activity of plants since it reduces the sunlight penetration, and thereby ecosystem is seriously affected. The dyes and colours cannot be removed by conventional wastewater treatment process, as it is fairly stable to light, heat and resist biodegradation on account of its complex molecular structures. In recent years several physico-chemical decolourisation process have been developed like membrane-filteration, reverse osmosis, flocculation, biological treatments and adsorption process (Ademorali *et al.*, 1992; Ajmal and Khan, 1985; Kothandaraman *et al.*, 1976; Shah, 1985).

Adsorption process offers most economical and effective treatment method for removal of dye and has edge over other methods due to its sludge – free clean operation. Though activated carbon is an ideal adsorbent for organic matter due to its organophilic character, it is uneconomical for wastewater treatment, owing to its high production and regeneration costs, and about 10 to 25 per cent loss during regeneration by chemical and thermal treatment. High cost of activated carbon in India has prompted search for cheaper natural substitutes, which are biodegradable, and abundant in nature. Natural materials that are available in large quantities, or certain waste products from industrial or agricultural operations, may have potential of inexpensive sorbents. Rice husk is a waste product from agricultural operation and have potential of inexpensive sorbents. Due to its low cost, after, it can be disposed off without expensive regeneration.

In present work efficiency of rice husk is compared with PAC. The study included batch pH/kinetics/isotherm studies for rice husk and subsequent comparison with PAC. The Freundlich and Lagergren models were tested for their applicability.

Material and Methods

Forosol Blue a commercial dye used by Raymond Ltd., Sausar, Chhindwada, M.P. was taken as adsorbate. Rice Husk collected from local rice mill of Nagpur was used as adsorbent. Chemical nature of forosol blue could not be obtained from Raymond Ltd., keeping it as trade secret. Rice husk was first washed frequently with water and than with distilled water. It was ground to powder and sieved through 85mesh sieve and again activated at 110°C for 3-4 hrs. Elemental analysis of adsorbents was carried out by XRF scan method (Make – Philips, Model No. – PW

2403 -MagiX).

Batch studies was conducted as follows. First of all pH studies were conducted by shaking 100 ml volume of 10 ppm concentration of Forosol blue with 50 mg amount of adsorbent for 60min on magnetic stirrer over a range of pH values from 2 to 10. For pH adjustment 0.1N NaOH or 0.1N HCl was used. Blank was run simultaneously, without any adsorbent to determine the impact of pH change on the dye solution. The treated dye solutions were filtered through laboratory grade filter paper and then colour was measured by spectrophotometer at its λ_{max} (590nm). Once the optimum pH was identified, kinetic studies were conducted by shaking 50 mg of adsorbent in 100 ml dye solution at the noted optimum pH. Contact time was slowly increased till maximum colour removal was observed. Isotherm studies were conducted – by shaking varied quantity of adsorbent ranging from 20 mg to 400 mg in 100ml of solution at noted optimum pH for the period of 1hr and by shaking 100 mg of adsorbent in 100 ml of varied concentrations of dye varying from 2 to 40 ppm at optimum pH for the period of 1hr. Blank study was carried out simultaneously at similar conditions with only adsorbent in 100 ml of distilled water, to account for any colour leaching by the adsorbent. Studies in above mentioned cases were repeated twice and mean values are presented in graphical form.

Results and Discussion

pH Studies

Figure 37.1 shows changes occurring in adsorption capacity of individual adsorbent with change in pH for forosol blue. Ionic dyes upon dissolution release coloured dye anion/cation in solution. The adsorption of these charged ion groups on the adsorbent surface is primarily influenced by the surface charge present on the adsorbent which is further influenced by the solution pH (Konduru and Viraraghavan, 1997).

Table 37.1: Elemental Studies by X-Ray Fluorescence Scan of Rice Husk

Sl.No.	*Major Elements*	*Minor Elements*	*Trace Elements*
1	Fe	Ca	As
2	K	Al	Pb
3	S	Mg	Zn
4	P		Cu
5	Si		Mn
6	O		Cr
7			Ti
8			Na

Lignocellulose forms the chief structural component of plant cell wall. The lignocellulose complex contains cellulose, hemicellulose and lignin (Majumdar and Chanda, 2001).

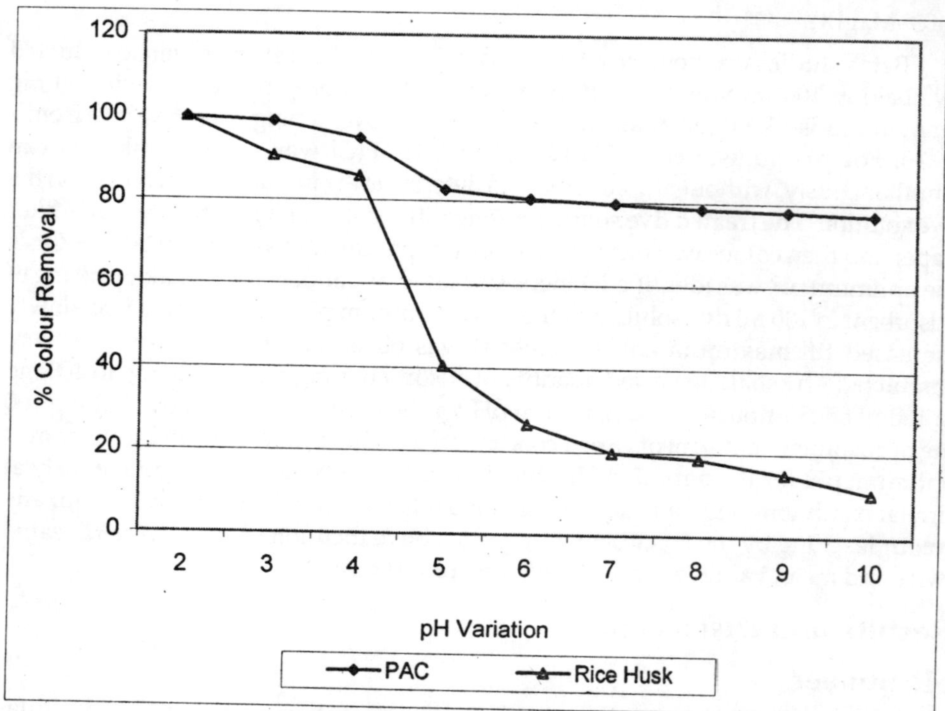

Figure 37.1: Showing Effect of pH on Colour Removal

The cellulose in contact with water is negatively charged (Vickerstaff, 1954). Reactive dyes belong to anionic dyes group, which suffer coulombic repulsion due to the presence of the strong anionic groups in adsorbent. The disperse dyes belong to cationic dye group which ionise to give the coloured cationic dye base and undergo attraction on approaching the anionic hemicellulose structure of natural adsorbents. Another main constituent of natural adsorbent is lignin, which consist of polar function groups *viz.* humic, and fulvic acid. These group make the natural adsorbents surface negatively charged (McKay *et al.*, 1981; Valentin, 1986). Hence, the structure of natural adsorbent has a high adsorption capacity for cationic (basic) dyes and show repulsion for anion (acidic) dyes.

Whenever the pH of dye solution increased, the percent colour removal was found decreasing. The result obtained may be due to the repulsion between dye and adsorbent surface in alkaline condition. The colour removal at pH 10.0 noted was 10.61 per cent by rice husk and 77.33 per cent by PAC While 100 per cent colour removal was observed by both at pH 2. In alkaline condition, effectiveness of PAC over rice husk can be explained on the basis of manufacturing process of PAC. Since PAC was washed with acid during its production, so its efficiency was improved (on the basis of literature supplied by PAC supplier).

Kinetic Studies

Figure 37.2 displays the adsorption kinetics of rice husk for forosol blue. It can be observed that the percent colour removal increased with increase in contact time and

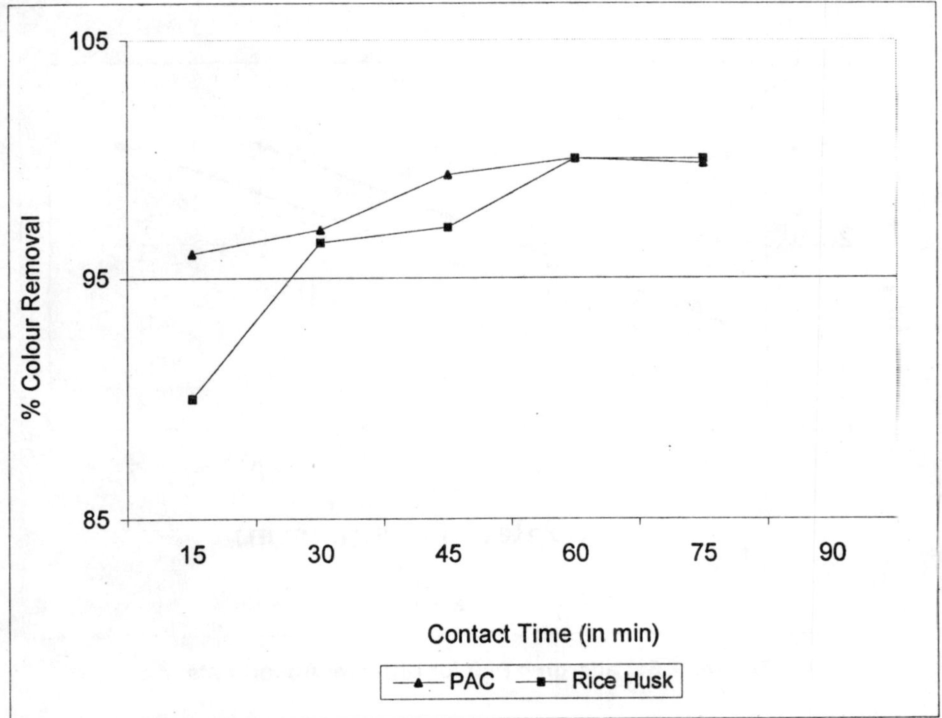

Figure 37.2: Showing Effect of Contact Time on Colour Removal

at some point of time it reaches a constant value beyond which meagre colour removal is noted. Maximum colour removal was noted in first 15 min and thereafter only gradual colour removal occurred.

In the first 15 min colour removed by PAC and rice husk was around 98.8 per cent and 90.03 per cent respectively. The equilibrium time for both was approximately of 60 min. Percent colour removal in 60 min was 100 per cent.

The rate constant for the adsorption of forosol blue by rice husk and PAC were studied by using Lagergren rate equation-

$$\log(q_e - q) = \log q_{e.} K_{ad.}\, t/2.303$$

where,

q_e and q are the amount of dye adsorbed (Mg/g) at equilibrium and at time t(min), respectively and K_{ad} is the rate constant of adsorption (Namasivayam and Arasi, 1997).

Lagergren plot is shown in Figure 37.3 Linear plot of log (q_e - q) Vs time was obtained for forosol blue in different dye concentration, which indicates that the adsorption process follow the first order rate expression.

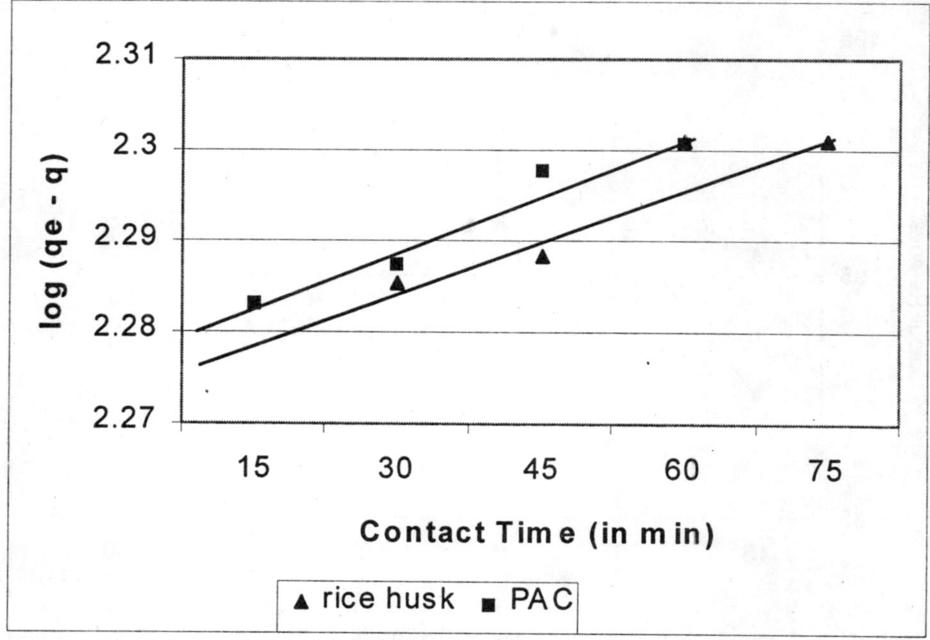

Figure 37.3: Lagergren Plot for Different Adsorbents

Isotherm Studies

Effect of Adsorbent Doses

Figure 37.4 displays the effect of adsorbent doses on colour removal. Around 68.49 per cent colour was found removed at 100 mg/lt of dose rice husk. With increase in doses more colour removal was evident. Hundred percent colour was found to remove at 500 mg/lt dose. The similar results were obtained by PAC only by using 200 mg/lt of dose.

Freundlich Isotherm

Freundlich isotherm shows multilayer adsorption and is expressed by the equation:

$$\log X/M = \log k + 1/n \log C_e$$

where,

x = weight of substance adsorbed

M = weight of adsorbent

C_e = concentration remaining in solution

K and n are constants, which primarily depends on temperature, adsorbent and adsorbate (Jain and Gupta, 1992).

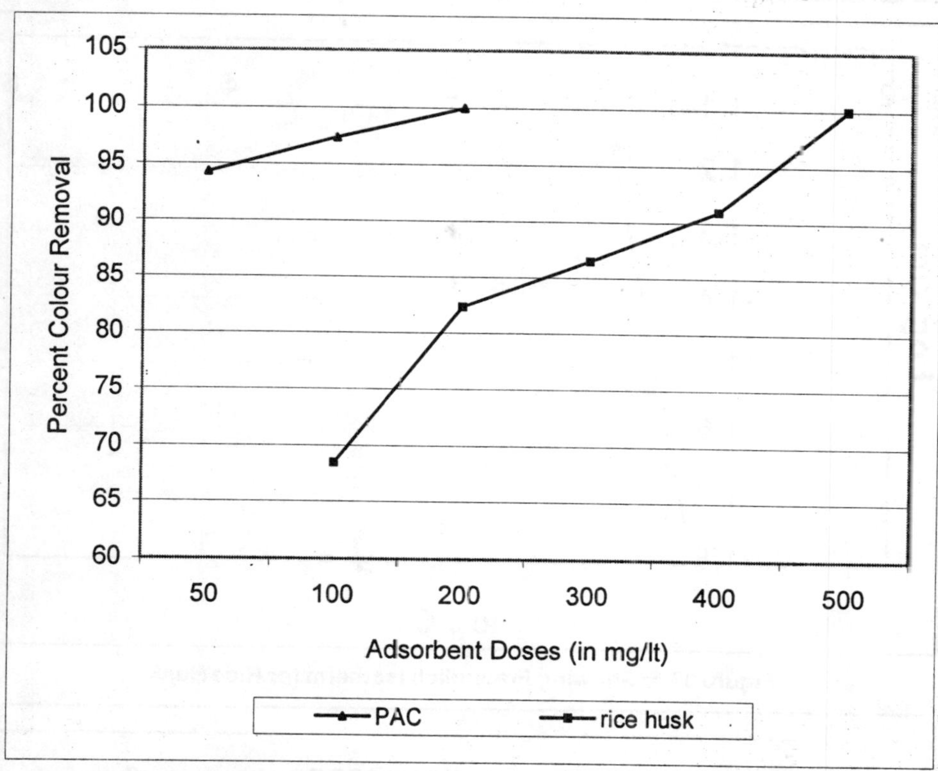

Figure 37.4: Showing Effect of Adsorbent Doses on Colour Removal

Figures 37.5 and 37.6 illustrate the Freundlich isotherm for rice husk and PAC. Linear plot of log X/M vs log Ce shows multi-layer adsorption of forosol blue following Freundlich isotherm. The values of constant K and n were calculated from intercept and slope of graph respectively and are presented in Table 37.2.

Table 37.2: Freundlich Constants

Name of Adsorbent	*Freundlich Constants*	
	k	*n*
Rice Husk	–1.12	0.08104
PAC	–0.46	1.1465

Effect of Initial Dye Concentration

Figure 37.7 shows the effect of forosol blue concentration on percent colour removal by different adsorbents. Percent colour removal were found to be higher in low concentration of dye solution while it decreases with the increase in initial dye concentration, showing effectiveness of rice husk for dilute solution of dye. Upto 10 ppm initial dye concentration of dye rice husk was found effective. In higher concentration of the dye, rice husk was not found effective to remove colour. For 40

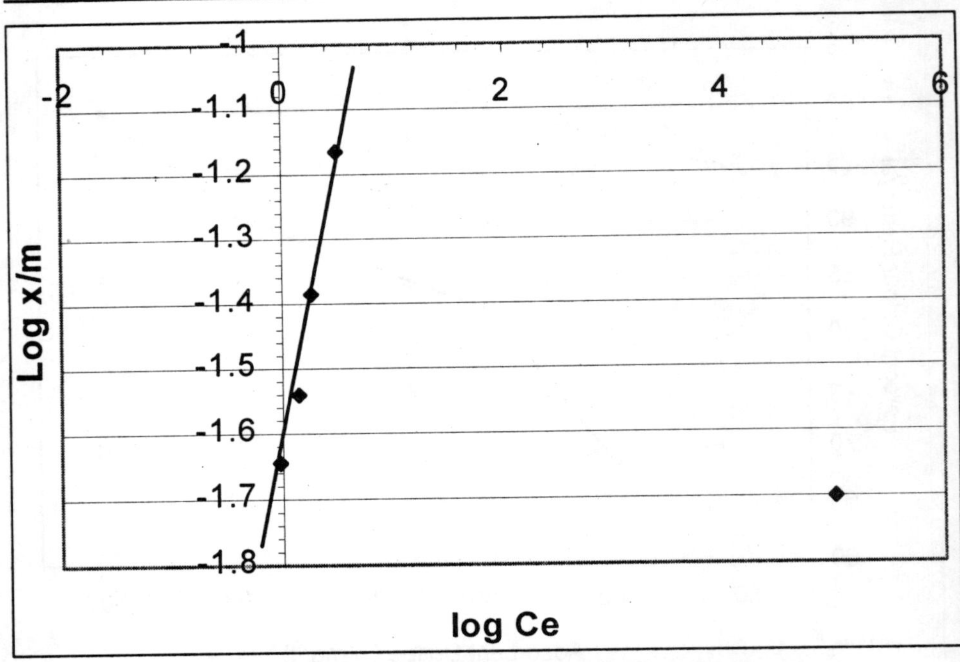

Figure 37.5: Showing Freundlich Isotherm for Rice Husk

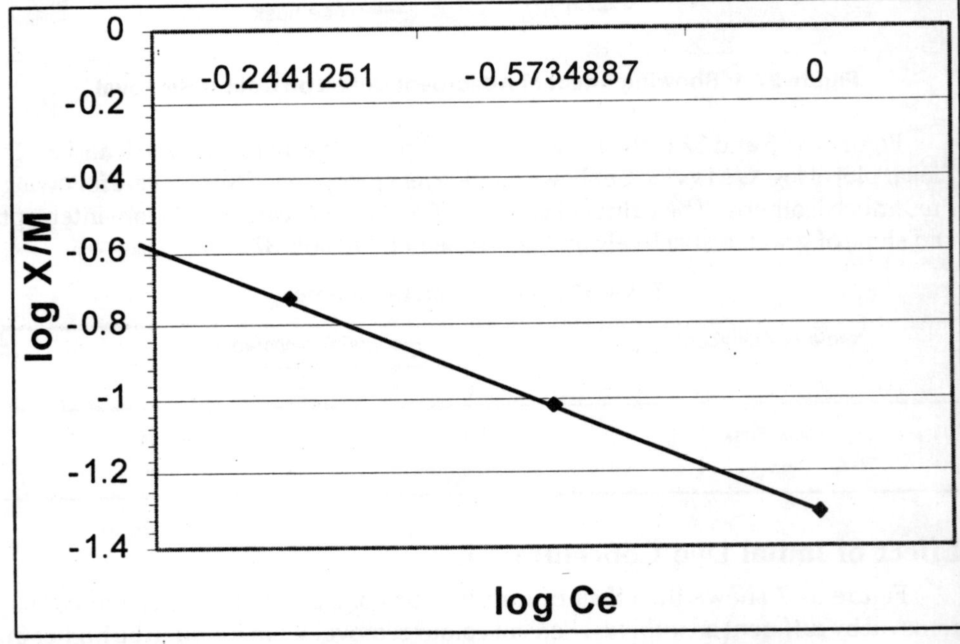

Figure 37.6: Showing Freundlich isotherm for PAC

ppm of initial colour concentration, only 58.03 per cent colour removal was obtained. While in case of PAC 100 per cent removal was obtained upto the 20 ppm dye solution and more than 95 per cent were obtained upto 40 ppm.

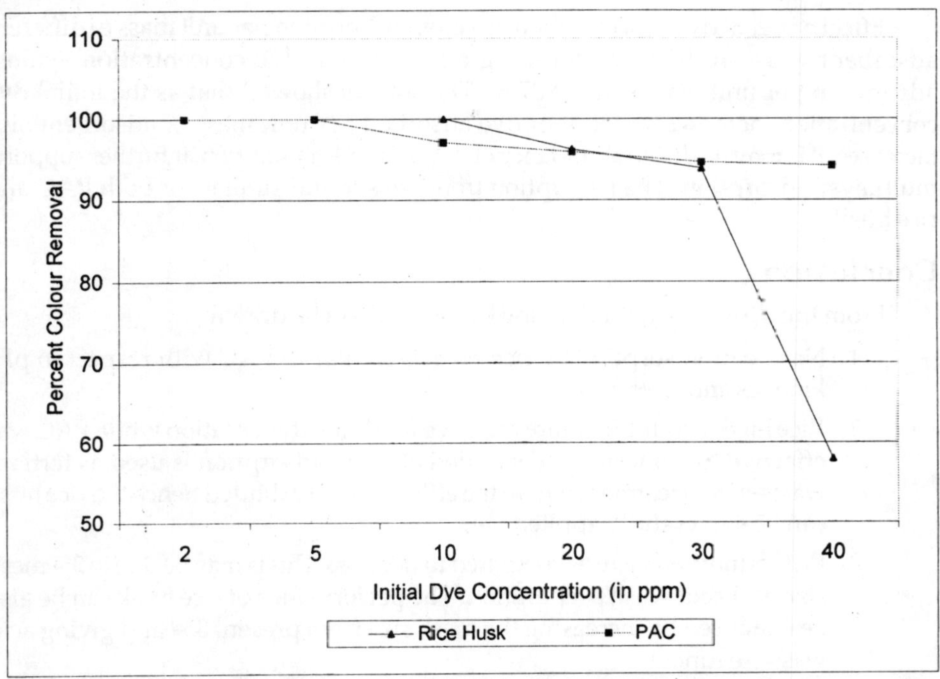

Figure 37.7: Showing Effect of Initial Dye Concentration on Colour Removal

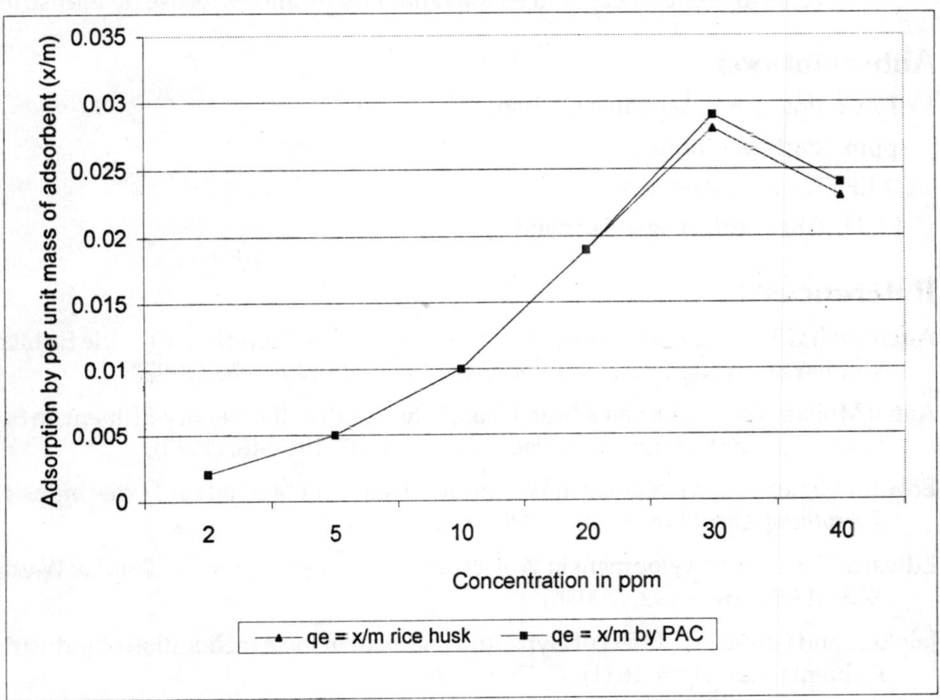

Figure 37.8: Nature of Adsorption by PAC and Rice Husk

Effect of initial dye concentration on colour adsorption per unit mass of different adsorbent were studied by plotting graph of initial dye concentration against adsorption per unit mass (Figure 37.8). The studies showed that as the initial dye concentration increases, amount of dye adsorbed per unit mass of adsorbent also increases. It is may be due to abundance of dye molecule in solution. It further supports multilayer adsorption. The adsorption trend was found similar for both PAC and rice husk.

Conclusion

From the above studies following conclusion can be drawn

1. Nature of adsorption by rice husk is similar to PAC with respect to pH, kinetics and isotherm.
2. Rice husk was found more effective for dilute dye solution while PAC was effective for concentrated solution also. As adsorption is used as tertiary wastewater treatment where the effluent is very diluted hence the rice husk can be successfully applied here.
3. PAC is more efficient as compared to rice husk This is may be due to 350mesh size and acid wash preparation. The performance of rice husk can be also be enhanced by increasing the mesh size from present 85# and giving acid wash treatment.
4. Rice husk can also give similar results at higher doses. Further rice husk is an agro waste, cheap and easy availability promotes its use by industries.

Abbreviations

PAC: Powdered Activated Carbon

ppm: Parts per million

XRF: X-ray Fluorescence

COD: Chemical Oxygen Demand

References

Ademorali C.M.A., Ukponmwan D.O. and Omode A.A, "Studies of Textile Effluent Discharges in Nigeria". *Int. J. Environ. Stud.* 39 (5), 291-296, (1992).

Ajmal Mohammad and Khan Ahsan Ullah, Effect of a Textile Factory Effluent on Soil and Crop Plants, *Enviroment Pollut* (Series A) 37, 131-148, (1985).

Edward Menezes, Development in Wastewater Treatment Methods in Textile Industry, *Everything About Water*, Sep – Oct, 65-70, (2003a)

Edward Menezes, Development in Wastewater Treatment Methods, *Chemical Weekly*, XLVIII (48), 187 – 192, (2003b).

Jain B.L. and Gupta I.C., Effect of Gypsum Treatment in the Amelioration of Industrial Effluents. *Curr Agric* 16 (1), 75-77, (1992).

Konduru R.R. and Viraraghavan T., Dye Removal by Low Cost Adsorbents, *Wat. Sci. Tech.* 36 (2-3), 189-196, (1997).

Kothandaraman V., Aboo K.M. and saatray C.A., Charateristics of the Waste from a Textile Mill, *Indian J Envl Health*, 18 (2), 93, (1976).

Majumdar P. and Chanda S., Chemical Profile of some Lignocellulosic crop residues, *Indian J Agric Biochem*, 15 (1&2), 29-33, (2001).

McKay G., Allen S.J., McConvey I.F. and Otterburn M.S., Transport Processes in the Sorption of coloured ions by Peat Particles". *Journal of Colloid and Interface Science*, 80 (2), 323-339, (1981).

Namasivayam C. and Arasi D.J.S.E, Removal of Congo Red from Wastewater by Adsorption onto Waste Red Mud, *Chemosphere* 34 (2), 401-417, (1997).

Shah J.K. "Problem of Pollution in Textile Mills and Processing House, In: Seminar on Pollution Problems and Prospects in Textile Mills and Process Houses, Reg. Directorate, N.P.C., Ahamedabad, (1985).

Valentin F.H.H., Peat Beds For Odour Control: Recent Developments And Practical Details. *Filtration and Separation* 2, 224-226, (1986).

Vickerstaff T., The Physical Chemistry of Dyeing Imperial Chemical Industries, London, (1954).

2013, Environmental Biotechnology
Editors: D.R. Khanna, A.K. Chopra, Gagan Matta, Vikas Singh & Rakesh Bhutiani
Published by: BIOTECH BOOKS, NEW DELHI

Pages 315–321

Chapter 38

Dissipation of Chlorpyriphos on Okra (*Abelmoschus esculentus* L.)

Mosmi Raina and Anil K. Raina
Department of Environmental Sciences
University of Jammu, Jammu – 180 006

Indiscriminate use of insecticides to combat the insect pests has led to accumulation of residues in okra which are harmful to consumers. In the present study, an attempt has been made to study the dissipation of chlorpyriphos on Okra fruits under agro-climatic conditions of Jammu where no study has been carried out earlier. Field experiments have been conducted for two consecutive years 2004 and 2005 to work out the safe preharvest interval. The average initial deposits of 0.91 mg kg^{-1} and 1.46 mg kg^{-1} have been recorded on okra fruits treated with the recommended (500g a.i. ha^{-1}) and double the recommended dose of chlorpyriphos (1000g a.i. ha^{-1}) respectively, showing percentage dissipation of 97.80 and 98.63 per cent correspondingly. On the basis of dissipation and prescribed MRL of 0.20 mg kg^{-1} of chlorpyriphos for okra, the half-life for the recommended and double the recommended dose has been worked out to be 1.33 and 1.41 days, respectively. The safe waiting period of 2.92 and 4.06 days have been suggested for the recommended and double the recommended dose of chlorpyriphos, respectively.

Keywords: *Chlorpyriphos, Dissipation, Okra, Maximum residue limit, Pesticide, Safe waiting period.*

Introduction

Use of pesticides to control pests is unavoidable as pests cause heavy loss to yield and quality of the food items including the vegetables, which forms a very important component of agriculture in India. Various insect pests *viz.* shoot and fruit borer, leaf rollers, jassids, aphids, moths, mites, fruit flies, caterpillars, weevils and hoppers etc. cause considerable losses to the vegetables which along with fruits and spices have been estimated to be of Rs. 30,000 crores in India only (MOCF, 2002). Several kinds of pesticides are being applied to control the pests, sometime close to the harvest or picking time of vegetables, thereby, leaving little or no time for their adequate dissipation. The presence of pesticide residue or their metabolites is a matter of great concern as it is directly related to the health of the human beings. Use of pesticides cannot be avoided but their quality, quantity and safe waiting period after the spray can be regulated so that minimum level of residues are left on the vegetables at the time of consumption. Supervised trials provide useful information for determining the waiting period which are considered safe interval between the last application of pesticide and harvest of edible part of the vegetables. This also facilitates the growers to adjust their crop harvest intervals accordingly in the interest of vegetable consumers.

Okra, commonly called as "Bhendi", is an important vegetable crop grown all round the year and throughout the country for fresh market consumption as well as for preservation, as the fruit is very rich in fiber, proteins, vitamins and minerals (Pawar and Jadhav, 1993). Several studies have indicated the contamination of market samples of okra with different kinds of pesticides including chlorpyriphos.(Dahiya and Chauhan, 1982; Chauhan *et al.*, 1983; Saxena *et al.*, 1990; Chahal *et al.*, 1997; Agnihotri, 1999; Chahal *et al.*, 1999; Rao and Rao, 2000; Kole *et al.*, 2002 and Shah *et al.*, 2000).

Chlorpyriphos (0, 0-diethyl 0-3, 5, 6-trichloro-2-pyridyl phosphorothioate) is a broad-spectrum organophosphorous insecticide used against a number of important arthropod pests in the study area. Till date no studies have been conducted on dissipation pattern of chlorpyriphos on okra under agro climatic conditions of J&K. The present studies were, therefore, undertaken to determine the residues of chlorpyriphos on *Abelmoschus esculentus* L.(Variety: Pusa Sawani) following its application at the recommended and double the recommended dosages to find out the desirable waiting period between the last spraying and harvesting of vegetable.

Materials and Methods

The field experiments were conducted at village Jessore (32° 38′ N latitude and 74° 45′ E longitude; altitude: 271meters above mean sea level, located in Low- altitude subtropical agro climatic zone characterized by the monsoon concentration of precipitation, hot summers and relatively dry but pronounced winters and preponderance of alluvial soils) located in Tehsil R.S. Pura of District Jammu for the two consecutive years. The okra crop was raised by sowing the certified seeds procured from J and K Agriculture Department, directly into the prepared field in the form of ridges placed at a distance of 60cm with plant to plant spacing of 30cm (recommended by Directorate of Extension Education, SKUAST-Jammu) in starting April 2004 and 2005 and thereafter normal agronomic practices were followed.

Chlorpyriphos 20EC, purchased from local market was sprayed twice at the rate of 500g a.i. ha^{-1} (recommended dose) and 1000g a.i. ha^{-1} (double the recommended dose) with Knapsack sprayer. First spray was done at the onset of flowering followed by second spray at an interval of twenty days. The sampling was done on 0(2 hr), 1, 2, 3, 4, 5, 7, 9 and 15 after the second spray for both the years. Samples from each experiment plot consisted of about 8-10 okra fruits and were brought to the laboratory in polythene bags with minimal finger touching for residue analysis.

Extraction and Partitioning

Representative sample of chopped okra fruits (50g) after proper mixing and quartering was blended in an electric grinder and rinsed with Acetone AR. The macerated sample was immersed in 100-150 ml Acetone AR and kept over night. The sample was filtered into 1 liter separating funnel using whatman's filter paper No. 1. After adding 600ml of 5 per cent NaCl solution to the above extract partitioning was done twice with 100 ml of dichloromethane AR. The lower layer was collected over 30g of anhydrous sodium sulphate and concentrated to about 5ml using rotary vacuum evaporator.

Clean Up

The cleanup of the extract was performed by adsorption column chromatography. Glass column of 2.5 cm (i.d.) x 60 cm (length) was packed with mixture of 20g silica gel (60-120 mesh size) activated at 130°C for 2 hrs and activated animal charcoal (300mg) sandwiched between 5g of anhydrous sodium sulphate layers. The drip tip of the column was plugged with cotton. After pre washing the packed column with 40ml of dichloromethane AR the sample extract was added to it and eluted with 150 ml solvent mixture of Acetone: dichloromethane (2:1). The eluate was evaporated to dryness and the residue was dissolved in 5ml of Acetonitrile (HPLC grade) and the sample was filtered using membrane filter media.

Estimation

The residues of chlorpyriphos was estimated by using High Pressure Liquid Chromatography (Shimadzu HPLC model LC-10A) equipped with dual pump (LC-10 AT), auto injector (SIL-10 A), UV detector(SPD-10 A) set at a wavelength of 225 nm. The column used for the separation was Shim pack CLC-ODS (M) 4.6 X 25 cm. long stainless steel tube packed with totally porous, spheric silica particles (5 μ m particle diameter, 100°A pore diameter).

The solvent system used was acetonitrile: water (90:10) at a flow rate of 1ml/min and the retention time of chlorpyriphos was 6.1 minutes. The method offered a sensitivity of 0.5 ng and limit of detection of 0.01mg kg^{-1} (on 50g sample basis).

The average recoveries from okra fortified with concentrations ranging from 0.20 to 1.0 mg kg^{-1} were found to be in the range of 83 per cent -93 per cent with an average of 88.42±2-31 per cent. All the solvents used for extraction and cleanup were glass distilled and chemicals *i.e.* silica gel, anhydrous sodium sulphate etc. along with glassware were washed with distilled solvents before use.

The suitability of all the solvents for residual analysis was ensured by running reagent blanks.

Statistical Analysis

The data was analyzed to work out half life value (RL_{50}) and safe waiting period (T_{tol}) according to Hoskins formula (1961).

Results and Discussion

The quantitative estimates of residues of chlorpyriphos on okra fruits for two consecutive years *i.e.* 2004 and 2005 at various sampling intervals (0, 1, 2, 3, 4, 5, 7, 9 and 15 days after the second spray) have been presented in the Table 38.1. The residues of chlorpyriphos on okra fruits collected from control plots has been always found to be below the detectable limits (0.01mg kg^{-1}) during both the years of study period.

Table 38.1: Dissipation of Chlorpyriphos on Okra Fruits during 2004 and 2005

Days after Treatment	**Residues of Chlorpyriphos (mg kg^{-1})*			
	1st Year (2004)		*2nd Year (2005)*	
	R.D. (500g a.i. ha^{-1})	*Double R.D. (1000g a.i ha^{-1})*	*R.D. (500g a.i. ha^{-})*	*Double R.D. (1000g a.i ha^{-1})*
0 (2hr)	0.86	1.80	0.96	1.12
1	0.51 (40.69)	0.95 (47.22)	0.55 (42.71)	0.80 (28.57)
2	0.43 (50.0)	0.59 (67.22)	0.46 (52.08)	0.66 (41.07)
3	0.21 (75.58)	0.31 (82.78)	0.22 (77.08)	0.33 (70.53)
4	0.12 (86.05)	0.23 (87.22)	0.11 (88.54)	0.24 (78.57)
5	0.04 (95.35)	0.07 (96.11)	0.04 (95.83)	0.06 (94.64)
7	0.02 (97.67)	0.04 (97.78)	0.02 (96.87)	0.03 (97.32)
9	BDL	0.02 (98.89)	BDL	0.02 (98.21)
15	BDL	BDL	BDL	BDL
RL_{50}(days)	1.3413	1.4039	1.3254	1.4388
T_{tol} (days)	2.820	4.450	2.999	3.576
Regression Equation	Y=2.8782-0.2244x r = -0.9443	Y=3.1632-0.2144x r = -0.9946	Y=2.9023-0.2271x r = -0.9444	Y= 3.0948-.2092x r = -0.9905

Control Samples showed residue level = BDL; Figures in parenthesis represent per cent age dissipation.
* Average of three replicates; BDL = Below detectable limit (0.01mgkg^{-1}); R.D. = Recommended dos

The average initial deposits of chlorpyriphos on okra fruits after the second spray were found to be 0.86 and 1.80 mg kg^{-1} at the recommended 500g a.i. ha^{-1}) and double the recommended dose (1000g a.i. ha^{-1}), respectively, for the first year (*i.e.*

2004). About 98 per cent of initial deposits dissipated within 7 days after its application at recommended dose while at double the dose about 99 per cent of the initial deposits get dissipated within 9 days. The time required (T_{tol}) for dissipation of residues below the maximum residue limit (MRL) of 0.2 mg kg^{-1} were found to be 2.82 and 4.45 days, respectively, for lower and higher dosages while the corresponding values for half life (RL_{50}) were found to be 1.3413 and 1.4039 days (Table 38.1).

Similarly for the second year (*i.e.* 2005) average initial deposits of chlorpyriphos on okra fruits were found to be 0.96 and 1.12 mgkg^{-1} following treatments of recommended and double the recommended dosages. Dissipation of residues was found to be about 97 per cent within 7 days and about 98 per cent within 9 days respectively for lower and higher dosages. The safe waiting periods (T_{tol}) at recommended and double the recommended dosages were found to be 2.999 and 3.576 days, respectively, and for half life (RL_{50}) these values were 1.3254 and 1.4388 days correspondingly (Table 38.1).

Table 38.2: Dissipation of Chlorpyriphos on Okra Fruits (Average of 2004 and 2005)

Days after Treatment	*Average Chlorpyriphos Residues (mg kg^{-1}) for Two Years ± S.D.*	
	500g a.i. ha^{-1}R.D.	*1000g a.i. ha^{-1}Double R.D.*
0 (2hr)	0.91±0.07	1.46 ± 0.48
1	0.53±0.02 (41.75)	0.87±0.04 (40.41)
2	0.44±0.02 (51.64)	0.62±0.04 (57.53)
3	0.21±0.007 (76.92)	0.32±0.01 (78.08)
4	0.11±0.007 (87.91)	0.23±0.007 (84.24)
5	0.04±0 (95.60)	0.06±0.007 (95.89)
7	0.02±0 (97.80)	0.03±0.007 (97.94)
9	BDL	0.02±0 (98.63)
15	BDL	BDL
RL_{50}(days)	1.3348	1.4151
T_{tol} (days)	2.918	4.058
Regression Equation	Y = 2.885 - 0.2255x r = -0.9446	Y = 3.1227 – 0.2127x r = -0.9914

Control Samples showed residue level =BDL; Figures in parenthesis represent per cent age dissipation.
*Average of three replicates; BDL= Below detectable limit (0.01mgkg^{-1}); R.D. = Recommended dose.

On an average (of the two years) initial deposits of 0.91 mg kg^{-1} on okra fruits, treated with the recommended dose (500g a.i. ha^{-1}) of chlorpyriphos, dissipated to 0.53, 0.44, 0.21, 0.11, 0.04 and 0.02 mg kg^{-1} on 1, 2, 3, 4, 5 and 7th day after the spray, respectively showing corresponding percentage dissipation of 41.75, 51.64, 76.92, 87.91, 95.60 and 97.80 percent, whereas average (of the two years) initial deposits of 1.46 mg kg^{-1} recorded on okra fruits treated with double the recommended dose of chlorpyriphos (1000g a.i. ha^{-1}), dissipated to 0.87, 0.62, 0.32, 0.23, 0.06, 0.03 and 0.02 on 1, 2, 3, 4, 5, 7 and 9th day after the spray, respectively showing corresponding percentage dissipation of 40.41, 57.53, 78.08, 84.24, 95.89, 97.94 and 98.63 percent (Table 38.2).

On an average, the rate of dissipation on okra fruits has been observed to be rapid during initial days and is almost similar at both the doses of chlorpyriphos which is in close consonance with the works of Hinduja *et al.*, 1979 and Samant *et al.*, 1997 on dissipation of chlorpyriphos.

On the basis of present study, the safe waiting period of 2.92 and 4.06 days have been suggested for the recommended and double the recommended dosages of chlorpyriphos on okra, respectively. Various workers have reported different safe waiting periods for different pesticides used on okra fruits. Rajabaskar *et al.*, 2001 suggested waiting periods of 2.09 and 4.5 days for okra fruits treated with recommended and double the recommended doses of endosulphan, while Iiango and Devraj, 2003 suggested waiting periods of 3.51 - 5.73 days for okra fruits treated with imidacloprid for different concentrations. Dixshit, 2000 recommended waiting period of 5 days following treatments of both imidacloprid and beta-cyfluthrian. Patel *et al.*, 2001 have suggested waiting periods of only one day for okra fruits treated with lindane while Singh, 1999 recommended waiting periods ranging from 3.93 to 9.22 days for okra fruits sprayed with different doses of monocrotophos. Biswas *et al.*, 1991 recommended waiting periods ranging from 4.7 to 6.3 days to be observed between spraying of monocrotophos and harvesting of okra fruits. Thus, the safe waiting periods of 2.92 days (with recommended dose) and 4.06 days (with double the recommended dose) which have been suggested in the present study, are in fair agreement with the findings of some of the workers who have worked on the okra with some other pesticides.

Conclusion

On the basis of dissipation and prescribed Maximum Residue Limit of 0.20 mg kg^{-1} of chlorpyriphos for okra, the safe waiting period of 2.92 and 4.06 days have been suggested for the recommended and double the recommended dosages of chlorpyriphos, respectively.

References

Agnihotri, N.P. 1999. *Pesticide Safety Evaluation and Monitoring.* AICRP (Pesticide Residues), IARI, New Delhi. pp. 173.

Biswas, S.K., Shukla, P., Chakraborty, A., Bhattacharyya, A., Dar, A.K. and Pal, S. 1991. Studies on residues of fenvalerate and monocrotophos in okra (*Abelmoschus esculentus* Moench.) under West Bengal agro-climatic conditions. *Pesticides Res. J.* 3(2): 119-122.

Chahal, K.K., Singh, B., Battu, R.S. and Kang, B.K. 1999. Monitoring of farmgate vegetables for insecticide residues in Punjab. *India J. Ecol.* 26(1): 50-55

Chahal, K.K., Singh, B., Kang, B.K., Battu, R.S. and Joia, B.S. 1997. Insecticide residues in farmgate vegetable samples in Punjab. *Pesticide Res. J.* 9(2): 256-260

Chauhan, R., Singh, Z. and Singh, S. 1983. Determination of organochlorine insecticides residues in vegetable samples at Hissar (Haryana). *Indian J. Ecol.* 10(1): 16-19

Dahiya, B. and Chauhan, R. 1982. Organochlorine insecticide residues in vegetable samples from Hissar market. *Indian J. agric. Sci.* 52(8): 533- 535.

Dikshit, A.K., Lal, O.P. and Srivastava, Y.N. 2000. Persistence of pyrethroid and nicotinyl insecticides on okra fruits. *Pesticide Res. J.* 12(2): 227-231.

Hoskins, W.M. 1961. Mathematical treatments on rate of loss of pesticides residue. *Pl. Prot. Bull. FAO.* 91: 163-168

Iiango, K. and Devraj, H. 2003. Residue levels of imidacloprid in okra. *Pesticide Res. J.* 15(1): 47-49.

Kole, R.K., Banerjee, H. and Bhattacharyya, A. 2002. Monitoring of pesticide residues in farmgate vegetables samples in West Bengal. *Pesticides Res. J.* 14(1): 77-82.

Ministry of Chemicals and Fertilizers. 2002. *37th Report of the standing committee on petroleum and chemicals.*

Patel, B.A., Shah, P.G., Raj, M.F., Patel, B.K. and Patel, J.A. 2001. Dissipation of lindane in/on brinjal and okra fruits. *Pesticide Res. J.* 13(1): 58-56.

Pawar and Jadhav. 1993. Bio-efficacy of some synthetic pyrethroids and endosulfan in/on okra pests. *Pestology*. 17(5): 16-18.

Rao, A.S. and Rao, P.R. 2000. Study on pesticide residues in vegetable. *Poll. Res.* 19(4): 661-664.

Rajabaskar, D., Thangaraju, D. and Murogesah, N. 2001. Residues of endosulfan in bhendi (*Abelmoschus esculentus*) fruits and its removal by washing and cooking. *J. Ecotoxicol. Environ. Monit.* 11(3): 185-189.

Saxena, R.C., Srivastava, R.P., Singh R. and Srivastava, A.K. 1990. Studies on residues of insecticides in/on locally marketed fruits and vegetables. *Indian. J. Ent.* 52 (2): 258-264.

Shah, P.G., Raj, M.G., Patel, B.A., Patel, B.K., Diwan, K.D., Patel, J.A. and Talati, J.G. 2000. Pesticide contamination status in farmgate vegetables in Gujarat. *Pesticides Res. J.* 12(2): 195-199.

Singh, Y.P. 1999. Persistence and degradation of monocrotophos in/on okra (*Abelmaschus esculentus* Moench) at medium high altitude hills. *Pestology*, 23(9): 3-6.

2013, Environmental Biotechnology *Pages 323–331*
Editors: **D.R. Khanna, A.K. Chopra, Gagan Matta, Vikas Singh & Rakesh Bhutiani**
Published by: **BIOTECH BOOKS, NEW DELHI**

Chapter 39

The Significance of Antioxidative Enzyme-Catalase in Cyanobacteria

Archana Tiwari and Shivani
Department of Biotechnology, Guru Nanak Girls College, Ludhiana, Punjab

Reactive oxygen species arise because of photosynthesis and respiration leaving behind active oxygen species, superoxide ($O_2^{\bullet}$), hydrogen peroxide (H_2O_2) and hydroxyl radical ($OH^{\bullet}$). The active oxygen species (AOS) are highly reactive and have the potential to damage membranes, proteins, pigments and genetic material thereby disrupting the metabolism of the organism. Management of free radicals is a very important task executed by cyanobacteria. Antioxidative enzymes play a pioneer role in scavenging free radicals. The major hydrogen peroxide scavenging activities in cyanobacteria are reported to be catalyzed by catalase, Superoxide dismultase and ascorbate peroxidase. Catalase is an extraordinary enzyme and its activity has been found in all cyanobacteria species. It is one of the most potent catalysts known and the reaction it catalyses is crucial to life. Catalases catalyze the dismutation of H_2O_2 into dioxygen and water, are common to animals, plants and microorganisms.

Recent genome projects of cyanobacteria and plants confirm the occurrence of Antioxidative enzymes. Genes encoding key components that respond to oxidative stresses, whose roles have been established in other organisms have also been found in cyanobacteria. The information about occurrence of antioxidative enzymes in the cyanobacteria gives us ample information to understand the mechanism of survival of organisms in extremes of environmental factors and their applications in biotechnology.

Keywords: *Reactive oxygen species, Antioxidative enzymes, Catalase, Cyanobacteria.*

Introduction

Cyanobacteria are oxygenic phototropic prokaryotes combining mechanisms of plant type photosynthesis and cytochrome c based respiration in one cell (Peschek, *et al.*, 2004). Cyanobacteria were the first organisms that produced molecular oxygen; they had to devise ways to manage active oxygen species such as superoxide radicals, hydroxyl radicals and hydrogen peroxide. For the efficient removal of reactive oxygen species the cyanobacteria as well as other organisms developed quite a number of strategies like antioxidative enzymes. Scavenging of AOS is reported to follow the Halliwell-Asada cycle (Halliwell and Guttweidge, 1989; Asada, 1994). Activity of some of the enzymes of the cycle has been found absent in Cyanobacteria (Obinger *et al.*, 1998). The effective H_2O_2-scavenging systems consisting of catalase and peroxidase activities were required in cyanobacteria to protect the oxidative damage. The production of H_2O_2 is inevitable in cyanobacteria since superoxide dismultase has been reported (Miyake *et al.*, 1991). The primitive scavenging system of H_2O_2 appears to be catalase (catalase- peroxidase), because many aerobic prokaryotes contain only catalase (catalase- peroxidase) and lack peroxide scavenging peroxidases. The absence of ascorbate in prokaryotes and fungi can account for the lack of ascorbate peroxidase in these organisms (Asada, 1993).

Aerobic respiration and oxygenic photosynthesis leave behind active oxygen species, superoxide ($O_2^{\bullet}$), hydrogen peroxide (H_2O_2) and hydroxyl radical ($OH^{\bullet}$). The active oxygen species (AOS) are highly reactive and have the potential to damage membranes, proteins, pigments and genetic material thereby disrupting the metabolism of the organism (Shaaltiel and Gressel, 1986; Scadalios, 1993). Free radical management is thus very important for organisms. In order to overcome the oxidative stresses, cyanobacteria and plants have developed a number of antioxidant systems. Recent genome projects of cyanobacteria and plants confirm the fact. Genes encoding key components that respond to oxidative stresses, whose roles have been established in other organisms have also been found in cyanobacteria.

Early reports indicated the presence of two major hydrogen peroxide scavenging activities in cyanobacteria, catalase and ascorbate peroxidase (Tel-Or *et al.*, 1985, 1986). Catalase activity has been found in all cyanobacteria species (Miyake *et al.*, 1991). For the detoxification of H_2O_2, three classes of haem proteins are involved: catalase, catalase-peroxidase and electron donor specific peroxidases. Catalases catalyze the dismutation of H_2O_2 into dioxygen and water, are common to animals, plants and microorganisms (Ames *et al.*, 1993). Among the bacteria, an atypical hydroperoxidase, which differed from the typical catalase, was first purified from *Escherichia coli* by Claiborne and Fridovich (1979). Nadler *et al.* (1986) have proposed the definition of a new class of catalase and peroxidase activities and sharing characteristics with the typical catalase and peroxidases from higher organisms. These enzymes have been observed in photosynthetic bacteria, facultative anaerobes and strict anaerobes (Hochman and Goldberg, 1991). Catalytic activity is present in nearly all animal cells and organs and in aerobic microorganism. Studies on its biosynthesis have been reported by Higashi *et al.* (1974), Sakamoto and Higashi (1974) and Yasukochi *et al.* (1974). Yeast catalase has been characterized by Seah and Kaplan (1973) and Seah *et al.* (1973).

Catalase is an extraordinary enzyme and is one of the most potent catalysts known. Catalase catalyses conversion of hydrogen peroxide, a powerful and potentially harmful oxidizing agent to water and molecular oxygen. Catalases also use hydrogen peroxide to oxidize toxins including phenols, formic acid, formaldehyde, ethanol, methanol or elemental mercury. Catalases have four subunits. Each subunit contains a heme group. This heme group is responsible for carrying out catalase's activity. The following reaction is catalyzed by catalases:

$$2\ H_2O_2 \rightarrow 2\ H_2O_2 + O_2$$

Catalase is an enzyme that can function in two distinct modes. The catalytic mode is responsible for H_2O_2 breakdown, which is thought to occur in two stages (Halliwell and Gutteridge, 1999; Sichak and Dounce, 1986): -

$$H_2O_2 + \text{Fe (III)-CAT} \rightarrow H_2O + \text{O=Fe (V)-CAT}\ (k_1 = 1.7\quad 107\ M^{-1}\ s^{-1})$$

$$H_2O_2 + \text{O=Fe (V)-CAT Fe (III)} \rightarrow \text{CAT} + H_2O + O_2\ (k_2 = 2.6\quad 107\ M^{-1}\ s^{-1})$$

(III and V indicates formal oxidation states of Fe)

In the reaction Fe (III)-CAT represents the native catalase molecules and O=Fe (V)-CAT represents catalase compound I, which was first described by Chance *et al.* (1979). Peroxide upon entering the heme cavity, is severely sterically hindered and must interact with His74 and Asn147 (Fita and Rossmann, 1985). It is in this position that the first stage of catalysis takes place. Transfer of a proton from one oxygen of the peroxide to the other, *via* His74, elongates and polarizes the O-O bond, which eventually breaks hetrolytically as peroxide oxygen is coordinated to the iron center. This coordination displaces water and forms Fe (V)=O plus a heme radical. The radical quickly degrades in another one electron transfer to rid off the radical electron, leaving the heme ring unaltered. During the second stage, in a similar two electron transfer reaction, Fe (V)=O reacts with a second hydrogen peroxide to produce the original Fe (III)-CAT, another water and a mole of molecular oxygen. The heme reactivity is enhanced by the phenolate ligand of Tyr357 in the 5th iron ligand position (Fita and Rossmann, 1985), which may aid in the oxidation of Fe (III) to Fe (V) and the removal of an electron from the heme ring. The efficiency of catalase may be due to the interaction of His74 and Asn147 with reaction intermediates. This mechanism is supported by experimental evidence indicating modification of His74 with 3-amino-1, 2, 4-triazole, which inhibits the enzyme by hindering substrate binding (Darr and Fridovich, 1986).

Typical catalases (EC 1.11.1.6) are found in almost all aerobically respiring organisms, both prokaryotes and eukaryotes. Catalase was first crystallized from beef liver by Summer and Dounce in 1937. Under normal physiological conditions, catalase controls the hydrogen peroxide concentration so that this does not reach toxic levels that could bring about oxidative damage in the cells. The structure of catalase from many different species has been studied by X- ray diffraction. Although it is clear that all catalases share a general structure, some differ in the number and identity of domains. Catalases have the advantage to catalyze the dismutation of

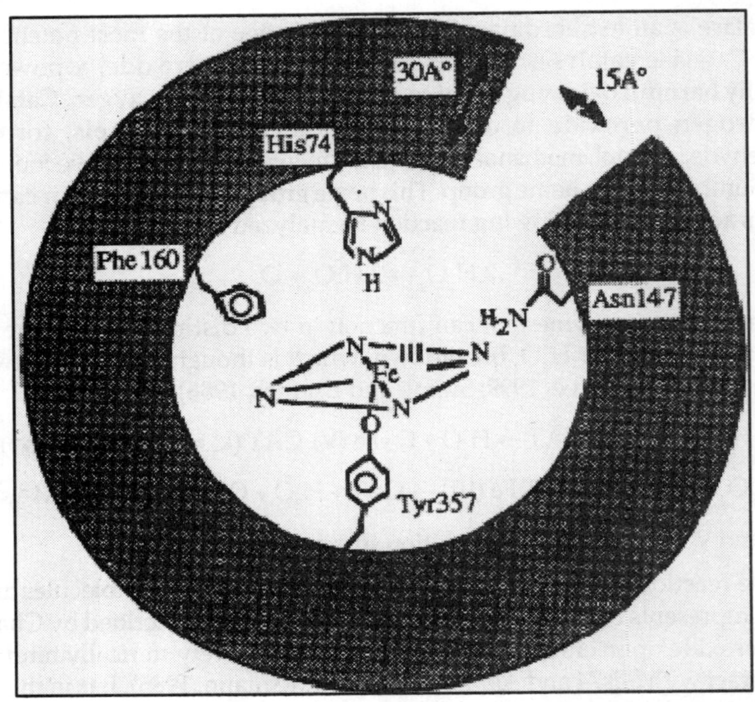

Figure 39.1: Schematic Drawing of Active Site of Catalase (Belal *et al.*, 1989)

hydrogen peroxide and oxygen without requiring an additional source of reducing power like all peroxidases.

Catalase fall into two main classes, the HP I and HP II catalases. HPII catalases catalyze just the disproportionation of H_2O_2 whereas HPI catalases have both the described activities. HPI catalases exist as two isozymes, HPI-A and HPI-B and they sediment at slightly different densities. Gregory and Fridovich (1974) reported a sensitivity activity stain protocol for catalase applicable to a polyacrylamide gel electrophoretogram. Haining and Legan (1972) described a polarographic assay utilizable in tissue homogenates. Kroll *et al.* (1989) worked out a rapid method for estimating the bacterial content of foods.

Catalase, like peroxidase is a hemoprotein and forms intermediate addition compounds with H_2O_2. The stepwise mechanism of its activity was first elucidated by Chance (1951), Oshino *et al.* (1975), Zidoni and Kremer (1974), Chance and Oshino (1973), Chance *et al.* (1973), Jones (1973), and Jones and Middlemiss (1972). Nichols and Schonbaum (1963) have reviewed catalase. Catalase uses H_2O_2 as a substrate as well as a hydrogen acceptor. Peroxidase requires a separate acceptor. Lanir and Schejter (1975) have reported on the difference in reactivity of the two enzymes. Nelson and Kiesow (1972) have studied the enthalpy of the catalase reaction. Corrall *et al.* (1974) and Oshino *et al.* (1973) have reported the oxidation of ethanol using catalase and Marklund (1972) studied its interaction with hydroxymethylhydroperoxide.

Catalase is of interest commercially wherever hydrogen peroxide is used as a germicide, for example, in the food industry for disposing of H_2O_2 used in pasteurizing milk prior to cheese-making (Chu *et al.*, 1975). Microencapsulated catalase is also of interest (Poznansky and Chang 1974). Miyake *et al.* (1991) have reported a scavenging system of H_2O_2 in cyanobacteria with respect to the acquisition of ascorbate peroxidase during the evolution of these organisms. Some species of cyanobacteria lack ascorbate peroxidase as a H_2O_2 –scavenging enzyme. The cyanobacteria can be divided into two groups: the first scavenges H_2O_2 with ascorbate peroxidase and catalase, and the second scavenges H_2O_2 only with catalase. Takeda *et al.* (1994) investigated the H_2O_2 removing enzymes in *Synechococcus* PCC 7942. They detected catalase activity but did not get ascorbate peroxidase activity. Mutsuda *et al.* (1996) reported the purification and characterization of catalase protein from *Synechococcus* PCC 7942 and the gene coding for its presence. They also found that *Synechococcus* PCC 7942 contained a catalase-peroxidase. Tripathi and Srivastava (2001) discussed the presence of stable oxygen scavenging enzymes- superoxide dismutase, catalase and ascorbate peroxidase in a dessicant- tolerant cyanobacterium *Lyngbya arboricola* under dry state. They reported five isoforms of SOD of which one was Mn-SOD, one Fe-Mn SOD and three Fe SODs, two isoforms of APX, and three isoforms of catalase enzymes forms in the crude extract of the freshly collected mats on native polyacrylamide gel electrophoresis.

There are numerous assays for catalases. The subject has been reviewed by Maehly and Chance (1954) and Chance and Maehly (1955). Catalase activity can be measured by monitoring the disappearance of hydrogen peroxide, in the reaction system. For a given concentration of hydrogen peroxide, the initial rate of its removal is proportional to the concentration of catalase. Decomposition of hydrogen peroxide can be followed by the loss of its light absorbance at 240 nm (Beers and Sizer 1952; Rao *et al.*, 1996) or by measuring the release of oxygen by using an oxygen electrode (Halliwell and Gutteridge, 1999).

In Cyanobacteria, a few reports suggested inducation of SOD, Catalase or some peroxidases under conditions of oxidative stresses based on activity measurements (Mittler and Tel-Or, 1991, Abeliovich *et al.*, 1974). Recently, the effects of high light, dark and light, UV, electron transport inhibitors and salt stress have been analysed by DNA microarray (Gill *et al.*, 2002, Hihara *et al.*, 2001, Kanesaki *et al.*, 2002, Hihara *et al.*, 2003, Huang *et al.*, 2002). It was found that hundreds of genes and ORFs are induced or repressed under stress conditions compared with normal conditions. Some of them appear to overlap between different stress conditions, while others may be unique to certain stress. However, little is known about the direct mechanisms to recognize ROS or oxidative damages in cyanobacteria.

Conclusion

The antioxidant system is an important and necessary part of an organism's response to a changing environment. Most environmental stresses appear to have an oxidative component. In the case of plants and other photosynthetic organisms, antioxidants are needed to adapt to changes in light, temperature and water potential as well as the anthropogenic stresses imposed by herbicides and pollutants.

Antioxidants, in general, have garnered much attention in the medical field. The information about occurrence of antioxidative enzymes in the cyanobacteria gives us ample information to understand the mechanism of survival of organisms in extremes of environmental factors and their applications in biotechnology.

Cyanobacteria are credible models for mitochondria, and mitochondria may be primary sites of oxidative damage in some diseases (Shigenaga *et al.*, 1994b). Human cells, like cyanobacteria, are compartmentalized. If antioxidant therapy is to be fully realized, the same approach must be taken as outlined for plants. Hence, therapeutic genes for antioxidants must be targeted to the place of damage for the cure of diseases.

References

Abeliovich A., Kellenberg D. and Shilo M. 1974. Effect of photooxidative conditions on levels of superoxide dismutase in *Anacystis nidulans*. *Photochem.Photobiol.* 19: 379–282.

Ames B. N., Shigenaga M. K. and Hagen T. M. 1993. *Proc. Natl. Acad. Sci.* U.S.A.90, 7915-7922.

Asada, K. 1993. In *Active Oxygens, Lipid Peroxides, and Antioxidants* (Yagi, K., ed.), Japan Scientific Societies Press, Tokyo, and CRC Press, New York, 289-298.

Asada, K. 1994. Production and action of active oxygen species in photosynthetic tissues. In *Causes of Photooxidative Stress and Amelioration of Defence Systems in Plants* (eds Foyer, C. H. and Mullinaeaux, P. M.), CRC Press, Boca Raton, FL, 77–104.

Beers R and Sizer I. 1952. A Spectrophotometric Method for Measuring the Breakdown of Hydrogen Peroxide by Catalase, *J Biol Chem* 195, 133.

Belal R., Momenteau M. and Meunier B. 1989. Why an oxygen and not a nitrogen atom as proximal ligand in catalase? Hydrogen peroxide dismutationcatalayzed by synthetic iron and manganese porphyrins. *New J Chem*. 13:853-862.

Chance B. 1951. The Iron-Containing Enzymes. C. The Enzyme-Substrate Compounds and Mechanism of Action of the Hydroperoxidases, *The Enzymes 2, Pt. 1*, J. Sumner and K. Myrback, Academic Press, NY.

Chance B. and Maehley A. 1955. Assay of Catalases and Peroxidases, *Methods in Enzymology* 2, S. Colowick and N. Kaplan, Academic Press, NY, 764.

Chance B. and Oshino N. 1973. Analysis of the Catalase-Hydrogen Peroxide Intermediate in Complex Oxidations, *Biochem J*. 131, 564.

Chance B., Boveris A., Oshino N. and Loschen G. 1973. The Nature of the Catalase Intermediate in the Biological Function, *Oxidases and Related Redox Systems I*, T. King, H. Mason, and M. Morrison, University Park Press, Baltimore, USA, 350.

Chance B., Sies H. and Boveris A. 1979. Hydroperoxide metabolism in mammalian organs.*Physiol Rev*. 59 (3):527-625.

Chu H., Leeder J., and Gilbert S. 1975. Immobilized Catalase Reactor for Use in Peroxide Sterilization of Dairy Products, *J Food Sci* 40, 641.

Claiborne and Fridovich (1979). Claiborne A. and Fridovich I. 1979. *J. Biol. Chem.* 254, 4245-4252.

Corrall R., Rodman H., Margolis J., and Landau B. 1974. Stereospecificity of the Oxidation of Ethanol by Catalase, *J Biol Chem 249*, 3181.

Darr D, Fridovich I. (1986) Irreversible inactivation of catalase by 3-amino-1, 2, 4-triazole. *Biocheml Pharmacol.* 35 (20): 3642.

Fita I. and Rossmann M.G. 1985. The active center of catalase. *J Mol Biol.* 185 (1): 21-37.

Gill R.T., Katsoulakis E., Schmitt W., Taroncher-Oldenburg G., Misra J. and Stephanopoulos G. 2002. Genome-wide dynamic transcriptional profiling of the light-to-dark transition in *Synechocystis* sp. strain PCC 6803. *J. Bacteriol* 184: 3671–3681

Gregory E. and Fridovich I. 1974. Visualization of Catalase on Acrylamide Gels, *Anal Biochem* 58, 57.

Haining J. and Legan J. 1972. Improved Assay for Catalase Based Upon Steady-State Substrate Concentration, *Anal Biochem 45*, 469.

Halliwell and Gutteridge, 1999; Halliwell B, Gutteridge JMC. (1999) Free *Radicals in Biology and Medicine*. 3rd ed. New York: Oxford University Press. pp: 134-140.

Higashi *et al.* (1974), Higashi T., Kawamata F., and Sakamoto T.1974. Studies on Rat Liver Catalase. VII. Double-Labeling of Catalase by ^{14}C-Leucine and ^{3}H-d-Aminolevulinic Acid, *J Biochem* (Tokyo) 76, 703.

Hihara Y., Kamei A., Kanehisa M., Kaplan A. and Ikeuchi M. 2001. DNA microarray analysis of cyanobacterial gene expression during acclimation to high light. *Plant Cell* 13: 793–806.

Hihara Y., Sonoike K., Kanehisa M. and Ikeuchi M. 2003. DNA microarray analysis of redox-responsive genes in the genome of the cyanobacterium *Synechocystis* sp. PCC 6803. *J. Bacteriol.* 185: 1719–1725.

Hochman A. and Goldberg I. 1991 *Biochim. Biophys. Acta* 1077, 299-307.

Huang L., McCluskey M.P., Ni H. and LaRossa R.A. 2002. Global gene expression profiles of the cyanobacterium *Synechocystis* sp. strain PCC 6803 in response to irradiation with UV-B and white light. *J. Bacteriol.* 184: 6845–6858.

Jones P. and Middlemiss D. 1972. Formation of Compound I by the Reaction of Catalase with Peroxoacetic Acid, *Biochem J* 130, 411.

Kanesaki Y., Suzuki I., Allakhverdiev S.I., Mikami K. and Murata N. 2002. Salt stress and hyperosmotic stress regulate the expression of different sets of genes in *Synechocystis* sp. PCC 6803. *Biochem. Biophys. Res. Commun.* 290: 339–348.

Kroll R., Frears E. and Bayliss, A. 1989. An Oxygen Electrode-Based Assay of Catalase Activity as a Rapid Method for Estimating the Bacterial Content of Foods, *J Appl Bacteriol*, 66, 209.

Lanir A. and Schejter A. 1975. On the Sixth Coordination Position of Beef Liver Catalase, *FEBS Lett* 55, 254.

Maehly A. and Chance B. 1954. The Assay of Catalases and Peroxidases, in, *Methods of Biochemical Analysis*, 357.

Marklund S. 1972. Interaction between hydroxymethylhydroperoxide and Catalase, *Biochim Biophys Acta*. 289, 269.

Mittler R and Tel-Or E. 1991. Oxidative stress responses and shock proteins in the unicellular cyanobacterium *Synechococcus* R2 (PCC 7942). *Arch Microbiol* 155: 125–130.

Miyake C, Michihata F, and Asada K. 1991. Scavenging of hydrogen peroxide in prokaryotic and eukaryotic algae: acquisition of ascorbate peroxidase during the evolution of cyanobacteria. Plant and Cell Physiology 32, 33–43.

Mutsuda M., Ishikawa T., and Shigeoka S. 1996. The catalaseperoxidase of *Synechococcus* PCC 7942: purification, nucleotide sequence analysis and expression in *Escherichia coli*. *Biochem J*. 316: 251–257

Nadler *et al.* (1986) Nadler V., Goldberg I., and Hochman, A. 1986. Comparative Study of Bacterial Catalases, *Biochim Biophys Acta* 882, 234.

Nelson, D. and Niesow, L.1972. Enthalphy of Decomposition of Hydrogen Peroxide by Catalase at 25deg.C (With Molar Extinction Coefficients of H_2O_2 Solution in the UV, *Anal Biochem* 49, 474.

Nichols P. and Schonbaum G. 1963. Catalases, *The Enzymes, 2nd Ed. 7*, P. Boyer, H. Lardy, and K. Myrback, Academic Press, NY, 147.

Obinger C., Regelsberger G., Pircher A., Strasser G. and Peschek, G. A. 1998. *Physiol. Plant*, 140, 693–698.

Oshino N., Jamieson D., Sugano T., and Chance B. 1975. Optical Measurements of the Catalase-Hydrogen Peroxide Intermediate (Compound) in the Liver of Anaesthetized Rats and Its Implication to Hydrogen Peroxide Production, *Biochem J*. 146, 67.

Oshino N., Oshino R. and Chance B. 1973. The Characteristics of the "Peroxidatic" Reaction of Catalase in Ethanol Oxidation, *Biochem J*. 131, 555.

Peschek G.A., Obinger C., and Paumann M. 2004. The respiratory chain of blue-green algae (cyanobacteria) *Physiologia Plantarum*, 120: (3), 358.

Poznansky M. and Chang T. 1974. Comparison of the Enzyme Kinetics and Immunological Properties of Catalase Immobilized by Microencapsulation and Catalase in Free Solution for Enzyme Replacement, *Biochim Biophys Acta* 334, 103.

Rao M.V., Paliyath G. and Ormrod D.P. 1996. Ultraviolet-B- and ozone-induced biochemical changes in antioxidant enzymes of *Avabi-dopsis thaliana. Plant Physiol* 110: 125-136

Sakamoto and Higashi (1974) Sakamoto T. and Higashi T. 1974. Studies on Rat Liver Catalase. VIII. Isolation and Translation of Catalase Messenger RNA, *J Biochem* (Tokyo) 76, 1227.

Scandalios J.G. 1993. Oxygen stress and superoxide dismutases. *Plant Physiol* 101: 7-12.

Seah T. and Kaplan J. 1973. Purification and Properties of the Catalase of Bakers' Yeast, *J Biol Chem* 248, 2889.

Seah T., Bhatti A., and Kaplan J. 1973. Novel Catalytic Proteins of Bakers' Yeast. I. An Atypical Catalase, *Can J Biochem 51*, 1551..

Shaaltiel Y. and Gressel J. 1986. Multienzyme oxygen radical detoxifying system correlated with paraquat resistance in *Conyza bonaviensis. Pestic Biochem Physiol* 26: 22-28

Shigenaga M. K., Hagen T. M. and Ames B. N. 1994b. Oxidative Damage and Mitochondrial Decay in Aging. *Proc. Natl. Acad. Sci. USA* 91: 10771-10778.

Sichak SP, Dounce AL. (1986) Analysis of the peroxidatic mode of action of catalase. *Archives of Biochem and Biophys*. 249 (2): 286-295.

Sumner J.B. and Dounce A.L. 1937. Crystalline catalase. *J Biol Chem*. 121:417-424.

Takeda *et al.* (1994) Takeda T., Ishikawa T., and Shigeoka S. 1994. The H2O2-scavenging system and tolerance system to H2O2 in algae. *In* K Asada, T Yoshikawa, eds, *Frontiers of Reactive Oxygen Species in Biology and Medicine.* Excerpta Medica, Amsterdam, The Netherlands, 143–146.

Tel-Or E, Huflejt ME, Packer L. 1986 Hydroperoxide metabolism in cyanobacteria. Arch Biochem Biophys. 246(1): 396-402.

Tel-Or E., Huflejt M.E. and Packer L. 1985. The role of glutathione and ascorbate in hydroperoxide removal in cyanobacteria. *Biochem.Biophys. Res. Commun.* 132: 533-539.

Tripathi S. N. and Srivastava P. 2001. Presence of stable active oxygen scavenging enzymes superoxide dismutase, ascorbate peroxidase and catalase in a desiccation-tolerant cyanobacterium *Lyngbya arboricola* under dry state. *Current Science* 81: 2, 197-200.

Yasukochi Y., Nakamura M. and Minakami S. 1974. Effect of Cobalt on the Synthesis and Degradation of Hepatic Catalase *in vivo, Biochem J.* 144, 455.

Zidoni E. and Kremer M. 1974. Kinetics and Mechanism of Catalase Action. Formation of the Intermediate Complex, *Arch Biochem Biophys*. 161, 658.

2013, Environmental Biotechnology *Pages* **333–347**
Editors: **D.R. Khanna, A.K. Chopra, Gagan Matta, Vikas Singh & Rakesh Bhutiani**
Published by: **BIOTECH BOOKS, NEW DELHI**

Chapter 40

Assessment of Preliminary Phytochemical Analysis and Antimicrobial Activity of Aqueous and Solvent Extracts of Leaf of *Tecomella undulata*

Mukesh Kumar[1], Kiran Nehra[2] and J.S. Duhan[1]
[1]Department of Biotechnology, Chaudhary Devi Lal University, Sirsa – 125 055, Haryana
[2]Department of Biotechnology,
[2]Deenbandhu Chhotu Ram University of Science and Technology, Murthal, Sonipat – 131 039, Haryana

The present study reports the antimicrobial activity of aqueous and organic solvent extracts (hexane, chloroform, ethyl acetate, acetone and methanol) of leaf of *Tecomella undulata* prepared in increasing and decreasing order of polarity. The antimicrobial activity was determined against twenty pathogens including twelve bacteria (five Gram–positive and seven Gram–negative) and eight fungi. Preliminary phytochemical analysis of different extracts revealed the presence of anthraquinones, flavonoids, cardiac glycosides, proteins, tannins, sterols, sugars, and terpenoids. High antimicrobial activity was observed with chloroform, ethyl acetate and acetone extracts (in both increasing and decreasing order of polarity) against most of the tested pathogens, particularly against *S. epidermidis, M. luteus, E. faecalis* and *K. pneumoniae* (showing a zone of inhibition between 6.5

to 23mm). Minimum inhibitory concentration (MIC) values were determined using broth microdilution assay. The MIC value of the extracts against the tested microorganisms varied between 25-1000 μg/ml. When compared with the standard antibiotics (ampicillin, penicillin and streptomycin); ethyl acetate, methanol and acetone extracts showed a lower MIC value, against *A. hydrophila, E. coli, K. pneumoniae, M. luteus* and *S. epidermidis*. The overall results thus provide promising baseline information for the potential use of the crude extract of leaf of *Tecomella undulata* in the treatment of bacterial and fungal infections. Hence, the plant can be subjected to further isolation of the therapeutic antimicrobial compounds for pharmacological evaluations.

Keywords: *Tecomella undulata, Antimicrobial, Antifungal, Phytochemical activity.*

Introduction

Plants used in traditional medicine have made large contributions to human health and well-being. They constitute an important source of novel drugs as they have been proved to be a rich source of new biologically active compounds. Utilization of plants for medicinal purposes in India has been documented long back in ancient literature (Tulsidas, 1631, Dradhbala, 1996). However, organized studies in this direction were initiated in 1956 (Rao, 1996). In India, plants are widely used by all sections of the population either directly as folk medicine or indirectly in the pharmaceutical preparations of modern medicines. In recent years, the popularity of this complementary medicine has increased due to an alarming increase in the emergence of drug-resistant microorganisms. This has been further augmented by the fact that the efficacy of the antimicrobial agents currently in use has been reduced due to an increase in the evolutionary adaptations by pathogenic organisms to commonly used antimicrobial agents. Therefore, the search for new drugs from novel sources such as plants has gained a high momentum in the last decade. Currently, research in this direction is focused on the investigations of the ethnobotanical uses of plants prevailing among native people. There are numerous reports evidencing the antibacterial activity of plants against microorganisms (Indu *et al.*, 2006; Lino and Deogracious, 2006; Malabadi, 2005; Malabadi *et al.*, 2005; Malabadi *et al*,. 2007; Savithramma *et al.*, 2007; Sudharameshwari and Radhika, 2007; Vedavathy *et al.*, 1997; Warrier *et al.*, 1997).

In the present investigation, attempts have been made to examine *Tecomella undulata* for its antimicrobial properties. *Tecomella undulate*, commonly known as Rohira, Rohitaka and Pharphugh belongs to the family *Bignoniaceae*. It is an important agro-forestry tree, found in the western parts of India. Plant parts are used for the cure of syphilis and eczema, as antiwrinkle agent and skin nourisher; and the bark possesses mild relaxant, cardiotonic and chloretic activities. It is used for the treatment of various diseases like urinary troubles, spleen enlargement, liver and abdominal diseases and for leucorrhoea. Ahmed *et al.* (1998) have reported the analgesic potential of the whole plant methanol extract. Thus, keeping in view the importance of *Tecomella undulata* in our daily life and with an attempt to include this plant in our medicines, the present study has been designed to explore the antibacterial potential of this plant against a large number of pathogenic microorganisms and to conduct a preliminary phytochemical analysis of its various extracts.

Materials and Methods

Plant Material and Extract Preparation

Fresh leaves of *Tecomella undulata* were collected from regions adjoining Sirsa, Haryana during July, 2009. Immediately after collection, the leaves were washed thoroughly, initially with tap water and then with distilled water to remove any debris or dust particles and then allowed to dry in an oven at 35°C. The dried plant material was ground to a fine powder and stored at room temperature in airtight containers. Aqueous and organic solvent extracts (prepared in both increasing and decreasing order of solvent polarity) were prepared using distilled water and five organic solvents. To 500 g of *Tecomella undulata* leaf powder, 1500 ml of each solvent *viz.* hexane, chloroform, ethyl acetate, acetone, methanol and distilled water was added serially in increasing solvent polarity, and in reverse order for obtaining extracts in decreasing solvent polarity (flow chart-1). Extraction with each solvent was allowed to carry out for 24 h at room temperature. After 24 h, the supernatant was recovered by filtering through Whatman No. 1 filter paper. The filtrate was concentrated by evaporating the supernatant in a rotary vacuum evaporator to obtain the crude extract. The process was repeated thrice with each solvent before proceeding with the same procedure with the next solvent in the sequence. These extracts (prepared both in increasing and decreasing order of polarity) were stored at 4° C until used further for the evaluation of antimicrobial activity.

Bacterial Test Organisms

For evaluation of the antibacterial response, the following twelve bacterial (five Gram positive and seven Gram negative), and eight fungal strains were used.

Gram Positive Bacterial Strains

Bacillus subtilis (MTCC–1133), *Entercoccus faecalis* (MTCC–2729), *Micrococcus luteus* (MTCC–1809), *Staphylococcus aureus* (MTCC-3160), and *Staphylococcus epidermidis* (MTCC–3086).

Gram Negative Bacterial Strains

Aeromonas hydrophila (MTCC–1739), *Alcaligenes faecalis* (MTCC–126), *Enterobacter aerogenes* (MTCC–2823), *Escherichia coli* (MTCC–294), *Klebsiella pneumoniae* (MTCC–3384), *Pseudomonas aeruginosa* (MTCC–1035), and *Salmonella typhimurium* (MTCC–1253).

Fungal Strains

Aspergillus flavus (RKS-108), *Aspergillus niger* (RKS-104), *Aspergillus oryzae* (4655), *Aspergillus terreus* (RKS-124), *Alternaria alternata* (6276), *Alternaria brasicola* (1707), *Alternaria solani* (4632), and *Alternaria vitis* (4921).

All the bacterial strains were procured from Microbial Type Culture Collection (MTCC), Chandigarh; the fungal strains used were either the indigenous strains

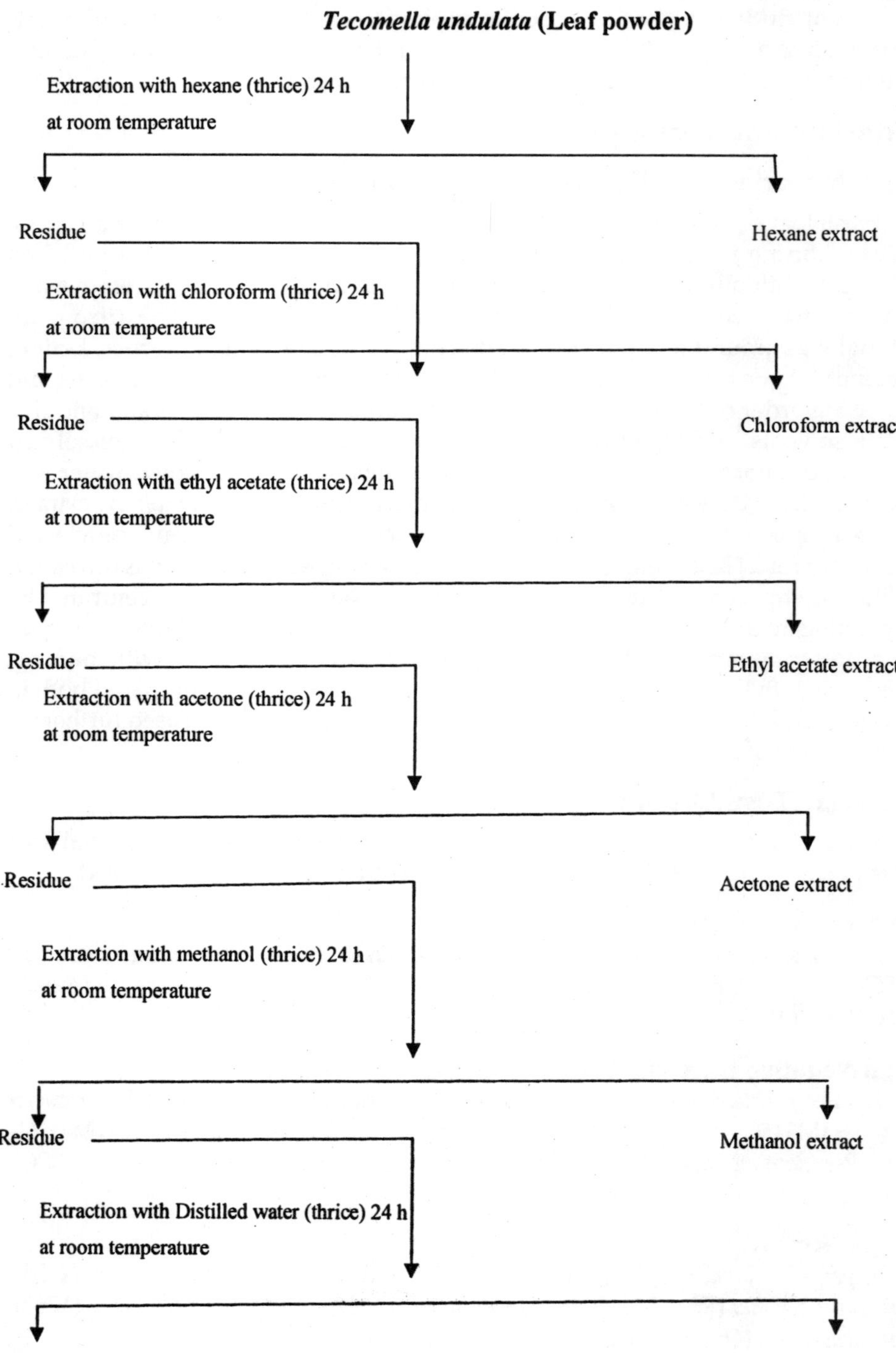

Flow Chart 41.1: Schematic Representation of the Extraction Procedure of Leaves of *Tecomella undulata* Prepared in Increasing and Decreasing Order of Solvent Polarity

maintained in the Department of Biotechnology, CDLU, Sirsa, Haryana, or were procured from Indian Agricultural Research Institute (IARI), Delhi.

Antimicrobial Assay

The evaluation of the antibacterial activity of the various plant extracts was carried out using the method as described by Perez *et al.* (1990). In this method, 100 µl of 24 h old culture of the test organism was inoculated on the agar plates and then spread on to the surface of the agar with the help of a sterilized glass spreader. After 30 minutes of inoculation of the test microorganism, wells (5mm diameter) were prepared with the help of a sterilized steel cork borer. Out of the five wells made in each plate, four were loaded with 60 µl of different concentrations of the test plant extract. Extraction solvent used as the negative control was loaded in the fifth well. Sixty µl each of standard antibiotics *viz.* ampicillin, penicillin, streptomycin, tetracycline, fluconazol, ketoconazole and miconazole were loaded in different wells in a separate plate and were used as a positive control. The plates were then aerobically incubated at 30°C for 24 h for bacterial and at 25°C for 48 h for fungal test microorganisms. Antimicrobial activity was determined by measuring the diameter (in mm) of zone of inhibition and comparing the results obtained with those from the standard antibiotics. The diameter was measured at cross angles and mean of three independent measurements was taken.

Minimum Inhibitory Concentration (MIC)

Defined as the lowest concentration of the test sample that results in a complete inhibition of visible growth, minimum inhibitory concentration (MIC) was determined by using the dilution method as recommended by the National Committee for Clinical Laboratory Standard (NCCLS, 1997). Different concentrations (ranging from 25 µg/ml to 1000 µg/ml) of all the extracts prepared both in increasing and decreasing order of solvent polarity were tested separately for each microorganism species. A stock solution of each active extract was serially diluted in 96-well microtiter plate with Mueller Hinton broth to obtain a concentration ranging from 25 µg/ml to 1000 µg/ml (with a gap interval of 25 µg/ml). A standardized inoculum for each bacterial strain was prepared so as to give an inoculum size of approximately 5 x 10^5 CFU/ml in each well. Microtiter plates were then kept at 37°C for an overnight incubation. Following incubation, the MIC was calculated as the lowest concentration of the extract inhibiting the visible growth of the bacterial strain. All experiments were carried out in triplicate.

Preliminary Phytochemical Analysis

Preliminary phytochemical analysis of the various extracts was conducted according to the standard methods (Sofowora, 1993; Trease and Evans, 2002). By this analysis, the presence of several phytochemicals like sterols, tannins, proteins, sugars, alkaloids, flavonoids, saponins, anthraquinones, terpenoids, and cardiac glycosides was evaluated.

Test for Sterols (Salkowaski Reaction)

Few milligram of the plant extract was dissolved in 2 ml chloroform and then 2 ml of conc. H_2SO_4 was added from the sides of the test tube. The test tube was shaken for a few minutes. Red colour development in the chloroform layer indicated the presence of sterols.

Test for Tannins (Ferric Chloride Reagent Test)

The test sample of each extract was taken separately in water, warmed and filtered. To a small volume of this filtrate, a few drops of 5 per cent w/v solution of ferric chloride solution prepared in 90 per cent alcohol were added. Appearance of a dark green or deep blue colour indicated the presence of tannins.

Test for Proteins (Xanthoproteic Test)

The extract (few mg) was dissolved in 2 ml water and then 0.5 ml of conc. HNO_3 was added in it. Yellow colour indicated the presence of proteins.

Test for Sugars (Fehling's Test for Free Reducing Sugar)

About 0.5 g of each extract was dissolved in distilled water and filtered. The filtrate was heated with 5 ml of equal volumes of Fehling's solution A and B. Formation of a red precipitate of cuprous oxide was an indication of the presence of reducing sugars.

Test for Alkaloids (Wagner's Test)

A small quantity of each extract was stirred with 5 ml of 1.5 per cent aqueous HCl and filtered. Wagner's reagent was prepared by dissolving 1.27g of iodine and 2 g of potassium iodide in 5 ml of water and diluting the solution to 100 ml with water. When few drops of this reagent were added to the test filtrate, a brown flocculent precipitate is formed which indicates the presence of alkaloids in the test sample.

Test for Flavonoids (Ferric Chloride Test)

About 0.5g of each extract was boiled with 5 ml of distilled water and then filtered. To 2 ml of the filtrate, a few drops of 10 per cent ferric chloride solution were added. A green-blue or violet colouration indicated the presence of a phenolic hydroxyl group.

Test for Saponins

One gram of each extract was boiled with 5 ml of distilled water and filtered. To the filtrate, about 3 ml of distilled water was further added and shaken vigorously for about 5 minutes. Frothing which persisted on warming was taken as an evidence for the presence of saponins.

Test for Anthraquinones

0.5 g of the extract was boiled with 10 ml of sulphuric acid (H_2SO_4) and filtered while hot. The filtrate was shaken with 5 ml of chloroform. The chloroform layer was pipetted into another test tube and 1 ml of dilute ammonia was added. The resulting solution was observed for colour changes.

Test for Terpenoids (Salkowski Test)

To 0.5 g of each of the extract, 2 ml of chloroform was added. Concentrated H_2SO_4

(3 ml) was carefully added to form a layer. A reddish brown colouration of the interface indicates the presence of terpenoids.

Test for Cardiac Glycosides (Keller-Killiani Test)

To 0.5 g of the extract diluted to 5 ml in water, 2 ml of glacial acetic acid containing one drop of ferric chloride solution was added. This was underlayed with 1 ml of concentrated sulphuric acid. A brown ring at the interface indicated the presence of a deoxysugar characteristic of cardenolides. A violet ring may appear below the brown ring, while in the acetic acid layer a greenish ring may form just above the brown ring and gradually spread throughout this layer.

Results and Discussion

Antimicrobial Assay

The results of antimicrobial response of the different extracts are presented in Figures 41.1–41.4. All the extracts prepared exhibited variable degree of antimicrobial activity against the tested microorganisms. In this study, the aqueous leaf extracts (prepared both in increasing and decreasing order of polarity) exhibited almost no antimicrobial activity against all the tested microorganisms. However, the organic solvent extracts showed a remarkable activity against most of the tested microorganisms. Among the five organic solvent extracts, ethyl acetate and chloroform fractions showed the highest zone of inhibition, followed by almost a comparable activity by the acetone and methanol fractions. Least activity was exhibited by the

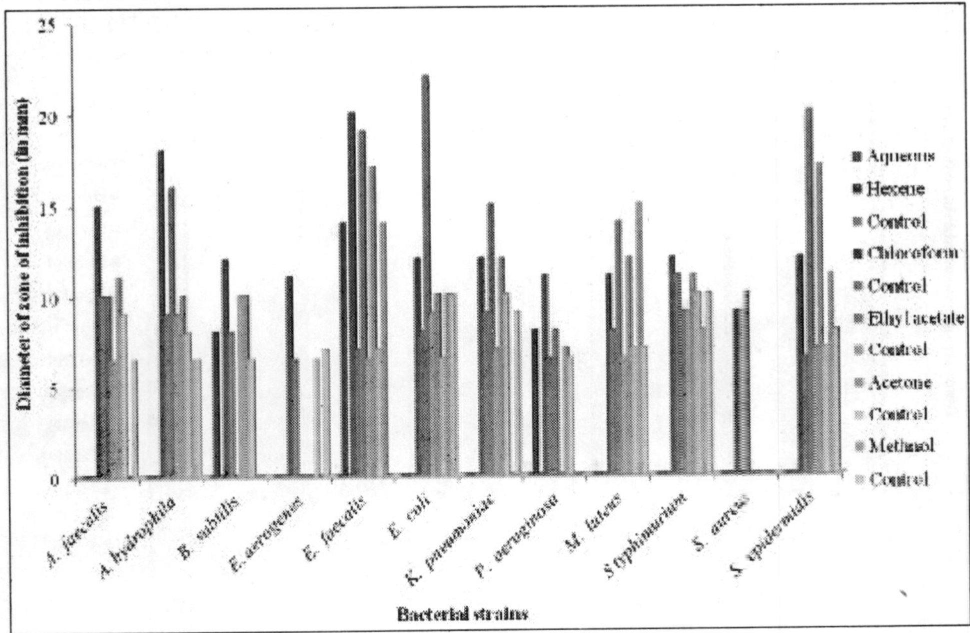

Figure 41.1: Antibacterial Activity of Various Leaf Extracts of *T. undulata* Prepared in the Order of Increasing Solvent Polarity

No activity was found in the aqueous extract.

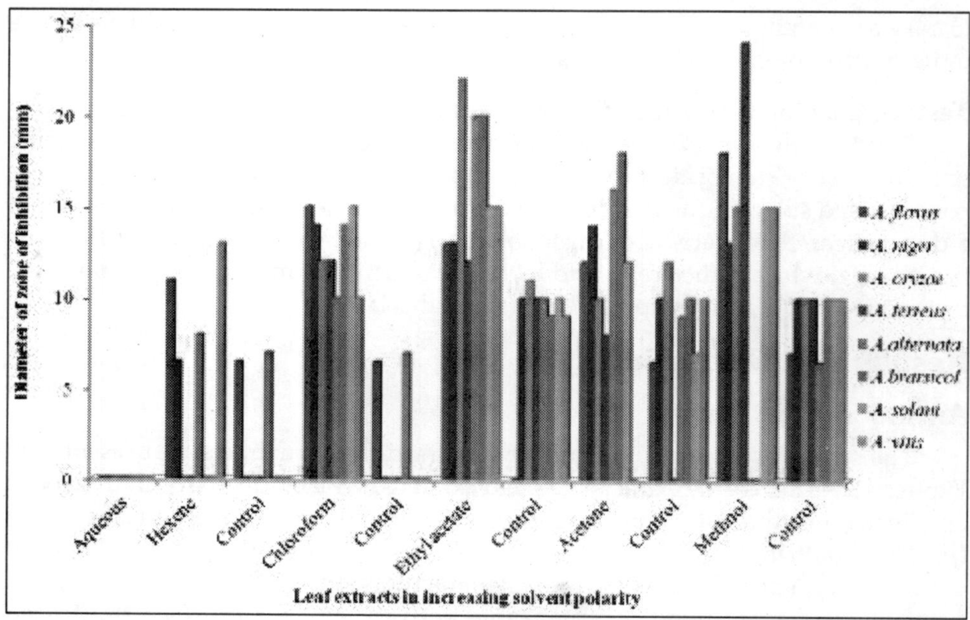

Figure 41.2: Antifungal Activity of Various Leaf Extracts of *T. undulata* Prepared in the Order of Increasing Solvent Polarity

No activity was found in aqueous extract.

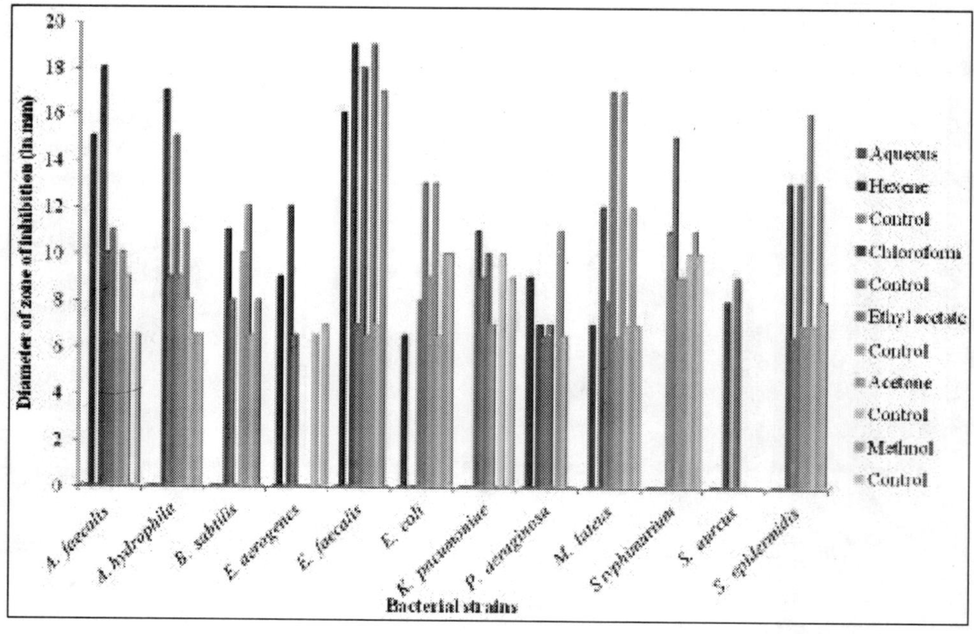

Figure 41.3: Antimicrobial Activity of Various Leaf Extracts of *T. undulata* Prepared in the Order of Decreasing Solvent Polarity

No activity was found in the aqueous extract.

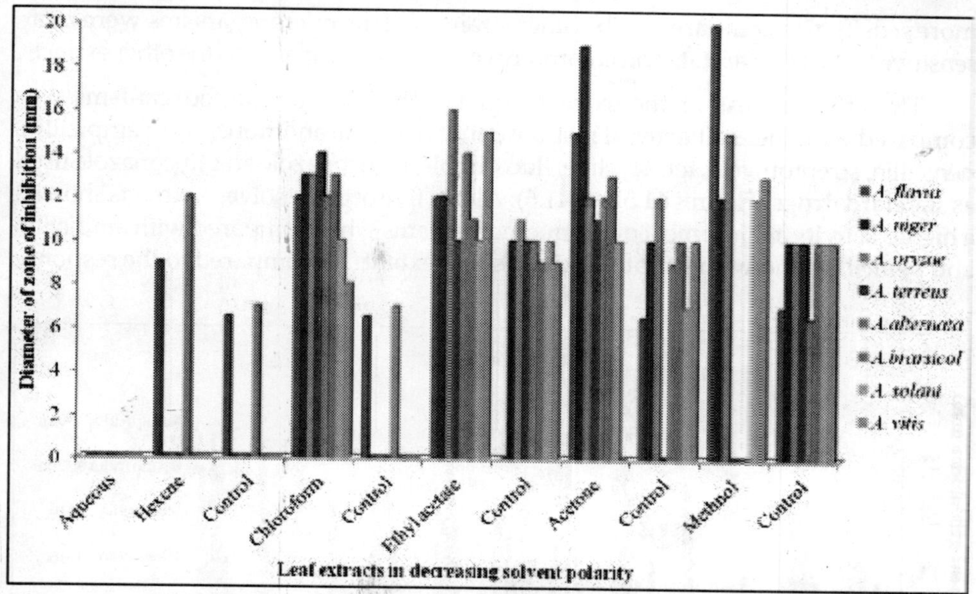

Figure 41.4: Antifungal Activity of Various Leaf Extracts *T. undulata* Prepared in Decreasing Order of Solvent Polarity

No activity was found in aqueous extract.

hexane fractions. For all the solvents, extracts prepared in increasing order of solvent polarity were more effective than those prepared in decreasing order of solvent polarity.

Ethyl acetate extracts exhibited a strong antimicrobial activity against all bacterial (except *E. aerogenes* and *B. subtilis*) and fungal strains tested. The zone of inhibition for the bacterial strains was in the range of 7-23 mm, and for the fungal strains, it was in the range between 10-23 mm. Chloroform leaf extracts also showed a significant activity against most of the bacterial (forming a zone of inhibition in the range of 10-20 mm), and fungal (10-15mm zone of inhibition) strains. Two bacterial strains, *viz.*, *E. coli* and *S. typhimurium* were found to be resistant to the chloroform extract prepared in decreasing order of solvent polarity. Acetone and methanol extracts exhibited almost a similar pattern of antimicrobial activity, showing zone of inhibitions in the range between 9-19 mm and 6-24 mm, respectively. A few microorganisms, *viz.*, *S. aureus*, *E. aerogenes*, *K. pneumoniae*, *A. brasicola*, A. *alternata* and *A. vitis* were found to be resistant to the two fractions. Compared to the other extracts, the hexane fractions were observed to be least effective (exhibiting a zone of inhibition in the range of 6-15 mm). Only 50 per cent of the microorganisms tested were found to be sensitive to the hexane extracts.

Among the bacterial strains tested, *S. aureus* was found to be the most resistant strain (forming a smaller zone of inhibition and against the least number of extracts), whereas, *E. faecalis* was the most sensitive strain (exhibiting a larger zone of inhibition against most of the extracts). Among the fungal strains, most of the fungi showed almost a similar inhibition pattern, however, *A. niger*, *A. oryzae* and *A. flavus* were more sensitive as compared to the other strains. All the microorganisms were more sensitive to the ethyl acetate and chloroform extracts as compared to the other extracts.

The effectiveness of the extracts against the various microorganisms was compared with the antibacterial response shown by the antibiotics, *viz.*, ampicillin, penicillin, streptomycin, tetracycline, fluconazole, ketoconazole and miconazole used as standard drugs (Figures 41.5 and 41.6). All the five organic solvent extracts showed a higher activity against most of the microorganisms when compared with ampicillin and penicillin, and with streptomycin to some extent. As compared to the response

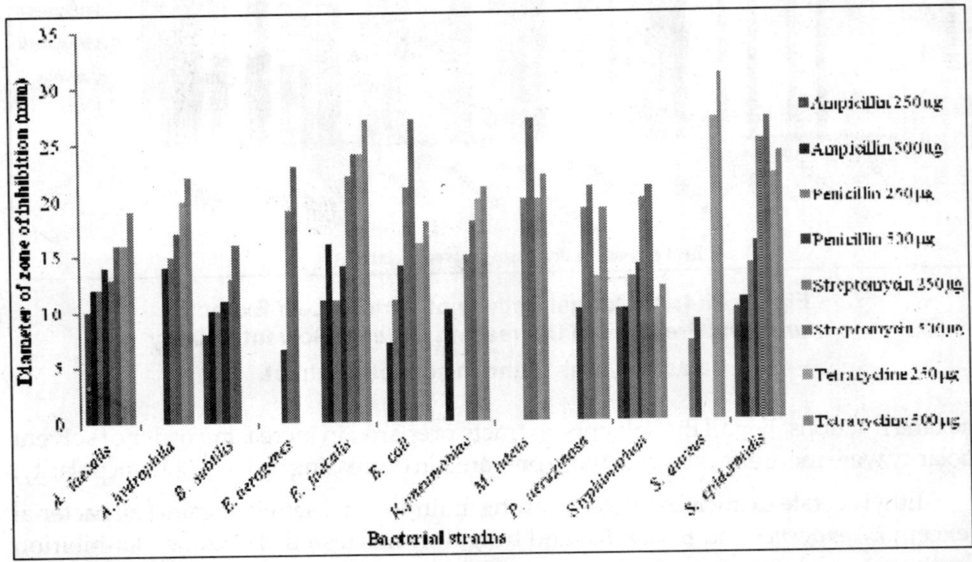

Figure 41.5: Antimicrobial Activity of Standard Antibiotics

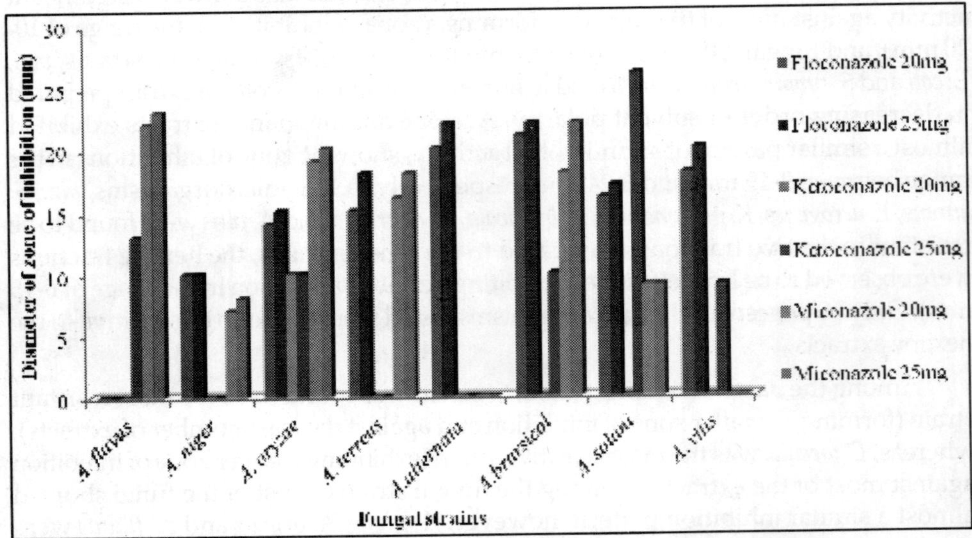

Figure 41.6: Antifungal Activity of Standard Antibiotics

exhibited by tetracycline, only the ethyl acetate and chloramphenicol extracts were found to be slightly better in case of some microorganisms. Among the three antifungal drugs tested, ketoconazole was found to be least effective, followed by a slightly higher response by fluconazole, and the maximum activity shown by miconazole. Compared with the zone of inhibition formed by the standard drugs, all the solvent extracts of *T. undulata* showed an appreciably greater zone than that formed by ketoconazole. The antifungal response exhibited by the ethyl acetate, chloroform, methanol and acetone extracts against some of the fungi was also comparable to that of the other two standard drugs (fluconazole and miconazole). Thus, the results in the present study show that *T. undulata* extracts exhibited significant activity against most of the tested microorganisms which was comparable to that of the standard drugs.

Minimum Inhibitory Concentration (MIC)

All the active extracts (leaf extracts prepared in organic solvents) were further subjected to determination of minimum inhibitory concentration, the results being shown in Table 41.1. The lowest MIC was shown by the chloroform extracts (200-400 µg/ml, except against *S. aureus* and *A. vitis* where it was slightly higher), followed by the acetone (between 175-500 µg/ml for majority of the microorganisms) and ethyl acetate (between 200-500 µg/ml for 50 per cent of the microorganisms) extracts. Methanol and hexane extracts exhibited comparatively higher MIC, indicating less effectiveness of these extracts. Among the various microorganisms tested, lowest MIC values were obtained for *S. epidermidis*, followed by *E. faecalis* and *E. coli*, indicating that these bacteria were most sensitive to the *T. undulata* leaf extracts. Among the various fungi, the most sensitive was *A. brasicola* (MIC values in the range of 175-400 µg/ml), followed by *A. oryzae* (MIC values in the range of 150-500 µg/ml). The results of MIC assay confirmed the findings of antimicrobial assays, wherein it was reported that ethyl acetate and chloroform extracts were more potent inhibitors of the microorganisms tested; and that *E. faecalis* and *A. oryzae* were among the most sensitive strains.

Phytochemical Analysis

Preliminary phytochemical analysis of leaf extracts (hexane, chloroform, ethyl acetate, acetone and methanol) showed the presence of alkaloids, anthraquinones, cardiac glycosides, proteins, tannins, terpenoids, sterols, sugar, and flavonoids; while flavonoids and saponins were found to be absent in all the tested extracts (Table 41.2). The relative antimicrobial activity of leaf extracts may not be easily correlated with any individual component but with a mixture of compounds present in these extracts. There are reports showing that alkaloids are responsible for the antifungal activity in higher plants (Cordell *et al.*, 2001). It has also been suggested that the antimicrobial activity is mainly due to the presence of alkaloids, terpenoids and other natural polyphenolic compounds or due to free hydroxyl groups (Rojas *et al.*, 1992). Moreover, secondary metabolites such as tannins and other compounds of phenolic nature are also classified as antimicrobial compounds. Therefore, the presence of these phytochemicals (alkaloids, anthraquinones, cardiac glycosides,

Table 41.1: Minimum Inhibitory Concentration (MIC) Values of Different Leaf Extracts of *T. undulata* Against the Tested Bacterial Strains

Sl.No.	Bacterial/ Fungal Strains	Minimum Inhibitory Concentration (µg/ml)									
		Leaf Extract of Tecomella undulata									
		Hexane (I)	Hexane (D)	Chloroform (I)	Chloroform (D)	Ethyl Acetate (I)	Ethyl Acetate (D)	Acetone (I)	Acetone (D)	Methanol (I)	Methanol (D)
	Bacterial Strains										
1.	*A. faecalis*	–	300	325	350	1000	225	400	500	–	–
2.	*A. hydrophila*	–	–	200	225	200	200	500	950	500	1000
3.	*B. subtilis*	925	–	375	400	–	–	500	250	–	1000
4.	*E. aerogenes*	–	525	225	200	–	–	–	–	–	–
5.	*E. faecalis*	275	300	250	250	375	250	250	350	275	225
6.	*E. coli*	–	450	375	–	200	200	250	200	250	450
7.	*K. pneumoniae*	–	–	400	425	200	500	225	–	–	–
8.	*M. luteus*	–	950	450	450	700	200	775	400	200	–
9.	*P. aeruginosa*	–	–	450	425	250	1000	1000	400	–	425
10.	*S. typhimurium*	–	–	400	425	1000	200	400	200	1000	200
11.	*S. aureus*	–	–	950	–	500	250	–	–	–	–
12	*S epidermidis*	–	–	200	250	200	200	200	200	200	200

Contd...

Table 41.1–Contd...

Sl.No.	Bacterial/ Fungal Strains	Minimum Inhibitory Concentration (µg/ml)									
		Leaf Extract of *Tecomella undulata*									
		Hexane (I)	Hexane (D)	Chloroform (I)	Chloroform (D)	Ethyl Acetate (I)	Ethyl Acetate (D)	Acetone (I)	Acetone (D)	Methanol (I)	Methanol (D)
		Fungal Strains									
1.	*A flavus*	175	900	175	200	350	250	200	200	400	200
2.	*A. niger*	975	–	225	200	200	250	200	200	850	400
3.	*A oryzae*	–	–	250	–	150	200	450	500	200	200
4.	*A. terreus*	–	–	200	–	250	1000	250	250	200	500
5.	*A.alternata*	325	400	250	250	825	550	200	425	–	–
6.	*A.brasicola*	–	–	350	200	175	250	175	400	–	–
7.	*A. solani*	–	–	325	250	250	250	200	700	200	250
8.	*A. vitis*	–	–	1000	–	450	200	–	–	200	200

All the values are an average of three determinations.

–: No activity; (I): Increasing solvent polarity; (D): Decreasing solvent polarity.

proteins, tannins, terpenoids, sterols and sugar) could to some extent justify the observed antimicrobial activity in the current study.

Table 41.2: Preliminary Phytochemical Analysis of Leaf Extracts of *T. undulata*

Phytochemicals Tested	*Test Performed*	*Leaf Extract*				
		Hexane	*Chloro-form*	*Ethyl Acetate*	*Acetone*	*Methanol*
Alkaloids	Wagner's test	++	++	++	++	++
Anthraquinones		+	+	+	+	+
Cardiac glycosides	Keller-Killiani test	–	++	++	–	–
Flavonoids	Shimoda test	–	–	–	–	–
Proteins	Xanthoproteic test	–	+++	+++	+	++
Tannins	Ferric chloride reagent test	–	–	–	++	++
Terpenoids	Salkowski test	–	++	++	+++	+++
Saponins	Foam test	–	–	–	–	–
Sterols	Salkowaski test	+	++	++	+	++
Sugars	Fehling's solution test	–	+++	++	–	–

Conclusion

The present study scientifically validates the antimicrobial potential of the traditionally important plant, *Tecomella undulata*. The results provide an important basis for the use of chloroform, and ethyl acetate extracts of the tested plant species for the treatment of infections associated with the pathogens used in this study, which could be useful for the development of new antimicrobial drugs. With an aim to identify the chemical nature of compounds responsible for the antimicrobial response, the study also involved a preliminary phytochemical analysis of the crude extracts. However, further studies related to the isolation and identification of the particular compounds responsible for the antimicrobial activity are underway. The antimicrobial mechanisms associated to each group of chemicals to which the isolated compounds belong, may explain the inhibition potency of the tested samples. Present results allow us to conclude that the crude extracts of *T. undulata* exhibited significant antimicrobial activity and properties that support folkloric use of this plant; corroborating the importance of ethno-pharmacological surveys in the selection of plants for bioactivity screening.

References

Ahmed, I., Mahmood, Z. and Mohammad, F. 1998. Screening of some Indian medicinal plants for their antimicrobial properties. *J. Ethno. Pharmacol.* 62: 188-93.

Cordell G.A., Quinn- Beattia M.L., and Farnsworth N.R. 2001. The potential of alkaloids in drug discovery. *Phytother. Res.* 15: 183- 205.

Dradhbala, C. 1996. The Charak Samhita. Sastri, K., Chaturvedi, G.N., Sastri R., Upadhayaya, Y., Pandeya, G.S., Gupta, B. and Mishra, B. (Eds). 22nd Revisd Edn. Chaukhaumba Bharti Academy, Varnasi.

Indu, M.N., Hatha, A.A.M., Abirosh, C., Harsha, U., Vivekanandan, G. 2006. Antimicrobial activity of some of the south-Indian spices against serotypes of *Escherichia coli, Salmonella, Listeria monocytogenes* and *Aeromonas hydrophila. Braz. J. Microbiol.* 37:153-158.

Lino, A. and Deogracious, O. 2006. The in-vitro antibacterial activity of *Annona senegalensis, Securidacca longipendiculata* and *Steganotaenia araliacea*- Ugandan medicinal plants. *Afr. Health Sci.* 6(1): 31-35.

Malabadi, R. B., Mulgund, G.S. and Nataraja, K. 2005. Screening of antibacterial activity in the extracts of *Clitoria ternatea. J. Med. Arom. Plant Sci.* 27: 26-29.

Malabadi, R. B., 2005. Antibacterial activity in the rhizome extracts of *Castus speciosus* (Koen). *J. Phytol. Res.*, 18 (1): 83-85.

Malabadi, R. B., Mulgund, G.S. and Nataraja, K. 2007. Ethnobotanical survey of medicinal plants from Belgaum distict, Karnataka, India. *J. Med. Arom. Plant Sci.* 29 (2): 70-77.

NCCLS–National Committee for Clinical Laboratory Standards. 1997. Methods for Dilution Antimicrobial Susceptibility Tests for Bacteria that Grow Aerobically. M7-A4. Wayne. Pa

Perez, C.; Paul, M. and Bazerque, P. 1990. Antibiotic assay by agar-well diffusion method. *Acta. Biol. Med. Exp.* 15: 113-5.

Rao, R. R., 1996. Traditional Knowledge and sustainable development key role of ethnobiologists. *Ethnobotany.* 8: 14-24.

Rojas, A.L., Hernanddez, R., Pereda- Miranda and Mata, R. 1992. Screening of antimicrobial activity of crude drug extracts and pure natural products from Mexican medicinal plants. *J. Ethnopharmacol.* 35: 275- 83.

Savithramma, N., Sulochana, C.H. and Rao, K.N. 2007. Ethnobotanical Survey of plants used to treat asthma in Andhra Pradesh, India. *J.Ethnopharmacol.* 113: 54-61.

Sofowora, A. (1993). Screening Plants for Bioactive Agents. In: Medicinal Plants and Traditional Medicinal in Africa. 2nd Ed. Spectrum Books Ltd, Sunshine House, Ibadan, Nigeria. 134-156.

Sudharameshwari, K., and Radhika, J. 2007. Antibacterial screening of *Aegele marmelos, Lawsonia inermis* and *Albizzia libbeck.* AJTCAM. 4(2): 199-204.

Tulsidas, 1631. Samavat, *Ramcharitmanas.*

Vedavathy, S., Mridula, V. and Sudhakar, A. 1997. The tribal medicine of Chittoor district. Herbal Folklore Research centre, Tirupati Andhra Pradesh, India, pp: 14-143.

Warrier, P. K., Nambiyar, N.P.K. and Ramankutty, G. 1997. Indian Medicinal plants. A. Compendium of 500 Species. Vol. I-V. Orient Longman Publications, Hyderabad, India, pp:14-143.

Index